Designing Mobile Interfaces

Steven Hoober
Eric Berkman

O'REILLY®
Beijing · Cambridge · Farnham · Köln · Sebastopol · Tokyo

Designing Mobile Interfaces

by Steven Hoober and Eric Berkman

Copyright © 2012 Steven Hoober and Eric Berkman. All rights reserved.
Printed in Canada.

Published by O'Reilly Media, Inc., 1005 Gravenstein Highway North, Sebastopol, CA 95472.

O'Reilly books may be purchased for educational, business, or sales promotional use. Online editions are also available for most titles (*safari.oreilly.com*). For more information, contact our corporate/institutional sales department: 800-998-9938 or *corporate@oreilly.com*.

Editor: Mary Treseler	**Indexer:** Lucie Haskins
Production Editor: Holly Bauer	**Cover Designer:** Karen Montgomery
Copyeditor: Audrey Doyle	**Interior Designer:** Ron Bilodeau
Proofreader: Jennifer Knight	**Illustrator:** Robert Romano and Steven Hoober

November 2011: First Edition.

Revision History for the First Edition:

2011-10-25   First release

See *http://oreilly.com/catalog/errata.csp?isbn=0636920013716* for release details.

ISBN: 978-1-449-39463-9

[TI]

Contents

Part II.   Components

Part III.   Widgets

Appendixes

Preface

The fact that you are reading this book means you don't need to be told how ubiquitous mobile is, how quickly the mobile market is growing and changing, and how much mobile computing is supplanting desktop computing as well as more traditional media such as film, television, radio, newspapers, and books.

Mobile is so huge and is growing so fast that astonishing growth numbers from just a few years ago pale in comparison to growth numbers today—so much so that we won't even bother quoting any figures, as they will be outdated long before the rest of the content loses its relevance.

One thing that has not happened yet is true standards for design. Movements are now underway to design for the mobile experience first, before focusing on other forms of computing. A good reason for this is that in many markets, many of your customers look at your website on mobile devices more than on desktops.

Yet, too much design is based on older paradigms for the desktop, or even for TV or print. Within mobile, too many design discussions are very narrowly focused. They pay special attention to applications on a single platform, or only to the mobile web—and almost always at the expense of every other platform. Certainly, almost no one discusses anything but smartphones, despite the huge market share and vast usage rates of feature phones.

Fragmentation is discussed as a bad thing for marketing, and sometimes for design, but designers themselves contribute to this fragmentation too often by focusing on pixel-based layouts and the specifics of their favorite OS. This does no one any good, and is especially pointless when you consider the user. Devices generally have many more features and methods of interaction in common than their differences might imply.

Serious mobile design now, and especially in the future, will require building for every user, and providing some solution on every platform.

This book offers a set of common patterns for interaction design on all types of mobile devices. A few patterns require specific hardware or form factors, but most are absolutely universal.

Most do not concern themselves at the top level with implementation details. The correct solution is correct even at the OS level, as an application or as a website.

Of course, there are notes to discuss alternatives, methods, and limitations to assist with decision making. And many of the specific patterns are coupled with alternatives or variations that allow similarly useful solutions to be achieved on any type of device.

Who This Book Is For

As with any good form of interactive design, we have kept a specific scope in mind from the moment we started writing this book. If this book was intended to be all things to all people, it would be much larger or we'd simply never have finished it. Our focus here has been on design. By this we mean information architecture; information, interaction, interface, and visual design; and copywriting.

If your job title, job description, or deliverables have names like those just mentioned, and you work in mobile, you need this book. Whether you are working on apps or websites for mobile (or any of many other things), this book addresses the common underlying principles in order to help you make better decisions and understand how to create better designs.

If you are moving from another field, such as desktop web design, or are switching from one narrowly focused mobile area to another, this book encompasses general patterns that can help you understand how to move from one type of device or one type of interaction to another.

If you work in a related job, this book still has something for you. Human factors engineers and HCI experts will find numerous discussions of why these solutions have become patterns, and references to cognitive psychology- and physiology-based reasons these are true.

Development is not addressed as such, but the book is organized so that you can use it to find specific solutions to any form of mobile interaction. If you don't have a dedicated design team, you can use the patterns to find and focus on solutions, confirm they are technically possible, and avoid common implementation pitfalls.

Hardware designers, or anyone who can influence hardware design, will find specific guidelines to best practices in interactive, such as key labels and the use of sensors. Though these are included primarily for use by interaction designers—to understand how the hardware influences their on-screen behaviors—they are also specific enough to be used for design of the interactive portions of the hardware itself.

What We Mean by "Mobile"

Over the years, the reaction to my job title, "mobile interaction designer," has migrated from blank stares to significant interest in this suddenly mainstream technology. Still, only about half the time do people have any idea what that title means. And if they do, they almost always assume my job is to design mobile phones or apps for them.

Figure P-1. *Traditional media, and desktop computing, require the user to make an effort to go to where the display is; even a laptop requires creating an ad hoc workplace. When the mobile device is always with you, everywhere is a place to do work, be entertained, or consume information.*

Occasionally, someone asks if we also design games for the Nintendo DS, or make maps for GPS navigation, or do work for some other sort of device. Has the definition of mobile changed? This is a list of the types of things we looked at to find and validate these patterns:

- Mobile smartphones
- Mobile feature phones
- Mobile network access points (aircards)
- Mobile Internet Devices (MIDs)
- Tablets
- eReaders
- Media players
- Image viewers and digital picture frames
- Portable game systems
- Remote controls
- Handheld navigation devices (see Figure P-5)
- Portable scanners
- Cameras and other capture devices
- Printers, scanners, copiers, and mopiers (multifunction devices or MFDs)
- Kiosks
- Wearable computers
- Telematics, and vehicle-mounted devices
- Industrial automation
- Portable surveying, measuring, and metering equipment

The preceding list is not exhaustive. And although the items in the list share many characteristics, many are not particularly mobile. Kiosks are, by definition, bolted to the ground, for example. And no one much thinks of a camera as being similar to a phone.

So our first answer is that *mobile* is not a useful word, and that this book addresses a lot of these devices. Their design can be informed by the mobile patterns in this book and elsewhere. The ubiquity of mobile devices also may mean that employing these as universal patterns is a good thing, as users may require less training when using interfaces to which they are accustomed. If you design cameras or printers, you should be paying attention to the state of the art in mobile.

Figure P-2. *Always-available devices encourage use at every idle moment, with an increasingly broad definition of idle time. Law, custom, or commonsense notwithstanding, devices will be used in every conceivable environment.*

We didn't come up with this answer out of the blue or just trust our instincts. Instead, years of work, discussion, writing, and arguing led to some principles of "what is mobile."

We like to think of the evolution of mobile telephony as having occurred in four eras:

- Voice
- Paging and text
- Pervasive network connectivity
- General computing devices

Yes, this leaves out a lot of interesting devices that are on the longer list shown earlier. The Game Boy and GPS receivers predate by several years what most of us would call mobile phones. But mobile telephony is what changed the world and ushered in all this, so it's a good anchor point.

If you consider a current mobile phone as being a "fourth-era" device, you find that it has the following characteristics:

Small
> It's small enough to carry with you all the time, preferably in a pocket (see Figure P-1).

Portable
> It's battery-powered and otherwise independent of the world, so it doesn't have to be plugged in or attended to regularly (see Figure P-2).

Connected
> It's wirelessly connected, not attached to the wall or connected only when the user makes special effort. Whenever possible, it is connected in multiple ways, to both voice and data networks (see Figure P-3).

Interactive
> It's inherently interactive. Unlike a watch, or even most MP3 players, which are limited to display, playback, and a small subset of interactions, a mobile phone allows more undirected actions to be taken, such as text entry and keyword search.

Contextually aware
> It (or services it is attached to) uses its ability to understand the network to which it is attached, and its other sensors, to help the user get things done, and preemptively gather information (see Figure P-4).

With this many facets, it's easy to disregard any one of them and still feel the device meets your needs. By strapping an iPad to a wall and calling it a kiosk, all you've done is simply remove its "portable" feature, so it's still a "mobile" device.

Consider the Wii or the Xbox Kinect instead. Though the display used on these devices is not at all portable, at its core it is aware of the user's position, it changes with the type of input being used, and the entire interface has been designed to support interaction via a game controller, or via the user simply waving his arms at the screen. These meet the interactive criteria to be a "mobile" device.

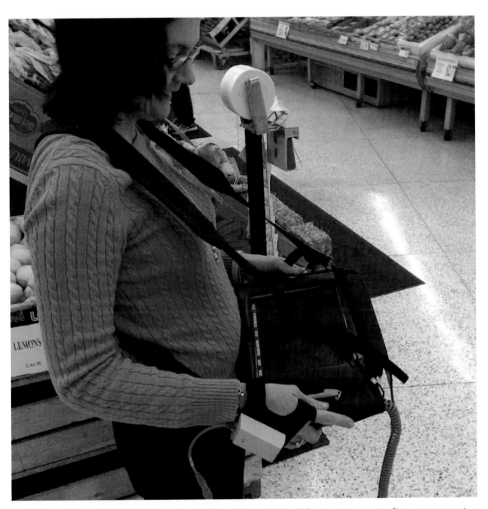

Figure P-3. *This wearable scanner, attached to a Tablet PC (used for inventory control), is representative of trends in workplace computing: contextually useful so that the operator has it with her constantly, and connected so that the enterprise can use the intelligence immediately.*

Now take a Windows Tablet PC. It has pen and touch input, can be quite small and portable, is networked, and has sensors. But we argue that it is not mobile. It's not really connected, because it connects like a desktop, so you have to open dialog boxes and press buttons. It isn't usefully interactive, because you cannot use it on the go, but rather have to stop to use it. It's not contextually aware, because the GPS, or camera, or accelerometers don't do much of anything by themselves.

Figure P-4. *A few years ago, I used this combination of a GPS and Windows Mobile device to record and track my location (in a snowstorm, no less). Today, the results of sensors in mobile devices can be seamless, automatic, and so intelligent as to risk violating privacy.*

What Type of Patterns We Will Cover

So, although this book does still focus on the mobile phone and, in particular, the smartphone, similar interactions from kiosks to game stations to telematics are also considered. In some cases, we'll even refer to these devices directly in the patterns.

The patterns are guidelines for implementing interaction design on these devices. So they talk about page-level components such as scroll bars, display components such as pop ups, widgets such as buttons, and input methods such as keyboards.

But they also talk about things such as the labels and lights for hardware keyboards. These are covered because you can influence them. You can fail to implement the keyboard correctly, and cause keyboard entry to fail. You can change scroll bar behavior, if there's a special case for your application. Increasingly, as HTML5 technologies roll out, mobile websites can take advantage of interesting interactive features.

Also, you might be working on a device operating system—likely the GUI layer on top of an existing OS. There are many, many devices, and new classes are still emerging. Overlays have migrated down to the point that end users may change the basic behavior of their handsets. You may have to work in this space sooner than you think.

If not, you still need to understand why certain OS-level behaviors are standard, or should not be, so that you can make informed decisions about your design.

What Is a Pattern?

There has always been a concept of reusing and reapplying known best cases in graphic design. There has always been a culture of sharing, borrowing, and building on the work of others. As graphic design (and designers) moved into and influenced interactive design, this philosophy of repurposing the best ideas, coupled with the principles of object-oriented development—which also develops by modules and reuses components—led naturally to the evolution of design patterns.

Patterns as applied to interactive design are much like patterns in software development. This has led to a conflict between design managers who push for repeatability and the use of templates, stencils, and patterns, and designers who want to be free to explore solutions. However, this is an artificial mismatch, arising from a misunderstanding of how patterns should be used.

The concept of a pattern was actually developed by the architect Christopher Alexander in the late 1970s. He argues that patterns are components of a language, and can be used to conduct a dialogue about building and organizing, and the nature of human existence in space. Although architecture and engineering are only so analogous to the fields of software and interactive design, the concepts did translate well.

Object-oriented software development applied the concept directly from Alexander's early work starting in the late 1980s. This was applied as a straightforward problem/ solution statement; select the pattern whose problem statement most closely matches the implementation technology problem. There is room to design the specific solution, and to modify it to meet the needs of the specific system, but they are still very plug-and-play.

While Alexander's arguments may be hard to follow—especially when he talks of concepts such as the "life" in spaces, or underlying "morphogenesis"—the core of his process is at the core of all design processes. Patterns are simply well-defined, well-researched best practices, but fundamental principles of design must always be followed, the user must always be kept in mind, and the purpose of the design must always be considered.

In mobile interactive design, we might summarize these core principles as *user-centered design*, context, and other principles. A set of more specific principles are listed at the end of this introductory section. You must always consider these principles—or others you may be more comfortable with—to ensure that the proper pattern for the situation is determined, and the correct application is created from the user's needs, his context, and by integrating the solution into the whole system.

Where Did These Patterns Come From?

Another way to think about patterns is that they are simply all the heuristics available, shoved together and placed into a form where you can simply look at the end result. In many ways, that is how this book was created. For years, we have been collecting implementations, trying to tease out behaviors and patterns, and gathering them up. In other formats, and at conferences, we've been distributing these to the mobile design community.

And when we went to write a book on them, we had a pretty good handle on the categories of information that had to be created, and the patterns that would occupy each one. And then almost immediately we ran into trouble. At this level of detail, gut checks and common practice and simply knowing how it works was not good enough. So we did research. A lot of it was simply observing devices.

Figure P-5. *A selection of the devices surveyed to discover, confirm, and understand the patterns that ended up in this book. This is just a small sampling. Add another 30 or more handsets, 10 tablets, 10 eReaders, numerous game controllers and portable game units, a handful of GPS navigation devices, and many others to the list. These devices are not just for show; many work, and are readily available to refer to during design (and when we were writing this book). Many have Velcro on the back, and are stuck to the wall of the design studio, to be in our faces every day when we come to work.*

Many are those we gathered over time, but we also acquired new ones, and asked people we know, work with, met at parties, or ran into at conferences to let us see their devices. We hit up stores when new devices came out. We noticed, and took photos of, the way PIN pads and kiosks work. We skulked around electronics recyclers to get old devices on the cheap and begged friends to let us have their dusty old phones. Figure P-5 shows a sample of these devices.

And then we compared the implementations. In many cases, the all-new, super-cool best practice was just a very minor change (or no change at all) to something on a 10-year-old PDA, or on many feature phones. In all too many cases, the newest technologies had lost common and best practices from the well-established methods of scroll-and-select devices.

We engaged with the users, too, in any environment. Whether at an airport, in a coffee shop, on a busy street, in the office, or in our family room, we recorded behavior. We observed these individual, social, and cultural interactions in varying contexts, paying close attention to how people use these devices in their everyday lives. We also interviewed them to gain rich, insightful, qualitative data about their needs, motivations, and attitudes about using mobile devices. All of this ethnography and contextual inquiry—whether formal research or ad hoc discovery—further validate the design recommendations presented throughout this book.

As a result of this research, we hope we have provided balanced and interlocking coverage of emerging technologies, the common practice of the most buzz-worthy current devices, and the best practice of the well-established "low-end" devices.

And whenever possible, we performed literature surveys to determine why those activities work the way they do, and to explain them—not just stating that they represent the right way, but *why* they do so. This sort of work helped us, and will help you, to understand the relationships among the different implementations of a pattern. Only understanding why lets us explore the edges without wasteful trial and error. Understanding human cognition, perception, and physiology lets you predict what will work and what will not before building it.

Art, Graphic Design, and Experience

We needed these patterns and heuristics before writing this book because we design mobile applications, websites, services, and OSes. Besides being paid to gather this information, we also have gained access to a lot of other information, such as user research and user behaviors with products in production. In addition, we have been able to perform our own research, and we have gotten firsthand experience with many of these device classes.

We also have other experience, in graphic design, art, human factors, industrial design, engineering, and education. And when working on real projects, to launch with real products, we also have to work with many other individuals, with dozens of other job functions. So these patterns are also grounded in the knowledge base of those skills, often backed by well-documented scientific research again.

Common Practice Versus Best Practice

A key overriding principle behind much of this work is the differentiation between common practice and best practice. Although not always explicitly stated, this is what drove activities such as the inclusion of antipatterns (or "worst practices") for each pattern. There are many, many design patterns that do not work, or do not work as well as alternatives.

This is a key reason so much effort went into researching the patterns. We didn't include something just because it was heavily used, or is a much-lauded feature of a new and well-covered device; if it was common or well known, but bad, we included it, but with warnings.

This also means, again, that we had some serious discussions about what qualified as a pattern. In general, a pattern must be a best practice, and common enough to be recognized or encountered.

Therefore, there may be some odd cases where an antipattern has general solutions listed, but no specific solutions in the body of the pattern. Though the problem is known, no single solution has emerged.

A best practice that is not implemented anywhere (or only very rarely) is not described, as it does not rise to the level of a pattern. Only real-world items are patterns by our thinking, not clever concepts, demonstrations, or videos of how the future might work. These are, however, sometimes mentioned as future technologies or options to look forward to.

Reading the Patterns

Almost before we even started to figure out which patterns we wanted, we started talking about how they should be internally arranged. There's a surprising amount of variation here, but primarily we decided that a single, totally consistent method was most important. That way, you have a fighting chance of taking two competing patterns and comparing them.

The most important section ended up being the one on variations. Something like 15 to 20 patterns disappeared over the course of writing this book, and ended up being better described as variations of existing patterns. And a few patterns we split into two or more patterns instead: after we started writing, the variations for some of the patterns were a bit too severe, so we split those patterns into two or more patterns, instead of just one with variations.

Names

Names are as short as practical while still being clear, and whenever possible they do not conflict with an existing concept. Some ended up being a bit of a mouthful as a result, but we did our best. In far more cases than expected, there was no name at all for a well-used design element, and we had to make something up.

We always used title case for the actual pattern names. If you run across a name that is capitalized and blue for no apparent reason while reading a sentence (e.g., "Input Method Indicator"), that means it is in reference to another pattern in the book, which you can refer to in order to make a comparison. Whenever possible, this is also made clear in the text, such as "see the Input Method Indicator pattern for an alternative method."

Problem

Some people get nervous when the word *problem* is used in a project or design sense. But problems foster solutions, so try not to worry about any history or bad implications that term may have to your organization.

The "Problem" section of each pattern is just a summary of why you'd want to use the pattern. Ideally, patterns are grouped with similar problems, and you can get to the right section and then compare the problem statements as a way to help identify which one you really have.

Solution

The "Solution" section provides a definition of what the pattern involves, which other patterns are key overlaps or provide key components, and (when relevant) the important technologies required to make it operate.

This is one of the sections that can vary widely, from a very brief introduction to comprising half of the pattern. If it is difficult to explain, difficult to implement, or often poorly implemented, this will get longer. Simple patterns are shorter.

Variations

Our patterns aren't stencils, so they aren't restrictive. All of these have variations that you can choose and that are defined so that you can choose the correct one based on the content to be delivered and the context in which you will use the pattern.

The length of this section varies widely depending on the number of and differences among the variations. Some have multifaceted variations, so more than one list may be encountered. In some cases, the variations are so pronounced that much of the interaction and presentation is covered in this section as well.

Interaction Details

This section explains how the user interacts with the item being described—including pressing buttons (or swiping the screen) and what the screen displays that the user can click on or type inside.

Presentation Details

This section explains things on the screen that you cannot click on, or details about the manner in which displayable items are presented which do not directly influence the interaction. A shadow on an interactive item might help with visibility, so this would matter but would not directly influence interaction, for example.

The difference between interaction and presentation can be a bit difficult to fathom sometimes, but breaking them up helps a lot when trying to seek the core truths of a function and separating what must be present and what is optional or additional.

Antipatterns

Specifics of the implementation you should watch out for are always listed. These cover both *antivariations* (methods that should never be used) and more minor pitfalls or edge case uses of proper variations to watch out for. These are not speculative, but are known to be bad because they violate heuristics, and often are verified by research.

These do not encompass all the possible antipatterns, but the key and most likely problems. Rest assured that there are many other ways to break a good pattern. Use design principles and heuristics, and carefully read the rest of the pattern to prevent poor implementations.

If you cannot avoid an antipattern for technical reasons, you should not use the pattern and instead you should find a technically feasible replacement. This is a common occurrence, and is a key reason the antipatterns are explicitly listed.

Examples and Illustrations

We deliberately chose not to include a lot of screenshots. In fact, we include hardly any. We did not arrive at this decision lightly; we gathered and extensively annotated screenshots for the first several patterns. But we decided to take this route for the purpose of practicality. It's very hard to find enough adequate examples, and often the best one is on a device that is difficult or impossible to capture. Some of the clearest examples are on feature phones, old PDAs, GPSes, and the like.

This leads to the key problem we encountered with screenshots: clarity. Patterns are the pure essence of an implementation. And almost every implementation layers its own style on top—or buries a pattern alongside others. Screenshots required explanation, and very often caveats about what not to do.

To solve the problems with screenshots, we used illustrations almost exclusively throughout the book. These are all of the same basic style, but vary widely in the detail level used, sometimes in adjacent drawings in the same pattern.

In each case, only the required amount of detail is used. Sometimes that detail is just boxes and lines, and the words and images are implied. Sometimes words and so on have to be in there to communicate the point. Sometimes actual raster icons or websites, drop shadows, and other effects are used.

As a general rule, large blank areas on a page do not mean there's nothing there. It just means we're not discussing that component, so we removed placeholder information for clarity. The Annunciator Row is almost always assumed, so space is provided, but is not displayed—again, for clarity and to reduce clutter.

Color, especially when clearly not naturalistic, generally has a meaning:

- Yellow usually refers to the displayed, interactive elements.
- Blue is for images, and graphical displays such as information visualizations. A different color is used so that it is clear that it's not just a box.
- Grays represent nonselectable items, like the parent when a child has popped up over it.
- Orange is used when the item is in focus, as when scrolling in a list, or to indicate the primary button that is going to be selected for a process.

This is not always adhered to, especially in the higher-fidelity drawings, but it is a good guideline.

That being said, there are a few photos and screenshots in the book. We used these when creating an illustration of sufficiently high fidelity would have been pointless, for clarity when describing certain hardware details, and as example implementations when introducing new categories of patterns.

Successfully Designing with Patterns and Heuristics

As we just discussed, patterns are often misapplied and used as rote answers to problems. Or they are specifically rejected because they are perceived as having to work in a certain way, and so stifle creativity.

While neither of these statements is true, the next problem with best practices is more insidious.

Avoiding the Heuristic Solution

Good, well-intentioned practitioners of interactive design, like you, apply well-established principles, procedures, and processes in an attempt to seek the right solution.

We all perform heuristic evaluations, apply patterns, copy best practices (or at least "common practices"), and—whenever possible—perform user research to confirm the design is good. Without inspiration or luck, all too often the end design is what could be called the "heuristic solution"—the rote, safe answer is the default result.

And very often this is fine, strictly speaking. There are no serious errors, and satisfaction is within norms. But have you ever found a design to be boring? Or that it only solves that specific issue, so you end up redesigning it for the next version?

A lot of demands are (and should be) placed on interactive designers to meet the brand, to find and meet the users' greater goals, and to differentiate from an ocean of interactive products competing for attention. Mobile especially is very competitive and has strong drivers for differentiation of your products in the market.

To develop interfaces that delight like this—or must entice users to revisit or share—requires using patterns and best practices as just one input into the design process. The product must be understood holistically, and design options must be developed tangentially in order to discover the multiple ways to approach the solution.

While there is no single method or movement to achieve better results, here are a few concepts that lead in that direction and are worth considering for your design process:

Conduct validation exercises

Before you even start, perform user interviews, ask the business what they want, and gather any other information about current and expected usage that you can gather. Develop measurable objectives, stick to them during design, and be sure to measure them after the product launches. Without feedback, you cannot learn.

Use studio methods

The best ideas come from individuals or small teams working independently. To get the greatest number of good, unique ideas, task those individuals or small teams to develop quick, independent designs and regularly share and regroup, iterating to a final solution.

Realize that every idea is unworthy

When working with design concepts, from competitors to the design teams just mentioned, remember to approach the design from a modular point of view, and evaluate the suitability of each element to the overall goals and process. Do not just accept (or reject) whole designs.

Embrace your constraints

Whether in conceptual exercises, during workshops, or as individual designers, only work within the domain, set preconditions, and remind everyone that the goals and objectives define the desired end state.

Collaborate

Design teams will, ideally, have a variety of individual skill sets, or at least multiple individuals that each has her own background and opinions. Use the individual skills of the team members to find solutions and explore concepts.

Seek outside opinions

Not everyone has all knowledge of the arbitrarily complex systems we work on all too much, so cross-functional collaboration can have great value in confirming concepts, getting input on the viability of concepts, and discovering tangential solutions already considered or in progress somewhere else.

Using User-Centric Execution Principles

The other key problem with interactive design is actually getting the product built. To us, this is the new, more critical "gulf of execution" and the most important problem across the practices of user experience and interactive design.

What are needed are principles of what we can call user-centric execution. They are not yet a process, or a series of fixed procedures. It is possible that they may never be. But like the principles, heuristics, and patterns of design, the idea should be followed and there are best practices. First, let's discuss the principles.

To encourage successful execution or implementation, UX teams should:

Never walk away

Always stick with the project through development, at least making yourself available for questions, rework, changes, and testing. Ideally, become integrated into the team, and attend daily meetings, test planning, and so on. Plan to do this from the start so that your budget accounts for it.

Ensure that goals are for everyone

The business and user goals you should have developed at the beginning of the project must be translated into actual, measurable metrics. Make sure the whole organization has these goals as their top drivers, instead of cost savings, efficiency of developers, or other internal measures. While "we're building for the end user" may not resonate, remind the team that they work for the larger company, not just their department. You may also have to push to include the analytical tools to make sure they get built and are not forgotten.

Use object-oriented principles when discussing and delivering

The efficiencies and enforcement of consistency that componentized, object-oriented practice emphasizes in design are just as valuable to software developers and the development process. Sometimes this is just called "modular reuse" or something similar, as "object-oriented" is a larger set of principles. But the core concept is the same. Instead of designing every detail for every state, and building by state or building hundreds of items to bolt together, a few dozen modules are built and reused over and over in common templates.

Design with polymorphism

This is a subset of the preceding item in the list, but it is harder for some organizations and designers to grasp, so we've broken it out. If there are several variations of an on-screen module you design, make sure you express them as variations of one another so that they are clear. Of course, if there is only one variant (*omnimorphism*), it should be explicitly stated as well. Always keep efficiency and reuse in mind.

You should not find any of these processes to be burdensome. They should instead make for a much more efficient method in which to develop, and ensure that everyone on the team works hand in hand, at every level.

It is also important to keep this in mind even if you are a developer. Make sure you do not fall into developer traps, and keep yourself true to the design principles.

Patterns are a reference, and a starting point for design. Use them carefully to avoid being overly constrained, and use the principles of modular design to efficiently communicate and build the end product.

Principles of Mobile Design

Principles exist at a higher level than any pattern. They can be considered patterns for the patterns, if you will. Each pattern, and each detail of interactive or presentational design, should adhere to each of these principles at all times.

Each section and chapter in the book will begin with a discussion of the core principles for their sections, as well as other helpful guidelines that apply to those patterns.

Each of these principles could be discussed in great detail, and in fact we could have organized the book differently so that each of them was a chapter, with patterns associated to it. In the interest of clarity, the discussion of each of these is limited. If you are interested in further details on the rationale, these are generally discussed in greater detail within the patterns they apply to, as well as the chapter introductions.

Respect User-Entered Data

Input is hard. Users slip. You have a new phone, or are borrowing someone else's, and someone jogs your arm: suddenly minutes of typing is gone. Do whatever it takes to preserve user data—from saving as the user types so that auto-complete can bring lost input back, to not clearing forms on error, to planning for a loss of connection. Consider contexts and plan for crises and real-world behaviors, not bench tops and labs.

Realize That Mobiles Are Personal

Although security is important, there is no longer the need to assume that maybe the website is being viewed on a library computer. Mobiles can be presumed to be "one device for one person," and no one wants to have to regularly tell his device his name, location, or favorite music. Only implement passwords and clear personal information when required by law or regulation, and take other types of reasonable and transparent precautions to prevent misuse of information.

Ensure That Lives Take Precedence

Mobiles are contextual, meaning they are used alongside people's actual lives. Desktops (and some other devices) can suck people in, so you can go ahead and issue alerts that blink in the corner of the screen and they will be noticed. Mobiles are glanced at, used in gaps between conversation and driving and watching TV. They are even used to enhance these other experiences. So make sure they don't interrupt unless they have to. And if they have to, make sure they interrupt in a way in which the interruption will be noticed. A blinking LED, for example, is easily missed when a device is glanced at for a fraction of a second.

Realize That Mobiles Must Work in All Contexts

Make the device behave appropriately, or allow users to make the device behave appropriately, to make it work where they are. Most devices are too bright at night, making it hard to read that last email before bedtime, or to tell what time it is when the alarm goes off first thing in the morning. If the phone doesn't have a good way to change brightness, your app can override it or your website can just have a dark/light switch. Think about the context in which the device will be used.

Use Your Sensors and Use Your Smarts

Whenever possible, perform actions for the user based on sensors and user data. Why should you have to silence your phone for a meeting, when the phone knows where you physically are and knows from your calendar that you have a meeting in that room right now? Mobiles can be better than computers, because of their personal nature and their sensors. Use them.

Realize That User Tasks Usually Take Precedence

If the user initiates a task, and especially if the user is in the middle of a task, do not interrupt the user so that the task is ruined. When the user is typing an SMS, feel free to beep, but do not change the focus so that the user is suddenly typing in another field. And never cancel the operation to take the user to another page, losing her information.

Ensure Consistency

Whatever the rest of the application does, do that. And the application standards should follow the edict: "whatever the OS does, do that." Even if the OS does something dumb, it's probably what the user expects, so changing the paradigm generally results in more problems than solutions.

Respect Information

Although presentation and visualization can be used to clarify information, or view it in different ways, do not modify the fundamental truth for saving space, or because you do not understand the value of it. More information than you might expect rises to life/health/safety levels with the ubiquity of mobiles. Weather, for example, must be presented perfectly accurately. Know the difference between precision and accuracy, and understand implications of meter types, relative values, off-scale errors, and more.

Naturally, these will change over time. Just in the past five years we have changed or expanded these several times. Be aware of the reasons these principles exist, and keep abreast of the industry so that you are aware of changes.

Although we feel these principles are universal today, you are very free to disagree. Many others do, and they have their own principles, or variations on the understanding of what these mean. Just be sure you develop a set of design principles or objectives for your work or your project, and then stick to them.

Safari® Books Online

 Safari Books Online is an on-demand digital library that lets you easily search more than 7,500 technology and creative reference books and videos to find the answers you need quickly.

With a subscription, you can read any page and watch any video from our library online. Read books on your cell phone and mobile devices. Access new titles before they are available for print, and get exclusive access to manuscripts in development and post feedback for the authors. Copy and paste code samples, organize your favorites, download chapters, bookmark key sections, create notes, print out pages, and benefit from tons of other time-saving features.

O'Reilly Media has uploaded this book to the Safari Books Online service. To have full digital access to this book and others on similar topics from O'Reilly and other publishers, sign up for free at *http://my.safaribooksonline.com*.

How to Contact Us

Please address comments and questions concerning this book to the publisher:

O'Reilly Media, Inc.
1005 Gravenstein Highway North
Sebastopol, CA 95472
(800) 998-9938 (in the United States or Canada)
(707) 829-0515 (international or local)
(707) 829-0104 (fax)

We have a web page for this book, where we list errata, examples, and any additional information. You can access this page at:

http://shop.oreilly.com/product/0636920013716.do

To comment or ask technical questions about this book, send email to:

bookquestions@oreilly.com

For more information about our books, courses, conferences, and news, see our website at *http://www.oreilly.com*.

Find us on Facebook: *http://facebook.com/oreilly*

Follow us on Twitter: *http://twitter.com/oreillymedia*

Watch us on YouTube: *http://www.youtube.com/oreillymedia*

If you have any thoughts or questions you'd like to share with the authors, feel free to contact us. We enjoy communicating with our audience.

Steven Hoober
Kansas City, Missouri
steven@4ourth.com
@shoobe01

Eric Berkman
Sydney, New South Wales, Australia
eric@4ourth.com
@ericberkman

For the latest lists of reference materials, visit:

www.4ourth.com/wiki

Acknowledgments

Though we have both written extensively before, and even self-published a book once, this was much more involved than we'd have guessed going in. We couldn't have done it—and certainly not this well—without the assistance of a number of others.

Mary Treseler, our editor, championed this whole project and showed great faith in both of us, especially during somewhat challenging times, as the scope of the book began to grow and as we stretched the bounds of what an O'Reilly technical book normally is.

The various members of the production team at O'Reilly have also been extremely helpful in working through our unique demands and sometimes-mediocre writing. Without them, this would be a much less readable and sensible book.

We'd also like to acknowledge the efforts of the technical editors. Steven also edits other books, and is very aware of how much work this can be. Josh Clark, Dan Saffer, Jennifer Tidwell, Bill Scott, and Christian Crumlish all gave us excellent feedback, if sometimes painful to hear and difficult to implement.

Similarly, there were innumerable small conversations on Twitter, in blogs, via email, on LinkedIn or Facebook, and face-to-face at work or industry events. Dozens of people gave us encouragement or useful feedback, or asked for a feature to be addressed that we might have forgotten otherwise.

Matthew Irish helped us with many technical aspects, such as setting up the wiki and taking some of the many screenshots we needed at the last minute.

Our device collection has been invaluable for research and perspective, so we want to thank everyone who donated an old device to us. We'd especially like to call out Ed Madigan, who donated his desk drawer collection on leaving Sprint. Though they do not know they helped, the Surplus Exchange in Kansas City—which has the best electronics recycling program, perhaps anywhere—has minimal interest in phones and PDAs, so we were able to get some neat old gear for terribly low prices.

We'd like to thank Allan Swayze, who provided better soldering and wiring gear, and a power supply to help Steven get many of these old devices running. He also exposed us to all sorts of interesting industrial automation devices, adding a lot of hidden technology to our knowledge base.

Thanks also go to Paige Miller, who has been growing up just down the street from Steven's house, and, along with her friends (especially Audrey and Lily), let me observe their phones, interview them, and perform free user tests. They have provided great insights into the youth market.

Thanks to Jesse Schifano and Mike LeDoux, for letting me put their exploration of an iPad kiosk in here, and to the rest of the Ai design team for only occasionally making Steven come out to dinner instead of writing or editing all night long.

And of course, thanks to our friends and families for putting up with almost a year of spending every bit of spare time writing, editing, photographing, drawing, and editing more in a seemingly endless cycle.

PART I

Page

The *page* is the area that you will spend your time designing for any application or website. A part of it is visible in the viewport of the mobile screen during its current state. There are states and modes and versions to be considered, as well as addressing what is fixed to the page, what can float, and what is locked to the viewport.

Based on cultural norms of reading conventions and how people process information, you have to design elements for the page, and place items on it in ways your users will understand. You also want to create information that is easy to access and easy to locate. Your users are not stationary, nor are they focused entirely on the screen. They're everywhere, and they want information quickly and to be able to manipulate it easily.

Unlike later parts of this book, which cover broader topics, the Page patterns that we will discuss here are contained in only a single chapter:

- Chapter 1, "Composition"

Helpful Knowledge for This Section

Before you dive into the patterns, we will provide you with some extra knowledge in the section introductions. This extra knowledge is in multidisciplinary areas of human factors, engineering, psychology, art, or whatever else we feel is relevant.

For this particular section, we will provide background knowledge for you in the following areas:

- Digital display page layout guidelines
- Page layout guidelines for mobile users

Digital Display Page Layout Principles

The composition of a page has to do with the assembly of components, concepts, content, and other elements to build up the final design. As we already know, consistency is important, so these elements are not placed arbitrarily on the page, or even just as the one page dictates, but on rules across the system, or even the whole OS.

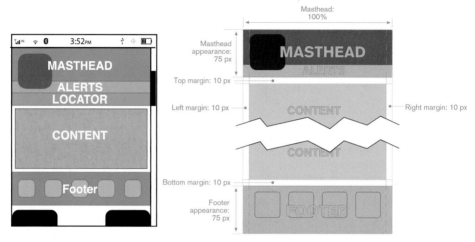

Figure I-1. *Developing a common grid and hierarchy of information for the entire application or site is key to a consistently usable and consistently branded experience. After a grid such as this is created and wrapper elements are defined, a series of templates can be created, and then individual pages and states.*

At the highest level is the *grid*. This is a regularized series of guides, defining the spacing and alignment of the main elements on all the pages in the system, as shown in Figure I-1. Some rules are inviolable, such as margins, and some are specifically designed to offer options for special page layouts, or unexpected future changes.

From these are developed a series of templates, from which each page in the application, site, or other process will be designed and built. This encourages a consistent user and brand experience that supports content organization and layout, advertising requirements, navigation, and message display characteristics such as legibility and readability.

Patterns within Chapter 1, such as Fixed Menu, Revealable Menu, Notifications, and Titles, are repeated at the same place on each and every page, and so reside in each template. A subsidiary concept is that of the *wrapper*, which defines these common components—and others, such as scroll bars—so that they are consistently designed and built for each and every page template.

Mobile users have specific tasks and goals. They require that the information be quickly located and effectively organized. Therefore, the page layouts need to reflect the mental models and schemas understood by users. If you ignore these, you will end up with situations such as that shown in Figure I-2. Your users can become frustrated and dissatisfied with their experience, create miscues and errors, and maybe even give up!

Furthermore, using page layouts wisely allows you to organize and place content effectively on valuable screen real estate, where every pixel is important.

Page Layout Guidelines for Mobile Users

Figure I-2. *If you don't follow the principles of a grid and templates, you will end up applying components in overly variable ways. Here, the title is far below the tab, separated by the banner ad. On the search page, there is a search dialog above the title, pushing it farther down. On the home page, the addition of the "more" icon makes it unclear whether "Top Stories" is the page title, section title, or something else. And in Settings, the new style of banner draws the eye, so it seems, for a moment, that this is the title.*

Here are some page layout guidelines to follow:

- Mobile screen real estate is valuable. Avoid the use of banners, bars, images, and graphics that take up space without any specific use. This use may be secondary, such as communicating the hierarchy or structure, but the designer should always be able to describe the reason.

- Lay out elements within a design hierarchy. There are optional versions, and some interactive types insist that time is another component, but for simplicity, the hierarchy is Position→Size→Shape→Contrast→Color→Form. The most important items are larger, higher, brighter, and so on.

- Consider the Gestalt laws of Closure, Continuity, Figure and Ground, Proximity, Relative Size, Similarity, and Symmetry. These are discussed further in Part II, "Components."

- Use consistent and simple navigation elements. People have limits to the amount of information they can store in their short-term memory. Therefore, they automatically filter information that is important and stands out. Information elements that are excessively displayed and irrelevant will be ignored and overlooked.

- Wayfinding is really rooted in real-world navigation, like getting around town or finding the right room in a building. Kevin Lynch, an environmental psychologist, established five wayfinding elements that people use to identify their position: Paths, Edges, Nodes, Landmarks, and Districts. These same environmental elements are also referenced when navigating digital content on websites or mobile devices. Page numbers, titles, headers and footers, tabs, links, and more provide a lot of help that we've inherited almost as a whole for interactive design.

- Consider how users will view your page when plotting content. Generally, users will look for high-priority information in the upper left of the content area (Nielsen 2010).

- Multicolumn text is not used to meet some design style, but to restrict line length. And line lengths are not based on fixed sizes, or even percentages of page width, but on character count. Long lines are harder to read. Around 60 to 65 characters (on average) is the maximum length you want to use. A definition for what is too short depends on how many long words you have.

- Titles describe pages, elements within a page, and content sections. Use them consistently and appropriately.

- Default text alignment is left. For right-to-left UI languages, default text alignment is right.

- Try to use bulleted information instead of a table.

- The term *false bottom*, or *false top*, is specific to interactive design and refers to users thinking they are at the end of the content and not continuing to scroll. If text flows from one column to the next or one page to the next, it must be designed so that the relationship between the columns or pages is clear. "Continued on page 86" is all but a hyperlink from the past, which has been inherited by interactive; "Read the rest of this blog post" is basically the same thing.

- Interactive systems have an additional challenge in that the page might be larger than the screen (or *viewport*, as we often call it).

Getting Started

You now have a general understanding that a page is an area that occupies the viewport of a mobile display. Pages can use a wrapper template to organize information consistently across the OS that will allow for a satisfying user experience. When making design decisions you must consider everything in this page section, even if you don't have control (i.e., you're not building an OS). You must know what it might do to your design. Consider your user's goals and cognitive abilities, page layout guidelines, and the importance of legibility and readability in message displays.

The following chapter will provide specific information on theory and tactics, and will illustrate examples of appropriate design patterns. Always remember to read the antipatterns, to make sure you don't misuse or overuse a pattern.

Composition

A Little Bit of History

To many people the year 1440 signifies a major shift in global communication. It was during this time in Mainz, Germany, that a goldsmith by the name of Johannes Gutenberg invented one of the most important industrial machines of the modern period: the printing press.

The printing press's use of movable type was inspired by earlier uses found in China and Japan as early as the 7th century. During this time, printers used a method of block printing, which involved a carved piece of wood used to print a specific piece of text.

Further advances took place in the 11th century. A Chinese alchemist, Bi Sheng, invented a process called *movable type*. His process consisted of having individual Chinese characters carved on blocks of clay and glue. These blocks were arranged on a preheated piece of iron plate where they were pressed on paper. Bi Sheng's process was not without limitations, however. The process was slow and was not advantageous for large-scale printing, and it relied on clay blocks, which created problems with the adhesion of ink.

A Revolution Has Begun

Gutenberg's invention advanced the process of movable type further, and consisted of individually cut or cast letters, sorted onto composing sticks, locked up into galleys, and then inked and impressed into paper. This, the invention of modern typography, marked the birth of mass printing. It allowed information to move from merely permanent and portable to the first mass-media product. It allowed for perfect replication, standardization, and affordable books for the middle class.

Composition Principles

Figure 1-1. *Just some of the variety—and similarity—of Annunciator Rows, Notifications, Menus, and other device-wide features built into mobile devices*

The invention of the letterpress not only allowed for mass production of existing content, but made it so easy that there was demand for more content. And with that content was a need for well-understood page composition principles, not just those handed down secretly by cloistered monks transcribing old works.

So composition became a process of assembling a layout that consistently arranged components and content on a page. These rules were repeated on all other pages, creating a recognizable system of component relationships that were understood by social reading norms.

This helped readers to understand why a composition element was arranged within a specific part of the page. Readers could then expect to find that same element on all other pages within the rest of the book.

These composition principles made books usable for the first time. Mass consumption meant the addition of scientific texts, and reading for entertainment, and portable books that could be read anywhere, by anyone. Literacy rapidly grew as well, from less than 30% to more than 90% in the 20th century. Users adapted to the technology as much as the technology adapted to them.

As type principles became standardized, so did binding, type and page sizes, and then margins and gutters. Page numbers, titles, and chapters on each page followed over time.

These standards became promulgated as best practices, and were implemented as grids (to which everything was aligned) and templates, which were used on every page to make volumes feel like single works.

Using templates is essential in mobile design. As designers, we want to create our layouts based on cultural norms of reading conventions and how people process information. We also want to create information that is easy to access and easy to locate. Our users are not stationary, nor are they focused entirely on the screen. They're everywhere, and they want information quickly and to be able to manipulate it easily.

The Concept of a Wrapper

Throughout this book, we discuss patterns from which you can make the very specific templates that you can use for any particular product or project.

The templates that are used across a product, on most every page of a website or application, we call a *wrapper* because they enclose (wrap around) all the other components and the content.

Considering design from the wrapper down allows:

- The designer to organize information within a consistent template across the OS

- Information to be organized hierarchically on a page

- The user to identify the organization structure, quickly increasing learnability while decreasing performance error

Grids are also important to consider in design, but they are unique to each project and are beyond the scope of this book. They are discussed in many general design books and web tutorials; if you are not already familiar with the principles, you can use print or desktop web principles too.

Context Is Key

Figure 1-2. *The lock screen on this device is as informative in presentation, and gestural in interaction, as the rest of the experience. Even notifying the user of an error on entering the code is organic to the design. Apply your interface and interaction paradigms as broadly as possible.*

Wrappers must be designed based on the content and the context of their use as much as any other part of the product. A wrapper for a mobile phone application will be quite different from that for a portable GPS, or a kiosk. When determining which information belongs in the wrapper, you must decide on a multitude of things regarding context of use:

- The technological, functional, and business requirements and constraints
- Where the context of use is occurring
- The goals of the users
- Which tasks are needed to achieve these goals
- What types of information must be displayed to achieve each goal or task

Patterns for Composition

Using appropriate and consistent wrappers will create mappings and affordances that will allow for positive user experiences. Figure 1-1 shows a selection of key components. Within this chapter, we will discuss the following patterns based on how the human mind processes patterns, objects, and information:

Scroll
> When information on a page exceeds the viewport, a scroll bar control may be required to access the additional information. Scrolling of information should almost always occur along one axis, except in rare cases.

Annunciator Row
> This displays the status of hardware features on the top of each page. The status of functions that may be displayed is radios, input and output features, and power levels.

Notifications
> When an alert requires user attention, a notification will occur in some form of visual, haptic, or audible feedback. These notification displays must allow for user interaction.

Titles
> Pages, content, and elements that require labels should use titles. These titles should be horizontal, be consistent in style, and follow guidelines of legibility and readability.

Revealable Menu
> This type of menu displays additional menus that are not immediately apparent. A gesture, soft key, or on-screen selection will cause these menus to immediately display on-screen.

Fixed Menu
> This type of menu presents an always-visible menu or control that is docked to one side of the viewport. This menu is consistently placed throughout the application. These interactive controls are most likely icons with textual coding.

Home & Idle Screens

These screens are used as display states when either a device is turned on or an application has exited, timed out, or returned to a device-level menu display.

Lock Screen

Mobile devices use this display state to save on power consumption. When necessary, the application's sleep state may become locked to protect the security of the data the user has input. Additional user interaction is required to exit out of the lock screen, as shown in Figure 1-2.

Interstitial Screen

This type of screen is used primarily as a loading process screen during device or application startup. Wait indicators may be used to show loading progress.

Advertising

When advertising is used within a mobile application, the advertisement must be distinct and must not affect the user experience. Obtrusive advertising could prohibit the user from achieving his task-based goals. Advertising must adhere to the specific guidelines set by the Mobile Marketing Association (MMA).

Scroll

Problem

More information is in the page or element than can fit in the viewport. You have to provide a method to access this information.

Usually, the OS provides this function. Certain behaviors will occur automatically, but in application design especially, you may need to customize your interaction and interface to work in the best possible manner.

Solution

Figure 1-3. *Scroll indicators may be complete bars, or simple indicators floating over the content.*

Scroll bars (Figure 1-3) have long been used in information display systems of all sorts. For mobile, you will display them to indicate the scrollable axes, and the relative position within the scrollable area.

Due to the scale of mobile devices, as a general rule you should not allow the scroll bar to be manipulated directly. Instead, allow the content to be grabbed directly by a gesture, or make the entire area movable via dedicated Directional Entry keys.

You will find that scroll behaviors are key to interaction with mobile devices. As with all the patterns around page Composition, Scroll will be mentioned in most of the patterns in the rest of the book. It is especially relevant to the list and list-like Display of Information patterns:

- Vertical List
- Infinite List
- Thumbnail List
- Fisheye List
- Carousel
- Grid
- Film Strip

Although scrolling does occur in other patterns, do not confuse it with other patterns, such as Infinite Area. That pattern does not use scroll bars due to the arbitrarily large data set presented.

Variations

Whenever possible, you should make sure scrolling takes place on a single axis. This is why the whole set of patterns based around Vertical Scroll exist. When the situation demands, such as for zooming in to content which otherwise fits the area, you can also offer a secondary scroll axis. Keep the secondary axis in mind throughout the design process; it may help you to understand how to avoid behaviors that may lead to confusion or cause the user to become lost when scrolling.

Figure 1-4. *A thumbnail may be a better way to indicate location within a large area, and may also be used to jump to other regions. Here, a full desktop-view website is in the corner, while a zoomed-in view of a small portion occupies the rest of the viewport. An indicator on the thumbnail indicates the current position and relative size.*

In rare cases, both axes of movement may be equally important. When zooming in on images, for example, the information is equally important in all directions.

Regardless of whether you are making a single-axis scroll or just selecting a primary axis, vertical scrolling is usually the easiest for users to understand and use. This is a result of language, where text uses horizontal space inefficiently, so additional items must be displayed in the vertical axis. See the Vertical List pattern for more discussion. This also means users are most familiar with vertical scrolling items, and so react best to them, though this may change over time.

Using the horizontal axis is occasionally useful for certain kinds of data, but is most useful when you must present a subsidiary scrolling area. If you have a vertically scrolling page of information, you can provide areas which allow for scrolling horizontally. This can provide additional clarity, and removes conflicts around in-focus scrolling that we have all encountered. (Think of what happens when you encounter a large form area in the middle of a web page.)

Whenever possible, you should display scroll indicators. Try to solve space and clutter problems with options presented in this pattern, and not by removing the indicators entirely.

For multiaxis scrolling especially, you can instead provide a thumbnail of the entire available area, as shown in Figure 1-4. This is usually, but not always, in addition to the scroll bars. The thumbnail shows the current viewport as a subset of the total area.

Only allow items that are in focus to scroll. For scroll-and-select devices especially, be sure this is clearly communicated and strictly enforced. See the Focus & Cursors pattern for a discussion of communicating focus.

For scroll bars on touch or pen devices, it is almost always best to not allow the scroll bar itself to be directly manipulated. Handset-size devices are too small, so this would encourage accidental input on the screen. Instead, the entire page is scrolled by gestural movement. Use care with detection of gesture to avoid selecting items on the page.

When designing for devices with larger screens, you may be able to add actionable scroll bars. Often, these still are distractingly large, so you may want to make them appear only conditionally. A similar function, suitable for many sizes, is discussed in the Location Jump pattern.

Scroll-and-select devices will use a scroll key pair, or more often a five-way pad of some sort. Scrolling may be by item (line by line, or jumping link to link) or by moving a pointer on the screen. When you use a pointer, do not allow it to get too close to the edge of the screen; scroll when one-third to one-fourth of the way across. Of course, at the limits of the displayable information, it may approach the edges in order to select all items in the screen.

For item scrolling, do not allow the user to jump past content. For example, when viewing a web page, if the primary method jumps link to link, when there is a large area of content with no links, temporarily suspend this and scroll a few lines at a time so that all content can be seen.

You can also provide item scrolling as a secondary method. If you have set the Up and Down scroll keys to move line by line, the unused Left and Right keys may be set to jump from link to link.

When using the five-way pad, users should always be able to press and hold in one direction for a short time to accelerate scrolling. The absolute speed depends heavily on the amount of content. Scrolling must be slow enough to allow the user to see his current position; otherwise, he will not know when to stop scrolling. This is extremely expected behavior, and so is noted when not included.

For touch and pen devices, inertia scrolling has also become expected behavior. If the user's finger (or pen) initiates a drag action, and departs the screen while still moving, the screen will continue scrolling at the departure speed until it is stopped by another form of input. It is usually best if you configure this to simulate friction so that the scrolling gradually slows over time, but do not overdo this deceleration if the list is very long.

If a thumbnail of the total content is provided, this may also be interactive. For touch and pen devices, tapping another area of the thumbnail will jump to that area. For scroll-and-select devices, Accesskeys may be provided to allow jumping to a region of the displayed area.

You can also use Kinesthetic Gestures, such as tilting the device, to scroll pages. These are not very common, so there are few standards regarding their implementation, and users must generally be instructed in their use.

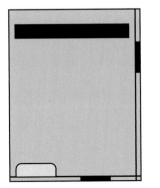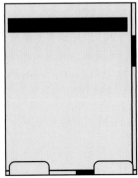

Figure 1-5. *Two types of two-axis scrolling. For items like images where both axes are equal, scroll bars must be equally easy to see and use. Make sure they are not obscured by options menus and other items, as shown to the left. For information oriented mostly along one axis, the other axis is secondary, and the scroll bar may be obscured as needed. On the right, the soft keys are not always visible, but they do sometimes occlude the horizontal scroll bar.*

Presentation details

Scroll indicators are not usually used in mobile devices to enable scrolling, but to:

- Provide an affordance (communicate the function) that the area is scrollable
- Convey the current location within the total content
- Indicate the relative amount of information the viewport displays, as a ratio of the total content

Always be sure to display an indicator of position. This may be hidden for full-screen views, when other information is hidden. But be sure to display the scroll position indicators whenever the user is interacting with the content, and especially whenever the user is scrolling.

Scroll bars may be made very small, or be obscured by other elements in some cases, as in Figure 1-5. Effective, visible scroll bars can be as small as two pixels wide: a single pixel line defines the outside of the bar, a single pixel line defines the inside, and a different colored area is for the current position. Naturally, this depends on the resolution of the display, so larger bars will be needed for very high-resolution displays due to the small pixels; for devices where the screen may overscan (such as for TV output) or the bezel is large enough, it may occlude the screen.

You can also eliminate the scroll bar itself, and just use the scroll indicator alone. A small tab can be anchored to the edge of the screen, protruding into the content a small amount. This must be larger (at least 5 to 10 mm wide), but since it floats on top of content, it doesn't take any actual room from the display area, unlike a traditional scroll bar.

The position of the scroll indicator within the viewport indicates the position of the viewport relative to the total scrollable area.

The height of the scroll indicator, in either case, should reflect the ratio of the viewport to the total scrollable area. For scroll bars, this should use the viewport as the 100% scale, so if very little is outside the viewport, the scroll indicator is almost as large as the viewport. For indicators without scroll bars, the size of the indicator must often remain relatively small to avoid obscuring information; a relative size change may still be possible, but will not be as effective at communicating the relative scale to users, or might obscure too much content.

When you have a scrolling area displayed, but it is not in focus so the scroll actions will have no effect, be sure to communicate this to the user. Generally, simply removing the scroll indicators will do, but often the secondary value of the indicator (position and size) is still useful. When displaying a Pop-Up layer, simply graying out the scroll bar in the parent window may or may not be sufficient. Additional changes to the scroll indicators may be needed.

Antipatterns

Do not allow users to become lost in the scrollable area. Especially be sure to not allow scrolling single-axis lists so far that no content is visible.

Consider anchoring secondary axis scrolling to an edge when scrolling on the primary axis. Two-dimensional scrolling is often very difficult to achieve with precision.

If you have to use multiaxis scrolling to show the information, do not assume it will be easily understood. Try it out, and add additional help or more prominent scroll indicators as needed.

Whenever possible, avoid vertically scrolling areas within other vertically scrolling areas. For example, a form text area within a page will, when it is in focus, stop the page scroll and grant all scrolling to itself until it has scrolled all the way to the bottom, then may allow the page to regain scroll focus. This is confusing, and often leads to errors. Avoid it instead.

For touch and pen devices, avoid drag-and-drop interfaces, or other interactions that require dragging an element within a scrollable area. If required, consider multifinger On-screen Gestures or Press-and-Hold interactions as a method to initiate a mode switch.

Be sure to support all the available input methods for the devices you will support. Touch and pen devices which also have five-way pads must be able to be entirely used (as far as scroll and selection goes) with the five-way pad. If there are multiple types of hardware for your target audience, be familiar with all of them, and do not just design for your handset.

When scroll indicators are used on an Infinite List, the location and relative size must reflect the total size of the available data. Do not allow the scroll indicator location to be based on the currently loaded data, and therefore to constantly change as new data is retrieved. You can do this by getting a count of the total number of items, and loading empty containers of the full size; only load data into them when demanded, instead of loading the container as well.

Annunciator Row

Problem

You must provide an easily discovered display of the status of important hardware features such as battery level and network connections.

The OS provides Annunciator Row features, but you may usually modify or suppress the Annunciator Row within applications.

Solution

Figure 1-6. *The Annunciator Row is a strip anchored across the top of the viewport. Note that the scroll bar stops at the Annunciator Row, as it does not scroll.*

Annunciators are lights, gauges, or sounds that indicate system status. Annunciator lights and panels date to the dawn of electrical devices.

The term was carried over to electronic design, and then to mobile OS design. Though the same feature is often referred to as a "status bar," this typically implies some overlap with the concept of Notifications, which we will discuss in the next pattern instead.

A bar, as shown in Figure 1-6, is displayed along the top of every screen with a series of iconic representations of the status of the device. You should always use common representations and placement of icons, so your users can understand the key indicators on any device without having to learn the specifics of every device.

Variations

The Annunciator Row is present on all screens. It may only disappear or be of lower prominence when other controls disappear as for full-screen game play or video playback.

Certain devices may not require using the space for constant reminders of system status messages. Appliances such as eReaders do not necessarily need an Annunciator Row as their battery life is very long, and network access is needed only intermittently. The status messages will still be needed, and you may be able to solve this with something like Notifications repurposed to display such information when it becomes critical instead. See that pattern for details on this functionality.

Kiosks and other devices where the user does not have full control of the device also may not need to display an Annunciator Row to general users.

Interaction details

Annunciators are notifiers only. You should generally not allow direct interaction with the items.

For touch and pen devices, it may be desirable to allow the user to select the Annunciator Row as a whole, in order to get more information, or to provide access to settings. You may also accomplish this by combining it with the Notifications area.

Presentation details

You should plan to display the default Annunciator Row on every screen. Make a careful, deliberate choice whenever hiding to regain screen real estate or to declutter the screen. Hiding the status icons is generally appropriate for video playback, most games, and many slideshows or similar interfaces. Browsers and readers of ebooks, PDFs, and the like may also benefit from hiding the Annunciator Row once reading or scrolling begins.

The Annunciator Row is generally displayed as a row of icons—as in Figure 1-7—laid out on a strip of color, gradient, or other background imagery to separate the icons from other, generally interactive display elements. Scroll bars do not intrude into the Annunciator Row as it does not move, but remains fixed at the top of the viewport.

Many devices allow the Annunciator Row to be modified. In general, you should restrict these modifications to display changes, such as switching out the background color, and changing icons to match. You can also use this modification as a half-measure, when completely hiding the status messages might not be the best option. For example, camera applications often use great amounts of battery power, so you could display the battery icon, but no others in order to leave more room for the live preview. If you do this, use the conventional position of the battery icon, simply leaving out all others.

Figure 1-7. *Common icons for the vast majority of conditions shown in the Annunciator Row. All items are enabled and at maximum graphical mode. This is an example; some are in conflict with one another, so this would never be seen. From left to right: Mobile network, WiFi, Bluetooth, NFC, Airplane mode, Audio level, Locked, Clock, Network activity and speed, Voicemail waiting, Synch, Location, USB connected, and Battery status.*

Within the row, the status messages are displayed as icons, with as few words or numbers as possible. Use common, universally understood or industry-standard icons whenever available.

Icons do not indicate the presence of a feature, but the status of that feature. No display means the icon is not functional, and displaying the icon means it is enabled. Optionally, disabled features may be displayed as grayed-out icons. This can be beneficial to communicate the availability of some features. Use caution to ensure that these are clearly disabled under all lighting conditions.

Whenever possible, you can add additional status messages to the icon, such as bars of signal or battery level, as shown in Figure 1-8. Use simple changes and well-understood signaling such as tall = more and red = bad.

Figure 1-8. *A series of exemplary statuses for the battery, from full to empty, then charging. Using the exclamation point in the icon is clearer than blinking the icon, and is a second code for users or conditions where red is not visible. The power plug icon is clearer to many users than the often-used lightning bolt.*

For certain features, whose presence is assumed, you must include explicit measures of their disabling or failure. For example, when no signal is available, the mobile network will have an error "X" in place of the signal bars.

Items are grouped by basic functionality. A conventional order has arisen, from left to right:

- Radios:
    - Mobile networks
    - WiFi
    - Bluetooth enabled, and active
    - NFC or contactless payment enabled
    - IrDA or other nonwired networking as available
    - Airplane mode
- Input and output:
    - Volume, vibrate, or silent mode
    - Screen or keyboard locks enabled
    - Network activity
    - Network speed
    - Message Waiting Indicator for voicemail, unless this is displayed by the notification area instead
    - Synch status or activity
    - Location services enabled; may or may not indicate when GPS is active
    - USB cable connected
- Power:
    - Usually a single item, which changes based on charge level and state (e.g., being charged)
    - A second battery indicator, as may be displayed on those now-rare devices with outboard (piggyback) or secondary batteries

The time of day (and sometimes the date) is also present, but may be in any of several places in the row. The most common is centered, followed by right-aligned. Time is always displayed, even on those few devices without an otherwise permanently visible Annunciator Row.

Naturally, features not included with the device are not given space in the display. Some items may share space, and the highest-priority feature or the one with the most important message is displayed.

Don't let the order or size of the row, or the details of the icons, change with different screens. Use one layout and one type of icon in all situations.

Don't reinvent the wheel. Reuse existing good design concepts so that users do not have to relearn your icons. How many of those in Figure 1-9 are immediately understandable?

Figure 1-9. *These are just some of the many ways battery charge level is depicted on mobile devices. Many are quite unreadable. Try to pick simple, easy-to-understand symbols, and reuse common icon styles from existing products and best-in-class examples.*

Except when notifying users of special conditions in places where the rest of the bar is suppressed (e.g., battery on a camera screen), do not pick and choose which items to show. Always show the same set in the same manner.

Avoid explanations that are jargon-laden. The percentage of usable battery is not nearly as useful as an estimate (even a bad one) of time remaining on a battery.

Avoid animations as sole explanations. Mobiles, and especially the status area, are often only glanced at. Blinks will instead be seen as solid on or off at a glance.

Notifications

You must provide a method to notify the user of any notifications, of any priority, without unduly interfering with existing processes.

The OS generally provides notification features, but they can sometimes be overridden. If you're building a notification-sending application, it must correctly interact with the system. Certain applications and sites can also call for their own notification processes that have nothing to do with the OS notification system.

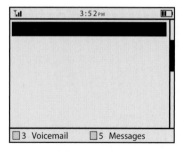 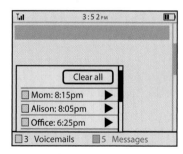

Figure 1-10. *Since mobiles generally have limited space, and notifications must usually be assumed to be secondary to the current process, even a dedicated notification area should be out of the way. When the notification area is selected, allow users to access more information, as on the right, where each item is described further.*

A single, consistent notification method should be provided across the entire OS. Make sure this method does not interfere with any processes the user is currently involved in. The user must be able to act on or dismiss the notification very easily.

When notifications are restricted to a specific subset of a device, such as an application, make sure they follow the same principles and do not conflict with the OS-wide notifications.

Multiple notifications must be able to all be displayed in a single view, so no notification obscures another.

Multiple variations of this pattern are available.

A dedicated notification strip or area may be used, with a portion of the viewport dedicated to notifications. This may be partly or completely hidden when no notifications are present, but will fade in, slide up, or otherwise appear when notifications are present. Figure 1-10 shows one example of this. The selection of any notification item will display the item in the application that hosted the notification. This may be difficult to use for scroll-and-select devices, so it will be most commonly encountered on touch and pen input devices.

Notifications may be combined with the Annunciator Row, as shown in Figure 1-11. This is, in fact, very common on basic and legacy devices, where the envelope icon means new voicemail or SMS messages have arrived. Additional notifications may be added to this area as well. These legacy utilizations are not Notifications as far as this pattern is concerned, as they are not generally interactive; the user must take separate action to dial into the voicemail system or open the text messaging application.

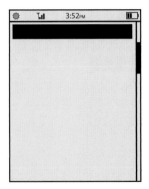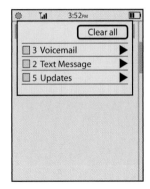

Figure 1-11. *The Annunciator Row commonly houses notification icons, as on the left, but may also be used as a method to access details or view the items. On the right, the user has tapped or pulled the Annunciator Row down to reveal a notification area.*

Especially for touch and pen devices, the Annunciator Row may include a notification area. You may place as many notification icons in the Annunciator Row as will fit. When it is opened, a complete list of all the current notifications will be revealed.

For scroll-and-select devices, and certain other cases (if there is no good place to put a notification area, or if the Annunciator Row is unsuitable for repurposing), a simple Pop-Up dialog box—as shown in Figure 1-12—may appear over the current context whenever a notification appears. This is a special case of the Confirmation dialog, which should be seen for behaviors and layout.

Currently, regardless of the other notification paradigms, incoming voice calls use a completely unique notification method, and automatically launch the phone application full-screen. This is a holdover, and may not be a permanent condition. Some OSes already seem to be addressing this by placing a "current call" icon in the Notifications area when in another application. Selecting the Notifications area allows the user to see certain details regarding the call, and switch to it rapidly. Using any of the aforementioned notification methods could provide a suitable a method of informing the user of the incoming call, and of accepting or declining it.

Interaction details

A key attribute of the Notifications pattern, as distinct from the Annunciator Row pattern, is the ability to directly interact with the alert. The user must be able to see each item individually within the notification method, and select it for viewing (or other suitable actions), or be able to dismiss it.

When more than one current notification is available, the method must support the ability to show all of them at once. If the Confirmation dialog style is used, for example, a special version must be made where a list of all the notifications is displayed instead, and selecting one will open a new dialog with the details and actions of that item alone. Within this list, batch operations, such as dismissing all notifications, should also be available.

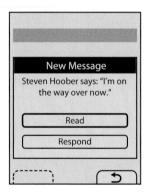

Figure 1-12. *Pop-Up notifications are used when other methods are inappropriate due to interaction limitations, or other spaces are occupied with too many other functions.*

When the Annunciator Row is used to access notifications, tapping or dragging the row will reveal the list of notifications. You should choose a method for the reveal gesture that is based on other gestures used in the OS, so it is discoverable and understandable when described and demonstrated.

Various actions must be available for any notification item, or for a single notification Pop-Up. See the Confirmation pattern for details on the dialog especially. Consider which actions might be helpful in reducing the number of steps the user must take. For example, instead of just "Read" and "Dismiss" buttons for a new SMS message, a "Reply" button could be added, if the message summary is enough to elicit a response. This will allow the user to act immediately. Use caution not to add too many options, cluttering the selections and reducing the size of displayed information and options.

When a notification is acted upon (e.g., the new text message is read), suspend any current operations or actions and save all user-entered data. When the action, application, or process initiated by acting on the notification is completed, return to this previous condition.

Presentation details

Fixed notification areas most commonly appear as a strip along the bottom of the viewport, to differentiate them from the Titles and Annunciator Row elements. The area should be at a fixed location within the viewport, and should not scroll. Scroll bars will not overlap the area when it is displayed. Individual notifications will be displayed as line items, or may be grouped by category, with counters indicating the number of notifications for each category.

Smaller devices, such as most mobile handsets, cannot dedicate this space, so they most often collapse the notification area to an icon or other small area within the Annunciator Row.

Labels for each notification item must be clear and comprehensible. Always state the service or application initiating the notification; usually, an icon will be sufficient for this. If a summary of the message cannot be displayed (such as for voicemail, or an MMS message with no text), do not display jargon or difficult-to-parse information such as the sender's phone number; use clear descriptions instead, such as "New voicemail." Whenever possible, display relevant information; look up senders of SMS/MMS messages in the address book and, when found, display their name instead of the phone number.

If summary information can be provided, such as the content of a text message, this may also be displayed. This is a rare case where *marquee* text, which scrolls within the available space, may be used to good effect. Text messages are short enough that such text display tends to be readable, and allows the user to read the entire message rapidly, without opening the messaging application.

Additional information may also be derived from the notifying application, or other handset services. For the case of a new SMS message, any avatar icon for the user may also be displayed, to assist in understanding what is being sent at a glance.

To keep the list of notifications from becoming too long, you should usually cluster them. Instead of a single line for each notification, display one line for all SMS messages, one for all email, and so on. Each line then has a counter of the number of items within it. The OS may limit the behavior you can implement with this. For example, you may not be able to reveal individual notifications, and you may have to load the application sending the alert.

For most notification types, you should also blink the LED and sound Tones and Haptic Output (vibration). These must be customizable so that the user can determine which types are critical enough, and they must follow the system-wide output settings to respect silencing. See the relevant patterns for additional details.

Antipatterns

Do not display notifications serially. If more than one is received at a time, use a multiple-notifications method, instead of showing one single notification after another.

Do not allow notifications to prevent access to other systems, even temporarily. The notification system must allow individual notifications as well as all current notifications to be dismissed.

Most media-centric activities, such as video playback, should not be interrupted by notifications. Very high-priority notifications may still interrupt, but must either pause playback or be very nondisruptive so that playback can continue during the notification.

Never display notifications to external display devices, such as TVs or projectors attached to the device.

Be sure you understand and follow the OS's method of notification and of marking messages or other notifications as having been read or accepted by the user. Dismissing a message from view in the notification area may or may not mark it as read in the application sending the notification. There is, as yet, no consistency in this regard, so no pattern can be determined at this time.

Titles

Problem

You should always label each key element to make context or process completion clear.

Titles are a key part of all OSes, applications, and web standards, but it is always up to you to include them in the design in the appropriate manner.

Solution

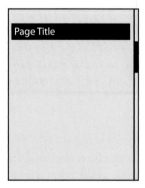 

Figure 1-13. *Titles should always be attached to freestanding elements, such as pages, windows, or pop ups. Follow the OS design guidelines for the use of title bars.*

Pages and elements or content sections within a page should almost always be labeled. Pop ups and other freestanding elements should have titles similar to the page-level title. Figure 1-13 shows both of these.

You should make a special point to use Titles in a consistent manner. Consider the size, location, content, and type style. The simplest method for the page level, especially for applications, is a straightforward title bar. We show this in almost all the diagrams in this book.

Titles are always horizontal, and any top-level title should be boxed, or otherwise separated out to make it clear that it is a key element.

Subsidiary titles are also text, but the text can be stylized as needed (bold, color, etc.) or additionally include boxes, rules, indents, or other graphical treatments to differentiate them from the remaining content, and to more clearly communicate their hierarchy.

If the OS calls for it, you should make the title of the running application display in a special title style. If so, there is no need to repeat the application name within the page title area. In some cases, the application title bar may disappear shortly after the application is loaded.

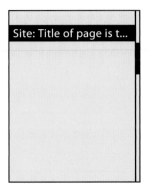 

Figure 1-14. *Page titles can be selected, by tap or drag usually, to reveal alternative information or functions. Here, within a web browser, the page title can be selected to show and edit the URL, and perform other browser functions.*

Titles are not required to have any interaction.

A typical interaction for titles of modules or sections is to make the title a link to another page.

The title bar of a full-viewport page can be used to reveal additional information (tap or drag down to reveal functions such as URL entry for a web browser, as shown in Figure 1-14) or as an anchoring element for very long pages (tap to return to the top of the page). These are generally only useful for touch and pen devices; for scroll-and-select input, having to scroll past the title is generally additional effort to be avoided for secondary functions such as this. If needed, place them within option menus.

Whenever possible, you should follow OS guidelines for title design. Though naturally this applies to applications, even web design should follow these guidelines when they can be targeted to the OS level. Even if significant changes are required for unique branding of your product, make title bars of similar typeface and style, size, shape, and position to the rest of the OS. For example, if the OS uses white title bars and black text, you can easily use a black bar and yellow type, and it will look consistent enough if the size, style, and position are identical.

Titles can include icons. Try not to be needlessly repetitive. When available, use a more specific icon instead. For example, in a Pop-Up, don't repeat the application icon, especially if the window behind it is clearly visible. Either display nothing, or use the space for icons that indicate the state, such as an error triangle if there's an error.

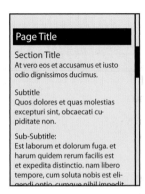 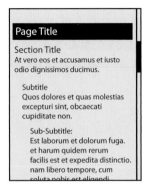

Figure 1-15. *Build all content with a hierarchy and title sections to follow, and express this to the end user.*

It is best to get professional writing resources. If this is not possible, obtain a style guide, define a communication style for the product, and stick to it. Use similar language for all descriptive titles.

- Use the same voice and, when practical, tense.

- Use a single name for your product, when it has to be referred to at all.

- Use consistent capitalization (sentence or title case). If parts of the product are considered proper names, make sure everyone has the list of these names.

Design the whole product around a simple hierarchy, and stick to it. Avoid going too deep; past about three or four levels usually becomes confusing, and it will be difficult to differentiate titles. Indenting is a common way to help express a hierarchical relationship. You might think that most mobile devices are too small to use this effectively, but don't worry, only a few pixels of indent can communicate this well. Compare this to the way a Hierarchical List is designed, and how it communicates the relationship between parent and child elements. See the examples in Figure 1-15.

Just like H-level elements in HTML (H1, H2, etc.), similar title and other display hierarchies are built into native OS development kits. Often, as with semantic web concepts, these have default attributes assigned, which can be useful when building the product and can add value to the user experience. Even if additional styles are imposed, you should try to use these basic attributes as the first definition level.

When titles are links, make this clear and follow conventions used in the rest of the application or site. Additional hints may be needed; even if color normally denotes a link, this may not be clear enough for titles which are often a different color from the content text anyway. Icons or underlines may be needed.

Antipatterns

Avoid jargon, or exposing internal processes. Avoid excessively harsh error messages, and other things which may confuse or annoy typical users. Even for certain special professions and hobbies, your customers will generally not understand your internal processes and organization. You have to explain this to them instead, in their language.

Do not repeat content. If the application is described adequately, do not keep restating the application name in subsidiary page or Pop-Up titles.

During testing, and periodically as maintenance, be sure to check all content. Often, only mockups or the primary path is inspected, but alternative paths and errors must be as clear, consistent, and well described as any other parts of the product.

Revealable Menu

Problem

You often will not be able to fit all functions for a page on the screen. A method must be provided to access these optional functions.

Often, the OS should dictate the general style of the menu structure, if only because users will become familiar with the style of interaction. However, there is generally much leeway in implementation, if variation from this is desired or called for.

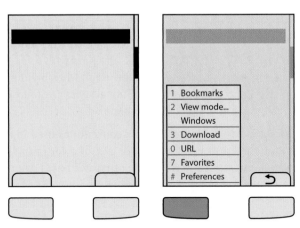

Figure 1-16. *Soft keys are very common still, revealing menus from tab labels based on pressing adjacent hardware keys. When one is open, the other provides a method to exit. The same paradigm can be used without the hardware keys on touch and pen devices by directly selecting the tab labels.*

When the user selects a key, selects a small on-screen element, or performs a gesture, an option menu is displayed with content relevant to the current state of the application.

Soft keys (hardware buttons tied to on-screen labels, typically on feature phones) are the archetypical version of this. Single actions that take place from a soft key (such as Cancel) are not a Revealable Menu since they are visible, and might be considered a Fixed Menu. There are many cases where the two menu access systems overlap, or switch back and forth depending on context.

Note that some devices use more than one menu scheme, for different purposes. For example, one may present options for the current application, and another may provide access to the running-applications list for the entire device.

Do not confuse display of this menu with the display of a Notifications list, and be sure the two do not conflict with each other.

Figure 1-17. *Menus can be of different types, here displaying fewer items, but also being more touch-friendly, and leaving more of the original context visible.*

Several variations of this pattern exist:

- The soft-key style uses one or more hardware buttons (or a portion of the touch/pen area outside the display) to reveal an option menu. When closed, these may or may not display a tab and label indicating their presence. Compare Figures 1-16 and 1-17.

- Soft-key-like on-screen displays always display a tab or button, usually along the bottom edge of the viewport. The closed state is always visible as this is the method used to access the function.

- Gestural menus, as shown in Figure 1-18, generally have no on-screen visibility. When the user swipes from an edge, the menu acts as though it accompanies the gesture, and moves into the viewport at the same speed. This is generally nonpersistent, and when the pen or finger leaves the screen, the menu collapses. Selections must be made in the same pen/finger-down gesture as the original reveal. Releasing while an action is in focus selects that action.

- A fourth variation combines gestural menu reveal methods with the on-screen button. When activated, the menu appears via another action such as sliding in from one side, or being revealed to be behind another component (such as by hiding an otherwise-present virtual keyboard). Other methods such as the Peel Away can also be used, but may be difficult to communicate to the user.

Items within the menu often reveal submenus, or lists of additional features. These may either follow the same principles as the top level of menu and appear as an attached subset, or appear as a freestanding menu, usually as a Select List or Grid of items in a Pop-Up dialog.

Treat opened menus as modal dialogs. For touch and pen devices, selection outside the menu area may, if desired, clear the menu instead of being ignored. This is especially true if the menu, when open, does not obscure the background. However, you should not allow users to select items in the parent window while the option menu remains open.

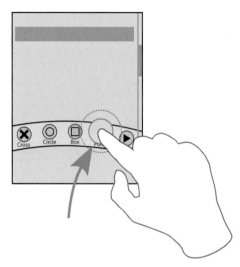

Figure 1-18. *Menus can be made to appear with a drag action, relying on no buttons or on-screen indicators. However, users will have to be taught this action somehow, so it is often unsuitable unless it is used across the OS, and there is supporting marketing or training can be tolerated.*

All opened menus must have a method to be exited. The typical methods are as follows:

- You can use the dedicated hardware "Back" key. When present, this is the preferred method, as the user will be accustomed to using it for similar functions.

- A spare soft key or on-screen tab (even if it is normally another key, or another selectable menu) will change to "Cancel." This key must be included in the items available for selection, and is not locked out due to the modality of the option menu itself.

- A close function may be added to the menu, either as a selectable menu option (usually the last one) or as a desktop-like close button in the corner. Both have certain pitfalls. The menu item must be clearly different, using a common icon. Also, a dedicated close button must be carefully placed to avoid accidental activation of it or adjacent items.

- The function that launched the menu may be used as a toggle. When it is available, and not obscured by the opened menu, selection when the menu is open will close the menu.

- Lastly, for touch or pen devices, you may make it so that selection outside the menu may dismiss the menu.

For devices with five-way pads, opened menus should be able to be scrolled through. Regardless of the input mechanism, selection of a single item will close the menu and initiate the action, which may change states, cause a different modal dialog to appear, load an entirely new page, or even exit the application. For vertical menus, see the Vertical List pattern for additional details on interaction and presentation.

If the selected item itself has options, a subsidiary menu will be opened, as shown in Figure 1-19. For vertically scrolling menus, this is usually presented as a visible child, adjacent to the primary menu, and is itself a modal dialog; the previous menu cannot be directly selected. For scroll-and-select devices, submenus can also be entered by scrolling to the right. For horizontally arranged icon menus, typically there is only one "More..." type of option which opens a Pop-Up dialog with a Vertical List, usually a Thumbnail List where the thumbnails are also icons.

To exit a submenu, scroll to the left or press the "Back" button or "Cancel" tab, whichever is available. This will only close the submenu, returning focus to the parent selection item, and not the entire option menu.

For any hardware menu keys or soft keys, pressing and holding the key can perform a different action. This action is usually a different kind of reveal menu, and may be associated with a soft key which normally does not open a Revealable Menu, and should have a relationship to the original key label to aid in discovery and recall. For example, within a browser, a soft key labeled "Back" could open a history menu when held down for a few seconds.

Presentation details

Figure 1-19. *Submenus or additional options can be displayed from any menu scheme. Icon menus, such as gesture menus or the icon bar menu on the left, usually should open a separate vertical dialog. Soft-key-type menus, such as that on the right, usually display them adjacent to or overlapping the main menu.*

In almost all cases, you should make the closed menu initiators or indicators, such as soft-key tabs, locked to the viewport. Do not allow them to scroll off the page. Visible tabs should always be visible, but may be hidden temporarily for media playback or otherwise to give more room to the primary content.

If your design has no on-screen menu visibility, and no dedicated button is provided, consider adding a visible component at page load, or during a training period when the device is first accessed to train the user that the feature exists.

Option menus, however they are initiated, must immediately appear. If animation is used to open the menu (e.g., it slides in), the beginning of the animation must start immediately. Since the menu may be partly obscured by the user, especially for touch and pen devices, Haptic Output (vibration feedback) of the action should also be considered. The menu should appear to emerge from the tab or button, overwriting the area.

Within the opened menu, display items as text, and use supporting icon elements only when there is room. See the Indicator pattern for more details on using icons along with text in lists. Most soft-key-like menus are vertical and designed for text density, so they do not work well when displaying an Icon for each item, but this is very suitable for other menu types.

Options that cannot be used should be grayed out to make them appear inaccessible. Leaving them in the list is helpful as it teaches the user what sorts of actions are available in the list, and preserves the position for additional consistency.

Display indicators for Accesskeys (keyboard shortcuts) for each item with an associated Accesskey. Do not display these for devices that do not have hardware keyboards or keypads.

You should place indicators of some sort next to items which can reveal additional menu items. Submenus may occlude the main menu if space is limited, or may hang off to the side. They should usually obscure the main menu in some manner to make the focus item clearer. Submenus should contain a Title, or the selected parent element must be highlighted and point to the submenu to serve as a label for the menu.

Antipatterns

Modal design, such as the entire principle of a Revealable Menu, can be difficult to communicate, especially if OS standard implementations are not used or the user is not accustomed to the platform. Carefully consider the whole concept, how to ensure that the menu can be launched, and how many controls can be modeless instead.

Within the menu, only allow access to options that can be used. If all options are selectable, error messages will be displayed which could otherwise be avoided.

Soft-key orientation, or which side gets which key, should follow the standards of the OS or the operator. These are usually not set at the device level, but are universal for all devices using the OS, or all devices the operator overlays. Most application frameworks allow specifying something like "primary" and "secondary" as soft-key identifiers; use these instead of "left" and "right" to ensure that your application complies with the standards of the device on which it resides.

Avoid too many menu levels. Generally, due to screen size and complexity, only one submenu level should be provided.

Fixed Menu

Problem

You have to provide access to options or controls across the application, but a Revealable Menu is already in use, or would be unsuitable due to lack of controls or conflicts with other key interactions.

Often, the OS will dictate the general style of the menu structure. There is usually much leeway in implementation, if variation from this is desired or called for.

Solution

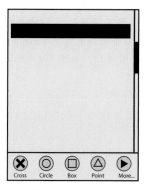

Figure 1-20. *The fixed menu is always visible, and does not restrict access to the rest of the page elements.*

An always-visible menu or control set is docked to one edge of the viewport, as shown in Figure 1-20.

Regardless of the OS-wide standard, you should often use a visible menu for media players, cameras, and other applications where a key set of controls or options should be visible at all times. Any time your users require immediate action, or you must avoid a learning curve (discovering a Revealable Menu), a Fixed Menu is a good solution.

When you can also use the menu to indicate position in the system, such as that an icon exists for each page, this is instead a set of Tabs, and not a menu.

Note that you should use this pattern only if it happens across the entire OS or an entire application. Controls on a single screen, such as zoom, pan, and playback functions for a weather radar image, are simply controls. Playback controls for a video player, which are the same in all modes of the application, are a Fixed Menu.

For game play, video playback or other full-screen displays such as readers may call for suppressing the Fixed Menu. It may be replaced with a Revealable Menu that opens in the same position.

Variations

Use the Fixed Menu to display a simple list of available options. These are generally displayed as a horizontal bar, so usually you should display each option as an individual Icon, each with an associated text label.

These may be used as the primary menu structure, but are often used in concert with a Revealable Menu, in which case they must be deconflicted. Both are best when placed along the bottom of the viewport. However, if the closed paradigm for the Revealable Menu has visibility, this may need to be changed, the visible tabs may need to be minimized in some manner, or the Fixed Menu may need to float higher up or be attached to a different side of the viewport. This may cause the Fixed Menu to be arranged vertically or horizontally.

On touch devices, the bottom of the viewport provides an additional advantage in that it may be easier for the user to reach, and for submenus to open, and so will not be obscured by the user's hand.

Pure text lists are also sometimes used, and generally follow the desktop metaphor (File, Edit, View, etc.), and so are always horizontally arranged. You can only use this successfully with a sufficiently large display. Avoid letting functions fall off the screen, as is shown in certain Tabs.

A Fixed Menu may include all options or controls required, or may offer subsidiary or additional lists of options.

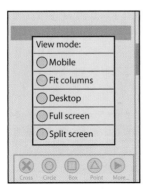 

Figure 1-21. *Additional menu options, or long lists of subsidiary options, generally are displayed in modal dialogs as on the left. Simple options, or single interactive elements, may simply reveal themselves as controls sliding off the main menu bar, as shown with the slider on the right.*

Fixed menus are either used for pen and touch devices, or used when the page content does not have to be interacted with directly, such as with video playback; the controls are the primary interaction method.

Fixed menus are not contained in a modal dialog, so you can offer interaction with the entire page's content, as much as needed. If it is not accessed as a matter of course, such as for video playback, this does not change the general state of the interaction. Other controls may also exist on the page, and are accessible at any time.

For devices with five-way pads, opened menus should be able to be scrolled through. Touch and pen devices use direct selection of the items in the menu. For controls, drag and other gestural actions (such as to change zoom level or jog to another portion of a video file) may also be supported.

Selection of any single item will initiate the action, which may load a new page, change states of the current content, cause a modal dialog of options to appear, or even exit the application. When the options in a subsidiary dialog are simple, they should be docked to the original selection. If you must prevent obscuring of the primary content, or if selection inherently disrupts other primary functionality, these subsidiary menus may be press-and-hold items; with touch or pen devices, press the menu item, then drag over to the selection and release when it is in focus.

If you must place more items in a menu than can be fit to the main menu, a subsidiary menu will be opened. This is typically a Pop-Up dialog with a Vertical List. When practical, to comply with the iconic presentation of the main menu, you may draw this as a Thumbnail List, where the thumbnails are also icons. Figure 1-21 shows two variations of this.

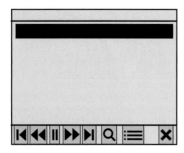 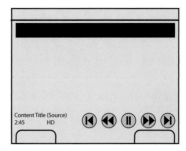

Figure 1-22. *A fixed menu may be the only menu visible on the screen, as on the left, or may need to be moved or modified to work with other menus, such as the soft-key-like Revealable Menu on the right.*

You should always lock the main menu to the viewport, and prevent it from scrolling off the screen. Do not allow the scroll bars to overlay the menu, to make this clear. Usually, to comply with the expectations for menus on mobile devices, the scroll bar is along the bottom edge, though it may be along one side to avoid conflicting with a Revealable Menu or along the top (below any Annunciator Row) to follow a desktop application paradigm. Keep in mind the issues of interaction conflict and obscuring described earlier when placing the menu anywhere except the bottom edge. See Figure 1-22 for two typical examples.

The menu is present in all screens and states of an application, but may be hidden temporarily for media playback or otherwise to give more room to the primary content. This hiding follows that of other fixed on-screen elements, such as scroll bars, Titles, or the Annunciator Row. Since they reappear by general interaction or after a time, and no explicit retrieval of the menu action is available, this does not convert the menu to the Revealable Menu pattern.

If hiding is routinely employed to retrieve on-screen space taken up by a Fixed Menu, reconsider the selection of this pattern. It is likely that a Revealable Menu is more suitable instead. Carefully consider the importance of consistency across the interface and the OS.

Do not stack multiple fixed menus on top of one another. Avoid having fixed menus immediately adjacent to other interactive bars of the same shape, such as a Notifications area. If this is the only suitable solution, prevent accidental activation through the use of gesture. You may make the items of the Fixed Menu tappable, but require a drag gesture to open the Notifications, for example.

Home & Idle Screens

Problem

You must display a default set of information and actions once the device has started, and to return to when all other user activities are exited or completed.

The device OS will provide this, but certain aspects can be modified, or widgets may be loaded by the end user which must integrate with the operation of the Home & Idle Screens correctly. Many of the same principles can be used for freestanding applications when the landing page offers enough functions or is used so often by the customer as to be functionally their home screen.

Solution

 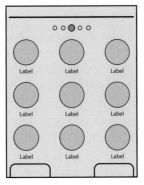

Figure 1-23. *Two key types of Idle Screens are the informational screen (generally associated with simple scroll-and-select devices) and the multipage iconic home screen, long used for PDAs and now associated with all touchscreen handsets.*

All mobile devices have an Idle Screen, originally used when the device was not doing anything (it is *idle*). This is used as a launching point or when the user is not specifically asking anything of the device. You can consider it to be similar to the Desktop on a computer, or to a web portal. Especially for smartphones and other, more capable devices, it provides a method to access all the applications, services, and information stored on the device, and can often be deliberately accessed by the user without exiting applications expressly for this purpose. Lately, it has been called a "home" screen quite often, and some OSes reinforce this to the end user by using the term, or a house icon, on the device or within the GUI.

If you are designing kiosks, or other, more constrained interfaces which present a smaller number of fixed options, the default screen is still considered an Idle Screen. It is just simplified due to the regular influx of new users and the relatively low number of options offered.

Do not confuse the Idle Screen with the Lock Screen or any other seemingly default screen. If the user must act to get to information or perform basic functions, it is not a Home or Idle Screen.

Variations

Most devices mix several design methods, in order to achieve all the needed goals.

The *Idle Screen* is the single screen that is loaded when the device is powered on, or when all applications are exited.

The *Home Screens*, often notably plural, encompass all the device-level menus that contain links to the applications. The Idle Screen is invariably one of these Home Screens.

Idle Screens generally follow one of two patterns, both of which are exemplified in Figure 1-23:

- The Idle Screen is largely occupied with status information and may have little or no direct access to applications.

- The Idle Screen is the center one of a series of related screens with icon-based representations of many or all of the applications loaded onto the device, generally displayed as a Grid with the Film Strip pattern used to move to and between other Home screens.

Status on the Idle Screen has traditionally used fixed elements, or those with only limited customization. Widgets are now supported on many devices, which may vary from an interactive Icon to display or interactive elements that occupy a large portion of the screen.

Some applications may appear to be continuous with the drilldown method of access. Settings, for example, should usually be considered an application, but the interface and interaction may be so seamless that the user is unaware she has left the Home Screen drilldown and entered the Settings application.

Additional features may be integrated into the Home & Idle Screens, such as lists of running applications, displayed as thumbnails of the user's current state or as a list of icons. Some of these additional uses of the Home Screens expand the interactivity to provide access via gestures perpendicular to the primary access. Rare or experimental versions use the features shown in the Simulated 3D Effects pattern to expand the home screen in yet another dimension. You can see that there is no clear limit on the variations that may emerge in the future.

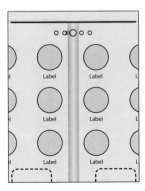

Figure 1-24. *Multipage home screens allow scrolling between each screen like a Film Strip. Position in the screens should be indicated and, while scrolling, should animate. Option menus, whether fixed or revealable, disappear during interscreen scrolling.*

Idle screens with status information are mostly for viewing. There may even be no direct interaction. If you are working with a simple scroll-and-select feature phone driven by a five-way Directional Entry pad, you will assign key directions to actions or to launch applications. The defaults are often printed on the face of the device hardware, but may be changed in settings. For these scroll-and-select devices, the five-way pad has no scrolling functions on the Idle Screen.

Other devices with this type of Idle Screen are generally arranged so that vertical scrolling will move between calendar or notification items, and horizontal scrolling will move between application shortcuts. If you add touch or pen control to these basic interfaces, these items may also be directly selected, and will launch a full application view.

Multipage home screens use the concept of a single page larger than the viewport. You can consider this to be a Film Strip pattern, as shown in Figure 1-24, and use it to access as many screens as desired and which the device can support. These are mostly suitable only for touch and pen devices. Scrolling between screens may not be clear or easily understood, if you try to make it available on scroll-and-select devices. However, when hardware Directional Entry is available, be sure to offer scrolling control, so users incapable of employing the touchscreen can still manage to use the device.

Additional information is almost always available via a list of all applications on the device. These are often represented as a vertical Thumbnail List, but may continue using icons as the primary label, in a Grid format. Items should be hierarchically ordered, ideally with user control over folder names and contents, to organize the information as needed. If a single list of all applications is shown, ensure that it is in an easily understandable order, such as alphabetical. See Figure 1-25.

Very often, contextually intelligent mobile devices should be presenting the last-used state to the user at all times. For example, if you are designing an eReader, you should present precisely the last reading state when the user returns to the device, even after a power cycle. While a "Home Screen" will still exist, it will be viewed much less often in this case and may be considered a Settings page instead. The same principles outlined here should still be used, however.

Regardless of your device, consider building interactive methods that avoid the Idle Screen, and allow continuous use of the device. Home & Idle Screens encourage "pogosticking" from one application, to the idle screen, to another application. Running application lists should be made available from within all applications, for example. You may even be able to use the same interaction method to access the list from all contexts.

Presentation details

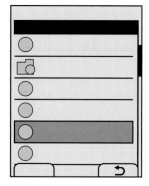

Figure 1-25. *A path must be provided to get from the Idle Screen (even if it is a multipage Home Screen) to the remaining Home Screens, to view all available applications and options. Within the list view, folders are displayed differently from individual applications or other leaf-level items.*

Home Screens should be distinctly different from application screens. It should be clear to the user when he is on a Home Screen, and especially clear when he is on the Idle Screen. One key method is to have no Title on the Home Screen and rigidly enforce the use of titles on all other screens.

When using a folder structure to organize items in the Home Screens or related drilldown lists, make the structure apparent. You should make sure all folders carry a folder-like icon, even if additional graphics are attached to it.

When on a drilldown menu Home Page screen, title all screens after the main screen. Usually, this will follow the title of the icon or link used to load it, and should be accompanied by the same icon. See the Title pattern for more details on the use of labels and icons.

Multipage Home Screens must have a Location Within widget to indicate which of the pages is currently in view. If you have more than three Home Screens, you should probably use a Location Jump widget, possibly integrated with the Location Within, to provide access to the far screens with greater speed.

For multipage Home Screens, the backdrop should scroll at a slower speed than the icons, labels, and widgets in the front layer. This simulation of parallax makes the screen appear to have depth, and the movement of the background helps to act as a wayfinding device for the user to better understand her position on the screen. Although the user can change the background, the default backgrounds should encourage these behaviors, and have depth built in, or appear to be slightly out of focus. They should not have truly repetitive elements, so users may become familiar with the various areas of the image.

Antipatterns

Ensure that users can understand the paradigm by which your Home & Idle Screens operate, without training. You should provide clear and easy access from the Idle Screen to the list of all applications, and to any menus of options.

Avoid violating device UI paradigms for the Idle or Home Screens. For example, the very common practice of disabling the scrolling function and making the directions correspond to shortcuts on the Idle Screen can be difficult to understand and learn. Users can experience a cognitive dissonance when switching between the two behaviors, which slows them down.

You must carefully design the method you use to add, remove, or move items from the Home & Idle Screens, to encourage user customization. The most common of the reasonably usable methods for smartphones involves press-and-hold to switch to an editing mode. However, users do not seem to recognize this as a universal feature yet; even when they are familiar with it on one platform, when they switch to another, they generally do not attempt to use the feature immediately. Hopefully, a standard will emerge or users will become more accustomed to exploring interfaces to discover interactions without clear affordance.

Lock Screen

Problem

Mobile devices must enter a lock/sleep state to reduce power consumption, to prevent accidental input, and sometimes to prevent unauthorized input. You must provide a screen to communicate this state clearly, provide key information, and assist with unlocking methods.

adding an Interstitial Screen to emphasize the power of the search taking place instead. Excessively fast searches can even be perceived by users as cheating, and the result viewed with some suspicion.

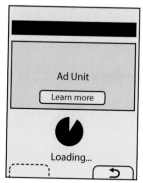

Figure 1-31. *If space allows, and the addition will not add confusion, advertising can be placed on the Interstitial Screen. Provide for interactivity with the ad whenever possible.*

The Interstitial Screen is usually described as a required step, and does not have any interaction itself. Key device features such as the ability to switch applications or exit back to the Idle Screen should not be disabled.

If you display the screen as part of a system update, or something else where interruption could cause irreparable harm, clearly communicate this, and set this expectation before initiating the process. For situations like this, whenever possible, prevent disallowed actions such as power-off by disabling the buttons, as well as describing the function.

When the process may be canceled, you should load an on-screen button or soft-key tab (depending on the device) with this function clearly labeled. When the user selects it, the user will return to the last stable screen. Ensure that this screen, if part of a process, has an easy method for changing the path or exiting the entire process or application.

When you choose to display advertising on an Interstitial Screen, as in Figure 1-31, it should be as interactive as possible. Allow the user to select the ad to view details, purchase the product, and so forth. Load the ad link into a new application (a new browser window, or the application store) so that the ongoing process is not interrupted. If possible, communicate this (possibly with another interstitial!) when the remote site is loading.

You should always display the Title of the application or service. Include the position in the process.

Branding should also be included. At least include an icon in the title bar, but the entire screen may be occupied with branding for loading screens, and in other cases.

Wait Indicators of one type or another should be used. See that pattern for details, and be as specific as possible.

Describe what is being loaded, and why. Much like "Submit" is too vague a button label, always say "Finding the best deals for you" instead of just "Loading." If the screen is being displayed due to a critical system process, and features such as application switching are disabled, clearly communicate this.

When you provide a cancel function, the button or soft key should be labeled "Cancel," or an equivalent label or icon should be used across the OS to denote the current screen or process is to be canceled. Although this may seem needlessly generic, and therefore in conflict with the labeling statements just made, this function is well understood by users when consistently implemented.

Additional spaces may be occupied with Advertising. However, ads must be clearly differentiated from the other messaging, so users are not confused by an apparent change in context, even if they glance away from the beginning of the screen being loaded. Simply labeling a message with the word *Advertisement* or similar is not an effective method of differentiation.

The advertising should not interfere with the user's understanding of the application. Advertising should almost never animate, while the loading indicator will, to imply activity. See the Advertising pattern for some discussion of display methods.

Antipatterns

Unless deliberately using a business model with a paywall, do not load interstitials purely to display Advertising.

Avoid using an Interstitial Screen for every loading condition. As much as possible, avoid locking during loading, use nonmodal Wait Indicators, use Pop-Up indicators, or load information fast enough so that it doesn't need any indicator.

Advertising

Problem

You must place advertising into an application, site, or other service.

Advertising is heavily used in all aspects of mobile, from SMS messaging all the way to being integrated with the device OS.

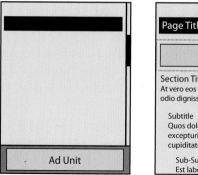
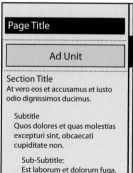

Figure 1-32. Advertising may be within the page, so it scrolls with content, or locked to the side of the viewport, and so does not scroll.

Many sites, applications, and even entire mobile services are ad-sponsored. If you must place advertising into a product, you must not try to hide the ads, nor make them so prominent that they damage the user experience. Generally, advertising is key to the business, and must be made as a necessary function of the product. Integrate advertising correctly, and well, or the product will be discontinued.

Advertising in mobile must be:

- Clearly differentiated from the content

- Clear, readable, legible, and able to be interacted with

- In the same place, and used in the same way, on each screen and in each state

- Unobtrusive enough to not interfere with the interaction of the actual product

- Easily actionable, so users can take advantage of the offer

While numerous graphic variations are possible, you really only need to consider two key methods:

- Ads may be loaded within the content of the page, so they scroll with the page. These are usually near the top and bottom of any particular page, associated with the masthead/title and any links or functions in the footer. For very large pages, there may be additional advertising at key breaks in the middle of the page as well.

- Ads may, alternatively, be docked, or fixed in the viewport. This is an especially useful solution for applications with multitier access, where payment removes the whole layer. The rest of the screen scrolls as normal, but this element is locked to the edge of the viewport, much like a Fixed Menu.

Figure 1-32 shows both of these options. Advertising in text, such as information appended to an SMS message, is not discussed here, but the basic principles still apply.

Interaction details

Your advertising should always be selectable so that the user may get more information or purchase the product. Make sure this link loads in such a way that it does not interfere with the existing process, such as by opening in a new browser window or launching the application store.

For tiered services, in which an ad is displayed for low-price or free access, you should provide a link immediately adjacent to the ad that allows the user to upgrade to the paid, ad-free version.

When designing for scroll-and-select devices, avoid placing advertising in such a way that the user must scroll past it to get to the actual functions or content.

Presentation details

You must become familiar with the standards of advertising. Banners should always display in standard sizes. Ad providers can then use existing ad units, and do not need excessive production details, as they can follow standards published elsewhere. In most cases, you can just follow the sizes and other specifications published by the Mobile Marketing Association (MMA). Figure 1-33 summarizes these standards; complete specifications are available at *http://www.mmaglobal.com/mobileadvertising.pdf*.

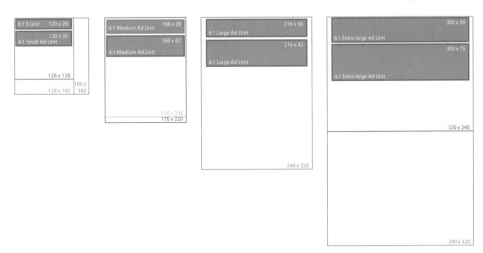

Figure 1-33. *The sizes of standard MMA banners and their suggested common screen sizes.*

Only for very large devices, especially tablet-size devices, will you need to use other sizes. These will often be provided by others (such as ad services) or the smaller common desktop sizes can be used instead. Of course, the MMA tries to stay on top of trends, so it will continue to expand its offerings over time.

Display the ad in a manner that clearly differentiates it from the actual page content. There are three basic ways to accomplish this, each illustrated in Figure 1-34:

- Place the ad on a shaded, tinted, or colored background. Usually, the background should extend the full width of the page.

- Separate the ad from the rest of the content with rules, which must run full-width, or in a closed box larger than the ad.

- If the ad is slightly smaller than the screen width (by using one size smaller than the suggested size for the screen), use different alignment than the content. If content is left-aligned, right-align or center the advertising.

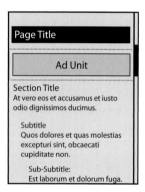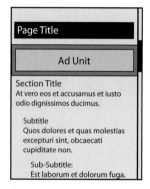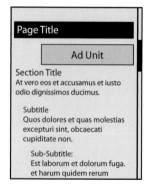

Figure 1-34. *The three methods of differentiating an ad from the surrounding content.*

When your ad is in a docked area, it should usually be docked at the bottom, and is considered to be one layer above the content. A Revealable Menu will overlap the advertising when open. The ad may be placed at the top when fixed menus are used, or when closed reveal menus would interfere with display. Advertising may be seen as a strip immediately above a Fixed Menu, and this can work, but it can become cluttered, and runs the risk of accidental activation.

Advertising may be integrated with the content, such as in a list of results. The advertising should not be counted as a line item, and so will not be numbered or included in result totals, nor should it use the guidelines covered in the rest of this pattern to visually differentiate it from the rest of the content. Typical uses of this are to have the first and last items in a page of results be sponsored.

When you display advertising in a list of other items, consider the available space and do not overload it with content. Although three sponsored links above a list may be acceptable on a desktop device or large tablet, it may be totally unsuitable on a small handset. The small viewport can also be used to add additional ad spaces; an extra ad can be provided in the middle of the same list. Interrupting about every one and a half screens can provide a useful break to the scrolling.

The ability to select the ad should be clear. Provide a link or button in the ad whenever possible. If there is insufficient room for this, use the current interactive standard method to indicate a link, such as underlined text or a border on the image.

Antipatterns

Mobile is generally so task-focused that failures in advertising—that make it too obvious or confuse it with the content—may cause an immediate loss of customers. Even casual activities such as gaming generally draw the user's focus, so an overly intrusive ad will result in dissatisfaction. Follow the guidelines here (and on the MMA website listed earlier).

When conducting user research, be sure to include real ads and do not carefully pick them to match the design. Borrow from competing sites if necessary, but get realistic advertising or your test will be invalid.

Do not make ads too large, making the content difficult to read. This is especially key for small screens using docked advertising space. If this could be a problem, switch to the inline style.

There is almost never a reason to place a text label such as "Advertising" next to the banners. The design should be able to communicate this. If it doesn't, fix the design instead of wasting space with labels.

You should almost never animate advertising. This will distract from the core content of the application or other process. If animation is absolutely required, use it very carefully and sparingly.

Make sure advertising in scrollable areas does not induce *false bottom* errors, making the user believe he has reached the end of the page content prematurely.

Do not use custom sizes for any banners. Advertisers will generally not go to the effort of making new sizes. All services you sign up for require that you abide by their guidelines, so your revenue base will be severely limited.

Summary

Wrap-Up

We just saw that a page is the area that occupies the entire viewport of the mobile screen during its current state. Within the page, a composition contains and organizes the appropriate components and content within that viewport. Mobile displays range in size, and on smaller devices, screen real estate is valuable and every pixel is important. Therefore, you should be aware of how little or how much room these components use when you are planning the UI layout. Mobile users want information access on the spot. So information that is displayed without hierarchy and structure will likely cause user tasks to be delayed or to fail.

When incorporating the page patterns into your design, consider:

- User needs, tasks, and reading goals based on context of use
- Principles of page layouts based on theories to ensure that the users can quickly identify the structure of the content

Pattern Reference Chart

The pattern reference chart in the following subsection lists all the patterns found in Chapter 1. Each pattern is accompanied by a general description of how it can apply to a design problem while offering a broad solution.

Cross-referencing patterns are common throughout this book. Design patterns often have variations in which other patterns can be used due to the common principles and guidelines they share. These cross-referenced patterns are listed in the following charts.

Chapter 1, "Composition"

This chapter described the composition as being a template, which contains or houses all the components and elements within a page. This allows the content to be consistently organized across the device's OS, allowing users to have an easier time navigating, searching, and accessing it. The ability for the user to quickly recognize the content's organizational structure will increase user learnability and satisfaction while decreasing performance errors.

Pattern	Design problem	Solution	Other patterns to reference
Scroll	More information is in the page or element than can fit in the viewport. A method of access to this information must be provided.	Scroll bars are displayed to indicate the scrollable axes, and the relative position within the scrollable area. Due to the scale of mobile devices, as a general rule scroll bars cannot be manipulated.	Infinite Area Vertical List Infinite List Accesskeys Directional Entry Pop-Up On-screen Gestures
Annunciator Row	The status of important hardware features such as battery level and network connections must be able to be discovered by the user with little or no effort.	A bar is displayed along the top of every screen with a series of iconic representations of the status of the device. Common representations and placements of icons are used so that users can understand the key indicators on any device without having to be familiar with the specific device.	Notifications Icon LED
Notifications	A method must be provided to notify the user of arbitrary notifications, of varying priority, without unduly interfering with existing processes.	A single, consistent notification method is provided across the entire OS. This method does not interfere with any processes the user is currently involved in, and can be acted upon or dismissed very easily.	Annunciator Row Pop-Up Confirmation Titles LED Tones Haptic Output Voice Notifications
Titles	Every key element should be labeled to make context or process completion clear.	Pages and elements or content sections within a page should almost always be labeled. Titles must be used in a consistent manner, in terms of size, location, content, and style. The simplest method for pages is a straightforward title bar.	Notifications Vertical List Infinite List Thumbnail List Fisheye List Carousel Grid Select List Confirmation Sign On Exit Guard Timeout Windowshade Pop-Up Hierarchical List Returned Results

Pattern	Design problem	Solution	Other patterns to reference
Revealable Menu	Not all functions for a page can or should be presented on the screen. A method must be provided to access these optional functions.	By selecting a key, selecting a small on-screen element, or performing a gesture, an option menu is displayed with content relevant to the current state of the application.	Fixed Menu Vertical List Pop-Up Thumbnail List Select List Accesskeys Icon Button
Fixed Menu	There is a need for access to options or controls across the application, but a Revealable Menu is already in use, or would be unsuitable due to lack of controls or conflicts with other key interactions.	An always-visible menu or control set is docked to one edge of the viewport. This is often used for media players, cameras, and other applications where a key set of controls or options should be visible at all times.	Revealable Menu Annunciator Row Titles Vertical List Pop-Up Icon Button
Home & Idle Screens	A default condition must be available for display once the device has started, and to return to when all other processes and applications have exited.	All mobile devices have an Idle Screen, originally when the device is not doing anything (it is idle). This is used as a launching point or when the user is not specifically asking anything of the device.	Lock Screen Grid Film Strip Thumbnail List Icon Link Pagination Location Within
Lock Screen	Mobile devices must enter a lock/sleep state to reduce power consumption, to prevent accidental input, and sometimes to prevent unauthorized input. The device must communicate this clearly, and assist with unlocking.	The Lock Screen is displayed immediately after a deliberate (user-initiated) lock. Key information such as events, alerts, time and date, and instructions on how to unlock the device is displayed.	Sign On Annunciator Row Home & Idle Screens Notifications Pop-Up On-screen Gestures
Interstitial Screen	A delay is encountered before a requested screen can be loaded. The information previously displayed cannot or should not be displayed.	The interstitial is primarily a loading process screen. Use it when there is a technical limit that prevents display of the previous context. This is commonly encountered when starting the device, or when applications are loaded.	Wait Indicator Confirmation Pop-Up Advertising
Advertising	Advertising must be placed in an application, site, or other service.	Sometimes a necessary function of the product, advertising must be integrated correctly, clearly differentiated from the content, readable, legible, and able to be interacted with.	Revealable Menu Fixed Menu Link Button

Additional Reading Material

If you would like to further explore the topics discussed in this chapter, check out the following appendix:

Appendix C, "Mobile Typography"
> This appendix provides additional information on appropriate use of message display characteristics, including typography, legibility, and readability guidelines, as well as further information on today's mobile display challenges and capabilities.

Components

Components, as described here, are a section or subsection of a designed interactive space. They take up a significant portion of the screen and may be as large as the viewport (or larger) or, when smaller, may appear to be in front of other displayed information.

Components must display a range of information types—images, ordered data, expandable lists, and notifications. They also allow the user to interact with the system in some significant, primary manner. Combining them with small, reusable, interactive, or display widgets gives the designer an almost unlimited number of options.

The components that we will discuss in Part II are subdivided into the following chapters:

- Chapter 2, "Display of Information"
- Chapter 3, "Control and Confirmation"
- Chapter 4, "Revealing More Information"

Types of Components

Components for Display of Information

When you display information, you need to reflect the user's mental model and mimic the way the user organizes and processes knowledge. Information displayed on mobile user interfaces that ignores this principle causes the user to become lost, confused, frustrated, and unwilling to keep using your product. Patterns found in Chapter 2, such as Vertical List and Infinite List, provide simple, well-understood solutions for displaying lots of information in a structured and contained manner that doesn't overtax the user's mental load. These patterns also provide solutions to arranging content where screen real estate is so valuable that every pixel counts. To prevent that, this chapter will explain research-based frameworks, tactical examples, and descriptive mobile patterns to use.

Components for Control and Confirmation

Humans make mistakes. As designers, we can create effective interfaces that can prevent costly human error resulting in loss of user-entered data. When costly human error is possible, we should create modal constraints and decision points to prevent errors before they happen. Patterns found in Chapter 3, such as Confirmation and Exit Guard, provide solutions that help prevent human error which can lead to the loss of important data the user has tediously typed on his tiny mobile device. You should always consider the context of the user's goals and current tasks when incorporating confirmation controls. But if you overuse these constraints and decision points during low-risk situations, you will frustrate the user by increasing his processing time and mental load, and delaying or stopping his task.

Components That Reveal More Information

When we design to reveal more information, we need to be conscious of the limitations of the devices, networks, and our human abilities. Screen size will limit the amount of information that can be displayed simultaneously. A device's OS, and the framework within which you have to work, will limit the types of interactions available. Hardware constrains the viewport size, input methods, processing speed, and loading times. Our memory limits cause us to filter, store, and process only relevant information over a period of time. You can use patterns such as Windowshade and Pop-Up to work around these limitations. They provide solutions you can use to reveal additional information when the user chooses, without having to saturate the screen full of content.

Getting Started

You now have a general sense of what components are and how they relate to information display. The component-related chapters will provide specific information on theory and tactics, and illustrate examples of appropriate design patterns you can apply to specific situations. And as always, remember to read the antipatterns, to make sure you don't misuse or overuse a pattern.

Chapter 2

Display of Information

Look Around

Take a moment and look around. Are you inside? Then you might come across books, a pile of mail, your computer, and your television. Or maybe you're outside, carrying your mobile device and checking your appointments. The world we live in is surrounded by ubiquitous information. Information that is visual, audible, and tactile. It is meant to inform, to entertain, to instruct, and to warn. Because we are constantly bombarded with this information in our daily lives, we must quickly collect, filter, store, and process which information is important to use for specific tasks.

Consider a busy intersection you are trying to cross. You are surrounded by the sights and sounds of pedestrians conversing, cars and trucks honking, birds flying, signage on billboards, and thousands of other types of stimuli. Our minds have an amazing ability to focus on the task at hand, filter the surrounding *noise*, and process, store, and allow us to act on only the relevant "signal" information.

When the crosswalk signal changes to "Walk," we identify the sign, interpret its meaning, determine an action to move our body forward, and carry out our actions by walking until we've crossed the street, achieving our goal.

Understanding how we process and filter visual information, or data, will help us to design effective displays of information on mobile devices. Let's first explore the types of information we will encounter.

Types of Visual Information

All humans have more or less the same visual processing system. However, without a standardized way to explain and notate our perceptions, our communication of this information becomes arbitrary and ineffective when designing to display information on mobile interfaces.

Ware (Ware 2000) introduces a modern way of dividing data into *entities* and *relationships*.

Entities are the objects that can be visualized, such as people, buildings, and signs. Relationships (sometimes called relations) define the structures and patterns that entities share with one another. Relationships can be structural and physical, conceptual, causal, and temporal.

These entities and relationships can be further described using *attributes*. These are properties of both the entity and the relationship, and cannot be considered independently. Some examples of attributes are:

- Color

- Duration

- Texture

- Weight, or thickness of a line

- Type size

For each of these we mean the attribute as it applies to a specific item. Not texture in general, or the texture of paper, but the texture of a specific type of paper (or even a specific sheet of paper).

Classifying Information

In addition to creating descriptions of our perceptions, we have also standardized a way to classify them. Common classifying schemes that we use are:

Nominal
    Uses labels and names to categorize data

Ordinal
    Uses numbers to order things in sequence

Ratio
    A fixed relationship between one object and another using a zero value as a reference

Interval
    The measurable gap between two data values

Alphabetical
    Uses the order of the alphabet to organize nominal data

Geographical
    Uses location, such as city, state, and/or country, to organize data

Topical
    Organizes data by topic or subject

Task
    Organizes data based on processes, tasks, functions, and goals

Audience

Organizes data by user type, such as interests, demographics, knowledge and experience levels, needs, and goals

Social

A collaboration of organizing data by users who share the same interests, such as tagging, adding to a wiki, and creating and following Twitter feeds

Metaphor

Organizes data based on a mental model that is familiar to the user, such as organizing computer files with folders, trash, and a recycle bin

Organizing with Information Architecture

Figure 2-1. *Even when given pen and paper, people will make lists, so it is not surprising that lists are the most common interactive element in mobile devices. Lists can be adapted almost infinitely, for viewing or selection, for any size, and for any type of interaction.*

Now that we can describe the data we perceive and knowledge types we store, we must understand how this information should be structured, organized, labeled, and identified on mobile user interfaces.

One of the most common organization structures humans have used through time is a *hierarchy*. A hierarchy organizes information based on divisions and parent-child relationships. When using hierarchies to organize information, Peter Morville explains rules to consider (Morville 2006): categories should be mutually exclusive to limit ambiguity. Consider the balance between breadth and depth. When determining the number of categories regarding breadth, you must consider the user's ability to visually scan the page as well as the amount of real estate on the screen. When considering depth, limit the scope to two to three levels down. Recognize the danger of providing users with too many options.

Another way to organize information is by *faceting*. In this, there are no parent-child relationships, just information attributes, such as tags, which may be sorted or filtered to display the most appropriate information. The tags do not have to be explicit, and faceting may be accomplished by searching text descriptions, or even through unusual methods such as searching for shapes, patterns, or colors directly within images.

Of course, these two methods, hierarchy and faceting, may be used in conjunction. A hierarchically ordered data set can also have tags attached to it, and the facet view may combine both strict and arbitrary ordering to display the information the user wants.

Date and location are essentially special cases, and depending on the data or needs, they may be approached either way, even though they are strictly defined. For example, location can be an arbitrary value, with filtering or sorting for distance around a single point. Or it may be considered as a hierarchy of continent→nation→state→county→city→address.

Information Design and Ordering Data

Figure 2-2. *Grids are used to display ordered data such as photos by date, or for user-organized information such as home pages, where they often become filmstrips as well. Grids can lend themselves to selection, reorganization, and even inclusion of larger items, as long as they fit in the grid.*

You can use the way in which people perceive attributes to communicate the relative importance and relationship of informational elements on the page. This design of pages or states, when it falls directly from the information architecture of the entire product, can be called *information design*.

Although many methods of considering these arrangements exist, an adequate grouping is from most to least important. Position is generally more critical to communicating importance than size, which is more important than shape, and so on:

Position

Although relative position is unarguably critical, it can easily be lost. The Annunciator Row, with battery level and so forth, is not the most important, because it is almost lost in the bezel. This is purposeful, and the size is adjusted to account for it. Each attribute works with others to function properly.

Size

Larger elements attract more attention, aside from providing more room for content. They can be too big, either obscuring other items or exceeding the expectation size for an element. Buttons can be so large that they are not recognized as buttons, for example.

Shape

At their simplest, pointed shapes attract attention. Warnings should be triangles, and helpful icons circles, for example. Rounded corners on some boxes, and square on others, will imply meaning. Make sure it's there, and not just a random design element.

Contrast

Contrast refers not to color, but the comparative value (darkness) between two different elements, discounting color. Contrasting elements are more easily read, less affected by lighting conditions, and not as subject to users with color deficit.

Color

High-visibility colors attract more attention, with significant caveats. The most important caveat is not always regarding colorblind users. Glare can make certain colors less prominent. Pervasive use of the color in the branding can also be problematic; a site with a lot of red cannot rely on red for warnings.

Form

The last thing to use is specific forms of an element. The most common is type treatments, such as bold and italics.

We will discuss these in detail in other chapters as well. Here, the concept is useful when determining how to relate the elements within a single informational item, and how to keep the elements and adjacent items from becoming mixed. Rules and bars of color are but some of the techniques. The preceding list covers six categories, with hundreds of design tactics included.

It is also useful to decide what information must be present. More can be said about a good, easy-to-understand interactive design by what is left out of any particular view, than what is included. For each of the information displays detailed in this section, only a portion is shown, and details, or alternative views, are available when the user takes action.

Naturally, make these decisions by following heuristics, standards, and styles of design that already exist, such as OS-level standards and universal hierarchies of visual communication. Most decisions for an existing platform can be easier to make by consulting the style guide. Only a few choices will exist, and these will be well understood by users.

Patterns for Displaying Information

One good way to think about the topic of interactive design is that it's all about displaying information. This chapter in particular is concerned with components whose sole task is to present ordered sets of information so that users may understand and act upon them.

These patterns have been developed and refined based on how the human mind processes patterns, objects, and information:

Vertical List

> Rather than using horizontal space inefficiently, this displays a set of information vertically using an entire allocated space. See Figure 2-1.

Infinite List

> This reveals small amounts of vertical information at a time because the information set is very large, and not locally stored.

Thumbnail List

> This uses a Vertical List with additional graphical information to assist in the user's understanding of items within the data set.

Fisheye List

> When a scroll-and-select device is targeted, and a Vertical List is called for, this can be used to reveal small amounts of additional information that can assist the user in his task.

Carousel

> This displays a set of selectable images, not all of which can fit in the available space but which can be scrolled through using many methods.
>
> Preserve the integrity of the image used. Do not use programmatic shortcuts visible to the user:
>
> - If the image is skewed to create perspective, skew the entire image. Don't clip off corners of the image to simulate the skew, as demonstrated in Figure 2-19.
>
> - Do not squash the image to change its aspect ratio. If necessary, add black bars or crop the image to fit a consistent space.
>
> - Do not flip the image when displaying the back of a card, as in three-dimensional carousels.
>
> Never use cheats, and come up with a different display method, or switch to another pattern such as a Grid or Slideshow.

Grid

> This presents information as a set of tiles, most or all of which are unique images, for selection. See Figure 2-2.

Film Strip

> This presents a set of information, which either is a series of screen-size items or can be grouped into screens, for viewing and selection.

Slideshow

> This presents a set of images or similar pieces of information using the full screen for viewing and selection.

Infinite Area

> This displays large, complex, and/or interactive visual information that must be routinely zoomed in on so that only a portion is visible in the viewport.

Select List

> This is usually a mode of another list pattern when selections, either individual or multiple, must be made from a large, ordered data set.

Vertical List

You must display a set of text-based information as efficiently and simply as possible.

The list is the key organization and presentation method used on all sorts of mobile devices.

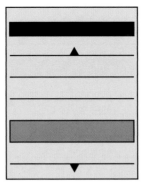

Figure 2-3. *The Vertical List simply stacks individual line items one above the other. Only a few are visible within the viewport, but many, many more may be in the data set.*

Text (at least in Western languages) is what we refer to as *horizontally inefficient*. It greedily occupies most of its space in the horizontal axis, but also uses a lot of that space to do it, and very often uses an unpredictable amount of horizontal space. Two labels for otherwise similar items may be of very different lengths. "Joe Smith" is just as much a person as "Narin Jaroensubphayanont," but the two labels could not be more different.

Text cannot be easily repurposed to be significantly narrower, at an angle or vertical. Vertical lists are lists of text, or text-centric lists with supporting information, such as a Thumbnail List.

A Vertical List allows inefficient use of horizontal space by permitting each item to use the entire line. The width of the list may be the entire viewport, or a portion of it. See a rudimentary example in Figure 2-3.

Items are then listed serially, using vertical space, also using either the entire viewport (minus a Fixed Menu, Annunciator Rows, the page and list Titles, etc.) or a subset of it.

Variations

Most variations of the vertical list have notable interactive variations, so you can find them discussed as separate patterns:

- Infinite List
- Thumbnail List
- Select List

Other variations are largely in visual display options, and are far too numerous to list.

You can add additional features to vertical lists to encourage their use, or increase the efficiency of the interaction. The most common are Location Jump, Search Within, and Sort & Filter.

Interaction details

Figure 2-4. *Details are key to making the Vertical List work. Scrolling must be smooth, information must be clear, and individual items must be well labeled.*

Selection of an item in the list will perform an action, such as viewing details of the selected item.

Scrolling acceleration, even on scroll-and-select devices, should be used to allow coverage of large amounts of ground without resorting to other methods. Even if other methods are provided, the user may prefer this or may not discover them.

In a *circular list*, when the end of the list is reached, another scroll action will jump back to the other end of the list, or just continue without interruption. Circular lists are strongly encouraged, but dead-end lists that do not cycle back to the top may be a better choice for certain data sets, or for specific tasks. If the OS standard is for dead-end lists, be sure to provide extra features such as Location Jump to allow easy return to the top.

Whether you use a circular or a dead-end list, when the end of the list is reached, acceleration should be canceled and the list stopped or slowed dramatically, to indicate a change has occurred.

Presentation details

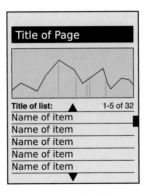

Figure 2-5. *Lists, like all components, do not need to be full-screen, if there is sufficient resolution. The list component must abide by the complete requirements outlined here, including titles, scrolling, and labeling.*

As with all Components, you may use the Vertical List in an area smaller than the viewport, as shown in Figure 2-5. Be sure the size of the scrolling area is clear.

You should usually align text labels to the left, to ensure that the list can be scanned. The order should be clear from the label; for example, if the list is in alphabetical order by last name, list the last name first, or use subsidiary alignment so that all last names are in a single column.

You may present additional information on a second (or third) line, as shown in Figure 2-4, but this information must be indented, smaller, and/or reduced in contrast so that the primary label line is clear.

Make sure the divisions between items are clear. You can do this by using divider lines, "greenbar" alternating colors, or simply ensuring that there is enough space between them. The text itself can often form a suitable visual block, allowing surprisingly little whitespace to yield a suitable break between lines.

Use the Location Within widget to indicate the position in the list and the total list size. Naturally, scroll bars should be represented correctly, to indicate current position and give a sense of the relative size of the list. See the Scroll pattern for more details.

A break indicator, or top-of-list label, should usually be presented at the first item in the list, to indicate circular scrolling. Especially on smaller lists, users may not understand the total size and scroll past it many times.

Antipatterns

Many lists, such as address books, may exist for the user with almost no content, or with many thousands of items. Be sure to consider all conditions when you make your design choices, and do not just pick one case. Also, be sure to consider the look and interaction of an empty list. Should it even be presented?

Lists should scroll smoothly, pixel by pixel. Do not scroll line by line unless this is a limitation of the equipment or OS in use. It rarely is anymore, so challenge your developers, and look at the documentation yourself, if they insist on doing it this way.

Similarly, never use page scroll. When the end of the viewport is reached, the next item in the line will appear. Do not jump and load an entire page (a viewport worth) of information, with the next selectable item at the top.

Use caution in selecting background and highlight colors. Make sure all users, in all lighting conditions, can both read the text and differentiate between selected and unselected items.

Infinite List

Problem

You need to display a Vertical List, but the information set is very large and may not be locally stored, so retrieval time is inconveniently long.

Increasing reliance on cloud services, and the availability of Ajax-like methods on the Web, makes the Infinite List a common answer to all sorts of applications and web interfaces.

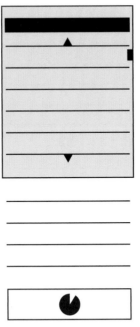

Figure 2-6. *The Infinite List will, ideally, load items off the screen, before the user even sees missing items or loading indicators. Lazy loading may be used on each line item in progress to make this clearer.*

Use an Infinite List to retrieve and display smaller amounts of information at a time. The first displayed set will fill the viewport.

Information displayed, and interactivity with the listed items, is the same as for any other list view. You can use all the variations of the Vertical List as an Infinite List.

You should preemptively load (prefetch) information outside the viewport whenever possible, to reduce the need for the user to be aware of the fetching and to reduce button pressing, as shown in Figure 2-6. Try out the interface, perform tests, and monitor usage in production to set this to the correct level. When detectable, give priority for prefetching to the existing direction of scrolling.

You may encounter resistance for implementing an infinite scrolling list on the grounds that it makes additional data requests. Be prepared to make your points about the data being needed by the user, and do what you can to reduce needless data both for the business and in consideration of the user's bill and device.

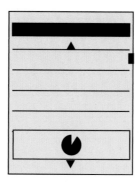

Figure 2-7. *Predictive retrieval loads user data before the user needs it. Occasionally, the user may catch up to the end of the list, and a loading indicator will be presented.*

The preferred method of interacting with the Infinite List is *predictive retrieval* and is largely transparent to the user. No explicit action is required to load additional information, although scrolling may be used as an implied action to retrieve information outside the viewport. To reduce network load, you will almost never try to download the entire data set.

For certain types of data, or if there are restrictions on the ability of the software to automatically act (such as browsers without full JavaScript support), the user may need to manually request additional information when the end of the displayed data set is reached. This is called *explicit retrieval* and is shown in Figure 2-8.

A useful subvariation of explicit retrieval is for new-only or otherwise filtered data sets (such as RSS feeds). The user must select which set of information to retrieve, not for systematic purposes but because that defines the additional information set in which the user is interested.

You can combine additional list types. You can use both the Thumbnail List and the Select List, as well as the conventional Vertical List, with the features of an Infinite List.

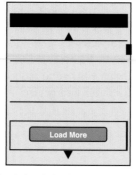

Figure 2-8. *When the end of an explicitly loaded Infinite List is reached, a message is presented and the user must choose to load additional data.*

For predictive retrieval, the user takes no explicit actions to load additional information. If an error occurs, a "refresh" or "reload" button may appear, and the list will temporarily function as an explicit retrieval list. You must carefully design the number of items loaded with each retrieval (especially the first), by considering the speed with which items will be browsed by the user. You should develop rules that react to a user scrolling repeatedly in the same direction by preloading more information than usual in the direction being scrolled.

For explicit retrieval, an informational box will be displayed as the last item in the list, with details on the required behavior, and a button to allow the loading of more items. It should be clear how many more items will be loaded, or some other definition of the range of items should be used, such as "Load next day" for calendar items.

For explicit retrieval where specific filters may be applied, the available filter criteria should be provided at the end of the list. A button may be offered to show "all" information, which will then load the remainder as an Infinite List—just a bit at a time. A series of selectors may be provided to allow selection of a specific limited set of information (such as all items within the past 30 days, 60 days, one year, etc.).

Some information sets may be browsed in arbitrary order. Compare an address book with an email client. Arbitrary-access lists will require Location Jump, Search Within, and/or Sort & Filter controls to provide access to the desired information without excessive scrolling (and excessive data retrieval). In these cases, a *Wait Indicator* will be more commonly seen, as predictive loading cannot be smart enough for all cases. The action of jumping or searching should always load the retrieved data set, and should not require an additional step to load the items.

Under absolutely ideal conditions, the operation of the Infinite List is transparent to the user, and the list appears to be a conventional Vertical List.

Practically, you cannot plan on the ideal state. If the user reaches the end of the retrieved information, a message area should become visible. Depending on the variation employed, this will be a Wait Indicator indicating that more information is being retrieved, as shown in Figure 2-7, or a selector to manually load more information. Error messages may also appear here if there were errors in retrieving the information. This informational area should be much larger than a single line item and be visually differentiated by color or style. A good practice is to have the information at the top of an area that is at least as tall as the viewport. The user can continue scrolling into the empty area, which implies that more information is available and adds to clarity when scrolling at high speeds.

Retrieve a count of the list length and display the scroll bar—or other scroll position indicators—as though the whole set has been loaded. For web pages, you can do this by loading the full number of elements needed (on a web page, the div and li elements), and ensuring that they are drawn to the right height but not populating them yet. Do not allow the scroll bar to readjust when additional content is loaded. For unlimited data sets (if scrolling decreases constraints, such as a geographical search sorted by distance), it is best to eliminate all location indicators such as the scroll bar, and to use counters and other such labels.

For all types of explicit lists, the end of the list may not be the bottom. If the user can encounter the end of the information when scrolling up, the explicit request should be at that point, above the list. Do not use a single location, or one that can be misleading.

There are almost no good design reasons to require explicit lists for arbitrarily ordered sets. Use predictive retrieval whenever it is technically feasible.

Data costs are going down, and more and more users have unlimited (or very large) data plans. Unless you know your users are on limited plans or bad networks, load much more information than they likely will need, to prevent them from running into the end of the list.

Thumbnail List

Problem

A Vertical List is called for, but you need to include additional graphical information to assist in the user's understanding of the items within the data set.

Almost every technology that allows for lists can simply and reliably include images adjacent to each text label.

Solution

Figure 2-9. *A Thumbnail List uses images, icons, or photos, usually to the left of the text, to help define and differentiate line items.*

The Thumbnail List is the simplest of concepts. You use any type of Vertical List, but in addition to text information, you add a small image, or thumbnail, next to each item, as shown in Figure 2-9.

Information displayed, and interactivity with the listed items, is the same as for any other list view. Variations of the Vertical List can be employed as a Thumbnail List.

Variations

Although the most common use of a Thumbnail List is where the thumbnail is expected to be the key identifier, and therefore it is the leftmost item, any arrangement is possible. Instead of assuming images are most important, try to lead—by position, size, or contrast—with the key element the user will use to index the list. This key element may be text or graphics.

If any one line item has no thumbnail image, it can be loaded without any thumbnail image at all, as shown in Figure 2-10, to place additional emphasis on the data which is relevant and specific to the item.

You can combine additional list types with this. The Thumbnail List may also be a Select List and can easily be loaded as an Infinite List.

Interaction details

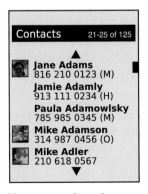

Figure 2-10. *Do not overuse placeholder icons. As shown here, gaps may be left and only actual custom images presented with an improvement in scanability and comprehension.*

The addition of the thumbnail to the list items affects no real changes in the interaction from any other list type.

For lists of information such as address books, additional options may be presented on-screen or within option menus to encourage setting up more contacts with custom avatars. These options should also be presented for individual items, either within the details for the list item (or even when an item is in focus but not yet selected) or within the detail or edit view for the item in question.

Naturally, you should make the thumbnail itself selectable. This may be as a result of the entire line being selectable, or you may choose to attach a special function to the thumbnail.

Figure 2-11. Any item that can be represented by an image can be displayed as a thumbnail list. In this example, websites use their favicon as a thumbnail.

Thumbnails are commonly cropped to a common size and shape, such as square. You must place sufficient room around the image to allow it to be differentiated from those on other lines, to be seen around the bezel, and for text adjacent to it to be readable.

Consider use of borders or shadows on thumbnails, so images masked to a color similar to the background, or that are in content similar to the background image or color, can still be detected as thumbnails. Favicons, as shown properly in Figure 2-11, often fall into this trap.

When no specific thumbnail image is available, you may replace it with a default icon or other image. However, do not repeat these too often. The value of thumbnails is in their ability to be identified at a glance. Too many placeholders can ruin this function for the real, custom images.

You may pay less attention to differentiating individual line items, as the thumbnails and their spacing (especially when used as the leading item) will solve this problem.

When the thumbnail performs an action that is different from the adjacent text, make sure this is clear. Preferably, make sure it is information carried by the thumbnail itself. For example, a call history log may only list in text the number and the date. Selecting those areas would reload the call details, allowing for callback, while the thumbnail would serve as a link to the Contact list itself.

Avoid using this pattern when there are not many unique thumbnails. Address books, for example, all seem to be Thumbnail Lists now, but are very often populated mostly or entirely with the default icon. This adds no value.

Even when only a few items have a default icon, carefully consider how to treat this. Images may be eliminated, a large set of default icons may be used instead of just one, or some other semi-unique visual identifier could be loaded based on other known attributes.

Do not crop images if the image itself is the key identifier. In a list of people, the thumbnail is an icon, but if an image list is being displayed, aspect ratio and full-frame images (even if thumbnails) may be crucial to identify the correct image.

Pay as much attention to type size, color, spacing, and weight as you would with a Vertical List. Do not assume that all text is equally secondary. Some users will rely on text, the image may not be helpful, or text may help differentiate similar images.

If text is truly not helpful, do not use a Thumbnail List. Use another pattern, such as Carousel or Grid, to make the selections.

Fisheye List

When designing for a scroll-and-select device, and a Vertical List is called for, you discover that displaying small amounts of additional information for each line would provide additional value.

This function relies on a focus or hover state, so for mobiles it is only really applicable to scroll-and-select devices. Many application frameworks support this feature natively. Web interaction will require scripting.

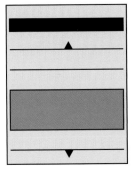

Figure 2-12. *In the Fisheye List, the item in focus expands to display additional information. All other items remain at their minimized state.*

In a Fisheye List, items not in focus appear exactly as they do in any other type of Vertical List.

When a list item is in focus but is not selected, it will expand vertically, as shown in Figure 2-12. This additional room will be used to display more information about the list item.

The term refers to the *fisheye* lens, which has severe spherical distortion. Items in the center of the viewing area are much larger than those at the edges.

Variations

As with other list items, details of arrangement are almost unlimited, and are not considered variations in principle.

This only works with selection methods that allow focus (or hover state) separate from selection. Current touch and pen devices do not support a hover or focus state, and so cannot use this. If you need something similar for those devices, or to preserve similarity of function on several device classes, you can apply the Windowshade pattern, or a Pop-Up, to each line item instead.

Do note that hover state indication is possible on many devices with touchscreens, if they also have a five-way-pad directional control. You would have to design the list on such a device to operate in both modes. If a Fisheye List is required, one of the touch actions may need to cause different behaviors on the list.

Interaction details

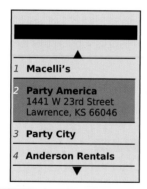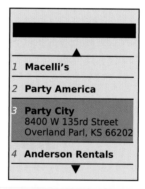

Figure 2-13. *As each item comes into focus, it is expanded and additional information is presented. This only works well on scroll-and-select devices, and generally one item will always be in focus, and expanded. Here, the second item is in focus, and then the user scrolls down one to make the third item in focus.*

On hover (when an item is in focus), the line item will expand, pushing down all items that are below it. See Figure 2-13. The new information will not overwrite other items in the list. Any Pagination or Location Within indicators will have to take this into account, as the number of items displayed may change.

When technically feasible, you should make the line item expand by transitioning to the new size, and revealing the new information as an animation.

Selection of an expanded item may perform any action desired, as with all other lists.

Presentation details

Figure 2-14. *When the list is expanded, additional information is presented, and this doesn't always have to be text. Here, an image of the contact uses alignment reminiscent of a Thumbnail List to expand on the information in the line. This is also one good solution for Thumbnail Lists where few of the list items have an associated image.*

An expanded list item will open downward, leaving the top of the list item in the same place.

You should leave unchanged and in place most or all of the layout and content that is visible before expansion. Add additional information in the space available below this information.

Additional information should generally be of lower priority, and so in smaller type or with lower contrast. Optionally, if truncation of the original list item information is a serious problem, such as with a book title, you may make the remaining information (at the same display type) wrap to another line. Rarely, when the point of the Fisheye List is to expand key information, the design may consider expansion to be from the middle, pushing any secondary information to the bottom of the list area and using the newly expanded middle space to display the longer title.

You can also display other types of information. For example, you can expand a text-only list to include a thumbnail, as shown in Figure 2-14, either because not all information has images or to provide for larger thumbnails than would otherwise be practical to display. An address book can expand to include call record information, for instance.

You should use care to ensure that the size of the expanded list item is clear and does not appear to be two items, or becomes confused with other list items.

Antipatterns

Images, when used as a Thumbnail List, should generally not expand or change size. This throws off the alignment of text and reduces the amount of text that is visible, significantly reducing the user's understanding of the state change and slowing her responses.

Do not confuse this pattern with others, such as Pop-Up. The expanded state pushes following items down the page, and never overlaps them.

Whenever possible, provide transition animations. Avoid jumping to the open state, and never refresh the entire screen. Users will not see the relationship between the two states and the value of the pattern will become lost. If you cannot do this for technical reasons, use another pattern.

Carousel

Problem

You need to present a set of information, most or all of which comprises unique images, for selection.

Carousels are graphically intensive, so they may not be easy to implement on devices without built-in support. Web implementations will need scripting or advanced HTML and CSS support.

Solution

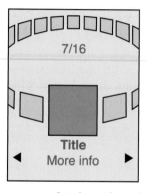

Figure 2-15. *Carousels may be laid out as actual circles, with much of the array visible to the user. Use of perspective and other 3D or simulated 3D effects assists with this.*

A Carousel displays a set of selectable images, not all of which can fit in the available space. It simulates a continuous strip, stack, pile, or other arrangement of images, as exemplified in Figure 2-15, only some of which can be seen at once due to the limited size of the device's viewport.

The images presented should be similar in size and aspect ratio. You can include a small amount of additional information about each image that can be displayed adjacent to the image as it comes into focus. In general, the in-focus item is larger, clearly has focus or actions attached to it, and so on.

The user can scroll through the set via any number of methods. Indicators should be present, but may not be based on OS standards. Scrolling brings more things in from the sides (or wherever), and changes the item in focus—it must always be live so that the items move instead of appearing in the middle of the page, or jumping from one position to the next.

Do not confuse this with a Slideshow, which only presents one image at a time, full-screen. The two are closely related, and users often switch between them within a single interaction. If you cannot implement a Carousel for technical reasons, use a Slideshow instead, or in combination with a simpler display method such as a Grid or Thumbnail List.

Variations

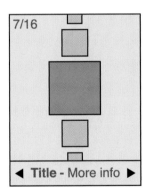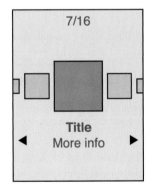

Figure 2-16. *Vertical or horizontal arrangements may be used, depending on the type of data and the principles upon which the device OS already operates, and sometimes simply on the amount of space available.*

You can design a Carousel to be presented either vertically or horizontally, as shown in Figure 2-16. Which design to use depends on the space available in the layout, device display or interaction paradigms, the nature of the items being displayed, OS conventions, and user expectation. Sometimes you may even wish to switch within a single implementation, due to device orientation changes.

Two-dimensional carousels use size differences to create a simple but false sense of perspective. Two-dimensional carousels may slightly overlap the items that are not in focus, to save space. Images should always be identifiable from the partially exposed image.

Three-dimensional carousels usually make the entire circle of the collection visible by using perspective. When oriented horizontally, the back of the circle is tipped toward the viewer. And contrary to the physical metaphor, the items along the back edge of the display are shown as though the image is on the back of the card, or as though the image is two-sided. Presenting the back of the image doesn't seem to work.

Note that three-dimensional carousels can present a more engaging interface, but you can rapidly run into issues with density and complexity of information. They tend to be more expensive in terms of screen real estate for the display and in processing needed for any image conversion or animation.

Interaction details

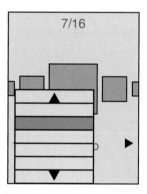

Figure 2-17. *The graphical nature of Carousels limits effect and actual on-screen space. Information and options will often need to be presented in informational bubbles, or as option menu items.*

The item in the front-middle of the Carousel is the item in focus. You will make this item subject to the conventional interaction criteria for any in-focus item. For example, when the user makes menu selections, as in Figure 2-17, or presses the OK/Enter key, the actions will affect this item.

For touch or pen devices, any visible image may be selected. You may choose whether the result of selecting an image not in focus is to bring it into focus, or to immediately carry out a selection action. This is dependent on the context and process in which the carousel is being used.

Scrolling may be by gesture, directional keys, or selection of scroll functions on the screen. Scrolling must always be "live" or show the items moving on the screen. If this cannot be done, do not use the pattern and select another method. Be sure to support multiple methods so that touchscreen devices with directional pads work with all input devices.

You can identify position in the list with a conventional Pagination or other type of Position Within widget. The total number of items should be easily obtainable or shown, but other features of these widgets may not apply to the particular data set or presentation.

If it makes sense for your information set, a Location Jump widget may be applied. For example, a photo gallery may use this to avoid putting images in folders, and allow jumping to dates that the images were taken instead.

In many cases, the action on first selecting an image in the Carousel is to display it full-screen, for a better look. You should always switch to a Slideshow at this point, and not require the user to go back to the carousel to select another image. The user may want to compare several images to select one for sending or deletion. Do not make anything difficult by excessively sticking to a specific interaction paradigm.

Presentation details

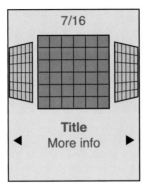

Figure 2-18. *Use all effects correctly, throughout the experience. Skewed images work well for presenting Carousels, but can be demanding to processors.*

The item in focus is centered in the screen (or in the array, if the array is off-center for dimensional simulation). You should also indicate this by size, by adding a border, or with other indicators. A common addition is the inclusion of text labels, such as the name, to only the item in focus.

Images not in focus should have attributes other than position to indicate they are in a secondary position in the list. Size and angle (skew to simulate not-facing) are the most used. See Figure 2-18 for a simple example.

The last image that is not in focus but is visible should imply that there are additional images outside the viewport. This may be via size (shrinking to infinity), via angle (tilting "up" to become perfectly thin), or by having part of the image not on the screen. Each of these should lead users to believe there is additional information if they scroll.

You should use additional scroll indicators whenever you can. Simply add arrow/triangle indicators to the sides, or immediately to either side of the labels for the item in focus. If they are selectable directly, the indicators should change when they are selected and while the carousel is moving to indicate their activity.

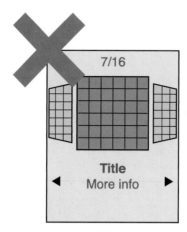

Figure 2-19. *Skew images correctly. Do not just crop flat images to a skewed shape; users can tell the difference.*

Preserve the integrity of the image used. Do not use programmatic shortcuts visible to the user:

- If the image is skewed to create perspective, skew the entire image. Don't clip off corners of the image to simulate the skew, as demonstrated in Figure 2-19.

- Do not squish the image to change its aspect ratio. If necessary, add black bars or crop the image to fit a consistent space.

- Do not flip the image when displaying the back of a card, as in three-dimensional carousels.

Never use cheats, and come up with a different display method, or switch to another pattern such as a Grid or Slideshow.

Grid

You must present a set of information, most or all of which comprises unique images, for selection.

A Grid display does not require unusual or special conditions in order to be implemented. It may be easily used for applications or websites on any device that can display the images adequately.

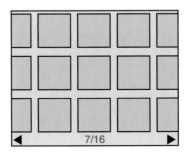

Figure 2-20. *Items may be displayed as a grid of images, with little or no text associated with each one.*

A Grid displays a set of selectable images, not all of which can fit in the available space. It simulates a continuous array of images, only some of which can be seen at once due to the limited size of the device's viewport.

The images presented should be similar in size and aspect ratio. You can display a small amount of additional information about each image as it comes into focus. The in-focus item will be indicated in some way, but does not change size.

The user can scroll through the set via any number of methods. Indicators should be present whenever acceptable to the OS standards. Scrolling brings more images in from the sides, and changes the item in focus. You must always use live scrolling, so the items move as part of the continuous list, instead of incongruously appearing in the middle of the page, or jumping from one position to the next.

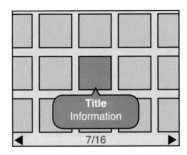

Figure 2-21. *Items in focus should clearly be indicated as such, and should display text labels such as titles and dates.*

You should only use a Grid when one axis of movement works well for the display and organization of your information. This one axis can be either vertical (Figure 2-23) or horizontal. Two-dimensional grids will generally scroll in the longest direction of the viewport, regardless of orientation. No movement should be allowed in the short direction.

All Grids simulate an array of images on a flat surface, such as a desktop. See Figure 2-20. But they do not have to do so literally; the backdrop may be anything, or a solid color. Today it is common to use the conventional backdrop image, which indicates that the images in the Grid are floating above a relatively distant surface.

Some 3D effects may be used, especially with accelerometer-equipped devices, as loading or scrolling animations, to allow viewing more images by tilting and so forth. See Figure 2-24.

A rare alternative type of Grid arranges the thumbnail images as though on a sheet or globe, allowing scrolling in any direction. This may be useful for emphasizing certain types of data relationships between images in the grid. This is very similar to the Infinite Area pattern, but the individual thumbnails align it more closely to the Grid.

Interaction details

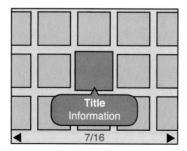

Figure 2-22. *Grid items may have additional effects applied when in focus, such as zooming slightly to display the image in more detail.*

Any visible item may be selected. Selection may result in any of a number of actions; you may display additional information, or perform drilldown methods such as a Pop-Up, a Slideshow, or simply displaying a new page of information.

If your information set can be applied to it well (such as for all images photographed on a certain day), a Location Jump widget may be useful.

You may offer scrolling by On-screen Gestures, Directional Entry keys, or selection of scroll functions directly on the screen. Scrolling must always be "live," or show the items moving on the screen pixel by pixel as the movement is made. If you cannot implement this due to technical constraints, do not use the pattern and select another method instead.

For most data sets, scrolling should be circular, where scrolling past the end of the grid will load the next end of the grid. This prevents the user from having to scroll large distances in many cases.

Array items so that lines are read in the short direction, and then jump to the opposite end on the next line. For example, for a photo album with a grid scrolling horizontally, the newest item is in the top-left corner. The next-oldest is below it, and so on to the end of that column. The next-oldest item is at the top of the next column to the right.

Presentation details

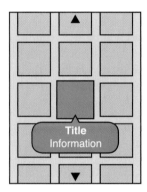

Figure 2-23. *Grids may also be displayed vertically.*

Position in the list should be indicated with a conventional Pagination or Location Within widget. The total number of items should be easily obtainable or shown, but other features of the Pagination widget may not apply to the particular data set or presentation.

The item in focus must be indicated, as shown in Figure 2-22 and detailed in the Focus & Cursors pattern. The in-focus item may display additional information, such as labels, or metadata. This may be presented below the image, or in another area reserved for this information, such as a bar across the bottom of the entire image area. This is a good place to use a Tooltip or Annotation, as shown in Figure 2-21, to hold this information. A small Pop-Up dialog may also serve this need, and may be easier to implement.

For pen and touch devices, without explicit in-focus or hover states, the focus paradigm is usually eliminated and no attempt is made to simulate it. This is exactly the same as it is with position in the Carousel pattern. However, be sure to support scroll-and-select actions for devices that have Directional Entry controls, even if they are secondary to a touchscreen.

For circular scrolling grids, a gap or other indicator should be displayed to mark the start and end of the grid. This may be as large as the viewport, so a very deliberate scrolling action is required in order to scroll to the other end of the grid.

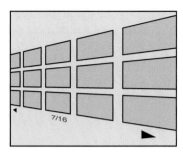

Figure 2-24. *Effects may also be applied to the entire grid. Tilt effects such as this may also be coupled with sensors, and related to the tilt angle of the device.*

Images at the end of the viewport, along the scrolling axis, should imply that there are additional images outside the viewport by being cropped and not fitting entirely within the viewport. This is to lead the user to believe there is additional information if she scrolls; if she just tries to scroll enough to see the rest of the image, she will also become aware of additional images.

Additional Scroll indicators are strongly encouraged. Simple arrow/triangle indicators to the sides, or immediately to either side of labels for the item in focus, are common. If they are selectable directly, the indicators should change when selected and while the carousel is moving to indicate their activity. Scroll bars or other such indicators may or may not work depending on the design of the page.

Antipatterns

Use caution when choosing the selection action. Do not make it too difficult to do alternative tasks, such as clicking on adjacent items or functions attached to the image, by making a single selection task too easy.

Avoid making the display overly complex. Despite the information density of the grid, a conventional one-axis scrolling grid is easy to control. Two-dimensional (spherical) grids present challenging mental models, and may be difficult to control.

Only use the Grid display if there is sufficient room and resolution for the thumbnail images to be viewed. If the screen is too small or is of insufficient resolution, use another solution instead, such as a Slideshow.

If live scrolling is not available, avoid using this pattern. Jumping screens or refreshing will not present the items—or their relationship to one another as a set—as clearly to the user.

Be sure to provide plenty of space between images. Do not butt them against one another, as it is difficult to discern individual images when they are similar in content. Also try to account for common OS elements such as the Annunciator Row, and the Fixed Menu.

Film Strip

You must present a set of information, which either is a series of screen-size items or can be grouped into screens, for viewing and selection.

Several mobile OSes and programming platforms support the Film Strip directly. It can be simulated effectively on many others. It may be difficult to use this in certain platforms, such as within web pages, due to scrolling conditions.

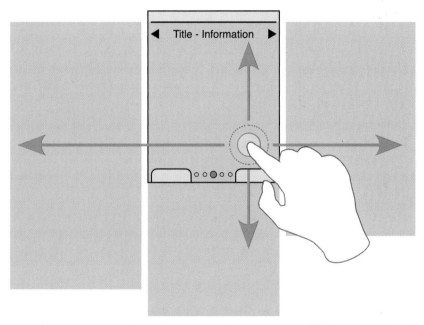

Figure 2-25. *When any one panel in a Film Strip is in focus, it is presented full-screen. In this condition, it cannot be differentiated from a slideshow.*

As with many of these terms and concepts, the filmstrip is a real machine-era display device, a sort of cheaper way to display a slideshow. Instead of the slide film being cut up and placed into slides, however, the film was kept on a continuous roll and fed into a very simple machine that displayed an area as large as one exposed frame. The transition was not hidden but out in plain sight for all viewers to see. One frame slides up while another takes its place, separated by a small gap. Clever presentations embraced this paradigm and had images span the frames, or discussed processes as continuous vertical flows, with arrows leading from one frame to the next.

In mobile devices, a Film Strip is similarly a series of screens, displayed as a continuous strip, with small spaces or markers between them. When a particular screen is centered it fills the entire viewport. During a scroll, part of two screens and the divider can be seen at the same time.

For viewing individual items, such as images, you may fit the entire image to the viewport. For collections or groups of information such as articles, lists, or groups of icons, the individual screen may also scroll up and down.

Variations

The Film Strip is used in two distinctly different ways.

When displaying full-frame items, such as images, this essentially is a Slideshow where the relationship between images is more clearly defined. You might find this to be helpful to your users when transitioning to this view from a Carousel or other list, where the relationship was explicitly stated.

When displaying complex items, such as a grid of icons or a page of text content, the individual frames are considered pages, either in a process or in a display of equal information. A common use is for a multipage Home & Idle Screen, as shown in Figure 2-26. The basic interaction of pages sliding in from the side may also form the basic interaction paradigm for the entire application or OS.

Rarely, you may wish to implement the Film Strip as a large grid, with scrolling between frames in both axes, or possibly even at angles. The full-screen nature of the individual panes makes this different from the rare version of the Grid or Infinite Area pattern.

Interaction details

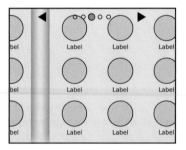

Figure 2-26. When scrolling or dragging from one frame to another, a gap between frames will be displayed. Shadows or some other effects should be used to make the distinction between frames clear.

When you use the Film Strip to display full-screen content, such as images in a Slideshow, only allow scrolling to reveal more images. Scrolling is most effective when performed side to side, or horizontal to the ground. This means you have to account for Orientation changes and make the display and interaction work in both portrait and landscape modes.

If you use the Film Strip to display complex, page-level information, the pages do not need to be restricted to the size of the viewport. Scrolling sideways via touch or with hardware Directional Entry keys will perform the usual action to reveal a new page, but conventional vertical Scroll may also be used within the page currently in focus. This is explained with gesture scroll in Figure 2-25.

Other methods of accessing the adjacent pages, such as clicking on a Link to reveal it, can be designed to look and feel related but are outside the scope of this pattern.

Scrolling must always be "live," or show the items moving on the screen. If you cannot do this for technical reasons, the point of the pattern will be lost. Users will not understand the relationship between the frames. Do not use the pattern and select another method.

When designing for scroll-and-select devices, or touch devices with hardware Directional Entry keys, make a single click load the entire next frame, by animating a sideways slide. For gestural scrolling, enable dragging so that when the frame divider is about halfway across the screen, it will "snap" to the frame in the gestured direction. If the user's touch input is stopped at this point, the next frame will slide into view anyway. Any movements that are smaller than this will revert back to the starting frame. For all of these, including snapping back to the starting point, the action should be animated.

For full-frame displays, such as Slideshows, include acceleration or key repeat functions to allow the user to scroll across several frames quickly. This should work for both touch and scroll-and-select devices. Unlike high-speed scroll through Vertical Lists, at each frame in a Film Strip, the scroll should pause for a fraction of a second. The time to transition may be greatly increased.

When not suppressed for full-screen display, the Annunciator Row should be fixed to the viewport and should not scroll.

When on any single screen or page, you can use conventional design elements, including scroll bars, Titles and Fixed Menus, or Revealable Menus. Or you may hide these if the frame is a full-page display and they are not appropriate. When the scrolling action is taking place, these items should generally be suppressed, to prevent users from becoming confused as to what is a selectable item.

For certain types of data, or certain interaction models on the device (e.g., especially scroll-and-select devices), you will want to add Pagination. This may be visible at all times, made to be visible via an option, or made to be visible only when scroll has otherwise been initiated. Unlike the other widgets, Pagination is generally available during the scroll actions.

Circular scrolling is encouraged when practical, but may not apply to certain types of data. It may be important, for example, if you have a list of date-ordered information that has a distinct starting point, and proceeds from there.

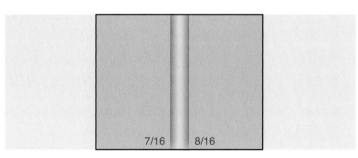

Figure 2-27. *When scrolling, the panels are drawn at full resolution, but only part of each one is visible.*

You must make a clear delineation between the individual frames. The best solution is to display a gap, as though each frame is a physical object, floating above another surface, as in Figure 2-27. If this is not practical, you can sometimes get by with divider lines, or by creating a "gutter" from the margins along the edge of each frame. Without some indication of the edge, the relationship between elements while moving may not be clear.

A Location Within widget should be displayed if there is no Pagination widget as described earlier. Depending on the type of data used, this may be visible only when scrolling, or at all times.

When on any single screen or page, you should follow conventional design methods for page design, and include common elements such as Titles.

Do not allow horizontal scrolling within a single panel of the filmstrip. For example, if you are displaying a list of images, zooming in must disable the ability to scroll between frames, or must not be allowed. Without this, it will be unclear where the edge of the scrolling within the image is, and users will unexpectedly jump, losing their place.

Make sure each vertically scrolling area is entirely independent. Scrolling to the bottom of a page and then moving to the right will load the top of that page.

If live scrolling (animating the transition between screens) cannot be implemented for technical reasons, the pattern being used is not a Film Strip.

Slideshow

You have a set of images, or similar pieces of graphical information, which must be viewed in detail.

Large-area display of graphics is easy and common, though there may be limits to the ability of some platforms such as web pages to display these at truly full-screen.

Figure 2-28. *In most modes of a Slideshow, there is nothing to see but the content, which is presented full-frame, with no scroll bars or labels.*

The core of the Slideshow is that each image is presented full-screen, as boringly shown in Figure 2-28, with a function to transition to the previous or next image in the series.

You will find this very often used exactly as its namesake, the slideshow. Much like presenting a series of vacation photos in the past, this allows the user a way to view a series of photos or other images. It can also be used to view other types of information summaries, such as consolidated sports scores.

Transitions between slides may be of any type, such as a cross-fade. This makes the Slideshow suitable for most devices, which may have technical difficulties with other types of patterns. If the images presented as though the images are on a continuous strip, this pattern becomes the Film Strip instead.

Do not confuse this with a Carousel, which presents all images in a continuous and visible series. The two are closely related, and users can often switch between them within a single interaction.

Variability in Slideshows really concerns only the degree of control that may be imparted to each slide. Some will be simple viewers, and no individual control is available. Some will allow, for example, editing of the images. Controls may be presented on-screen, or under option menus.

You should make sure the ability to switch between slides complies with the application or OS standards surrounding it. Gesture, directional keys, or selection of "Next" and "Back" functions on the screen are all suitable.

For image slideshows and some other types of data, additional options should be available, such as randomization, or ordering by various types of information. These may, of course, be available only under menus.

For certain types of data, or certain interaction models on the device (e.g., especially scroll-and-select devices), you will want to include a Pagination widget to allow the user to jump to specific locations in the list. This may be visible at all times, made to be visible via an option, or made to be visible only when scroll has otherwise been initiated.

A Revealable Menu should be made available via the expected (or an easy-to-discover) method such as that shown in Figure 2-29, to offer additional information or functions for the entire slideshow, or for individual slides. Even if the OS paradigm is for a Fixed Menu, the menu should be hidden by default to grant as much space in the viewport as possible to the slide information.

Circular scrolling is encouraged, but may not apply to certain types of data. For example, it may be important with date-ordered information that the set have a distinct starting point and proceeds from there.

Presentation details

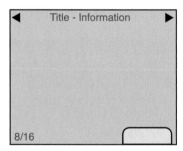

Figure 2-29. *If needed, or in certain conditions, information about the image should be presented on the screen. Soft keys or soft-key-like indicators may also be shown on the screen to allow access to the option menus.*

The amount of extraneous information presented depends on the intended use of the slide information. Pure slideshows will specifically have no text, icons, or other items on the screen. You may even wish to suppress most or all of the Annunciator Row. This will become especially important as more devices have high-quality video-out capabilities, and mobile devices may be used as the source for large, projected presentations.

Design supporting information, such as filename, dates, and the links to options, should be nonintrusive and should obscure as little of the slide image as possible. Although specific implementations will vary based on user needs, you may find it useful to restrict the information presentation to when some selection is made (such as a tap on the screen) or after a pause of several seconds on a single image.

OS-level Notifications should usually be hidden by default, to give as much space in the viewport as possible to the slide image. When other information is presented (such as option menus, or image metadata) these may be presented also, in their normal location and style.

If any additional information is presented as an on-screen overlay, a Location Within widget should be employed.

Antipatterns

Slideshows are not suitable for significant interaction with items within the individual slide. Although charts, graphs, text, or tabular data can be displayed as a slide, you will find that allowing selection of a single component, such as a link from part of the data, is not easy with the slideshow. If this is required, use another pattern, such as Film Strip for interacting with individual screens, or Infinite Area for interacting with arbitrarily large sets of graphical data.

You must take care to find the correct balance between information presentation and clutter. When you start to add all the useful controls, metadata, and indicators of position, you will rapidly run out of space and begin obscuring the image displayed. Consider the use of the product and the users. Push less-used items that have to be retained into option menus.

Do not develop new types of OS-level displays in order to save space. If battery level, notifications, and so forth are possibly important information within the Slideshow, use the normal icons from the Annunciator Row and hide the rest, or hide everything that is noncritical, to save space and avoid distracting the user.

Infinite Area

Problem

You have complex and/or interactive visual information, which should be presented as a single image, but is so large it must be routinely zoomed into so that only a portion is routinely visible in the viewport.

Infinite Area is best used with information considered actually infinite, such as maps of the whole world, and therefore usually requires a reliable data connection or very large local data capacity. The dynamic loading and reloading is difficult to implement or not supported by all platforms, such as some older web browsers.

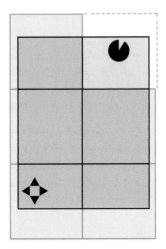

Figure 2-30. *Infinite areas break up the visible regions into rectangles of graphical data. When they are first loaded, or when the user scrolls too far, the user may briefly see lazy loading indicators, where data has not yet loaded.*

As we already mentioned, maps are the most common implementation of this, but there is no reason other information sets cannot borrow from it, and some demonstration projects already have used similar technology. You may find this to be a good way to display other photographic information (using very high-resolution, composite images) or to allow users to drill into complex infographics.

The entire breadth of the data set is generally available at every zoom level. Scrolling reveals more of the information. Zoom instead can be considered the depth of the data.

Whatever you are loading, the entire data set is considered to be a large, two-dimensional graphic, though these are generally actually broken into frames, as shown in Figure 2-30. Smaller subsets can be viewed as though "zoomed in" to portions of the larger image. Zooming reveals more information, generally with newly rendered images, which reveal additional layers of information. On a map, at the city level, you only see highways labeled, and street names only appear when they would be clearly useful and not clutter-inducing.

A key variation is to provide only two-dimensional scrolling of the data set, as shown in Figure 2-32. Consider a large graph of historical information, such as stock prices or key moments in history. At the far right is the current date, with past dates going off to the left, far off the screen. You can provide scales as well as all other controls to both pan and zoom in for further detail.

Two information-anchoring paradigms are also available. These may be encountered in the same application, and vary based on the immediate user task. However, they are different cases, so you should be aware which one you are using so that you can address them distinctly in design.

- In the first case, live data uses a real-world anchor point—or a simulated one, based on reviewing historical data; for example, the current position of the device, or the current time. As this changes, the data set moves to keep this dynamic point centered. Maps in navigation mode, keeping the current position centered as the map moves about it, are a typical example.

- In the second case, a fixed point or range is selected by the user and information is browsed around this point; for example, restaurants within 50 miles of a specific point (not where they are currently located), or a date range for historical graphs. Zooming of the data to fit the range selected, and other design features, will accompany this sort of behavior.

Don't be limited to the existing examples. You can also use predictive data, where range-limited data is presented for predicted future tracks, such as restaurants you will be passing on a planned route, or expected trends for other types of information.

The Infinite Area pattern, especially the two-dimensional version, can work well as a subset of a page, and does not need to take up the entire viewport. This makes it easier to implement on platforms that do not allow taking over the full screen.

Interaction details

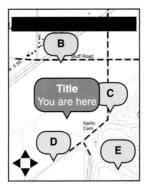

Figure 2-31. *Graphical data representing the real world, such as the map shown here, is very commonly used as an Infinite Area.*

You should, as with all of these Display of Information patterns, support scrolling with all the methods available on the device. In addition to the built-in gesture support or hardware keys, you might find it helpful to add on-screen buttons or other controls.

Zoom controls should be on the screen, or easily accessed. Zoom by gestures (e.g., pinch and unpinch) may not be discoverable, even if it is via a standard OS control, so an explicit backup to zoom control should be available.

Layers will be automatically controlled by the system to provide details as the user zooms. You can also offer optional layers, such as satellite versus map views, to the user as explicit controls. Optional methods of viewing data should be offered whenever they assist in understanding the data set. This can include manual control of the automatic layers, such as the ability to turn on and off road names at different zoom levels. You will have to design the system for the typical case, but not all data is typical, so user control can be helpful. In rural Kansas, roads are very far apart, and names can be useful at small scales; in downtown Boston (where every alley is named), even at maximum zoom it can be hard to see anything but the road names.

A key value of this pattern, or any information visualization like this, is that items within the data set may be selected and acted upon. For certain data sets, any point can be selected and actions taken. For a map, any point can display attributes, be saved as a favorite, and offer directions. You can offer such actions from option menus, or from more directly interactive features such as Annotations, as shown in Figure 2-31.

Presentation details

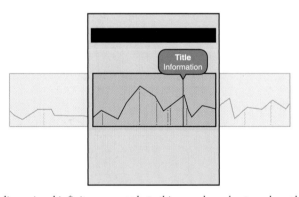

Figure 2-32. *Two-dimensional infinite areas apply to things such as charts and graphs. A single axis, such as time, is effectively infinite, and the other axis is of a very limited scale. Any infinite area may also be a subcomponent of a page, instead of a full page.*

Even if they are not used much, you may find that the presence of on-screen controls for zoom, scroll, and other functions is a great help in communicating to users that these functions exist. For complex interactive items, you should always consider the impact a visible control has, not just on the actual interaction, but on the perception of that interaction, and the indirect affordance, or communication, that an interaction is available.

In most view states—zoomed in to a portion of the image—user scrolling will require loading additional image data. These images are generally stored and loaded as panels, though when loaded the edges are seamless. You should always load at least one set of panels beyond the current viewport. When you can, develop simple predictive algorithms (based on scroll actions and other likely behaviors) to load additional panels in the direction the user is likely to scroll.

Panels that have not yet been loaded should display either a scaled (blurry) version of their previous zoom level or an explicit Wait Indicator.

You may receive great pressure from many sides to reduce prefetching as inducing too much overhead in data, processing, and so forth. Aside from arguments about making a compelling product, a well-designed system can, by the use of smaller-than-screen tiles, manage to use less data than a simple (but uninteresting) method of forcing the user to manually load the next whole panel. These require, among other things, overlaps in order to not let the user get lost, so pixels are loaded twice. Good Infinite Area design can be very data-efficient, and very engaging to the user.

Consider using 3D and other effects whenever it can add to the comprehension of the data. Tilting, or simulating elevation, may allow viewing relationships that are not clear in other ways (and not just for maps).

Scale must be indicated in some manner. This may be explicit, as marks for miles, but implied scale should also be used. For example, at close ranges on maps, roads can be depicted as actual width, and buildings can become visible. See the Zoom & Scale pattern for much discussion of this.

Decluttering is the process of determining how many layers of detail should be displayed for a particular zoom level. This includes not just the number of items, but how much detail is on a layer; a line on a graph, or a road, can be simplified to reduce the number of points drawn. The goal of decluttering is to provide as much relevant detail as possible without reducing readability and legibility.

Antipatterns

Use caution with clutter control methods. If you declutter too much, you may remove so much information that the data set is not valuable; too little and no information can be read. To add to the difficulty, portions of the data set may have more information, and single rules may not apply. For example, a declutter factor for a suburb works less well in a dense city.

Scrolling must be live. Do not refresh the entire page. When using scroll-and-select devices, or nongestural scroll controls, still show the data set moving from one part of focus to the other. Do not jump.

For maps, do not confuse GPS with "location." Many methods of finding location are available, and map applications should be available on devices without GPS, as well as on suitably equipped devices in which it is off, or cannot get a signal.

Do not assume all users will understand gestures, even those common to the OS. It may not be clear which methods apply to the particular application or element. Back this up with explicit controls, even if they are hidden in option menus.

Avoid the use of special effects, such as simulated 3D, that may cause data to be misrepresented or misperceived. Follow good infographics practices, such as those promulgated by Edward Tufte in his many books, lectures, and reports.

Select List

You must allow the user to make a selection—or several of them—from any of the large, ordered data sets discussed in this chapter.

The applications of this are essentially limitless, and can be applied to the simplest and the most complex of representations, and so can be used in one way or another for any type of implementation.

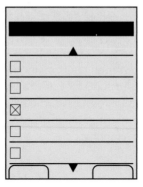

Figure 2-33. *The select list that is the most common and simplest to understand is the Vertical List, with selection checkboxes for each line item.*

You can turn any of the other patterns within this chapter, but especially the Vertical List, shown in Figure 2-33, and the Grid, shown in Figure 2-34, into a selection mechanism by simply adding a checkbox or some other explicit and visible selection component.

For many selection methods, simply tapping or pressing the OK/Enter key on hover will serve to meet the needs of single selection. The checkbox pattern is used for multiple selections (more than one item may be checked) or when the selection may involve investigation and confirmation before committing to the selection.

There are two variations to the select list. *Single-select* only allows one item to be selected, and then forbids subsequent selection, or deselects previous selections.

Multiselect allows you to check more than one item in the set. The limit is set by the application, or the requesting application, or may include the entire set.

Many select lists allow switching between these modes. This is useful when you cannot predict how many selections need to be made, but a typical number is just one, such as for selecting recipients of an email.

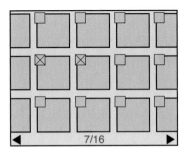

Figure 2-34. *You can make other methods of displaying large amounts of distinct data, such as grids, into select lists by simply adding selection mechanisms. A good use of this is to switch from a viewing mode to a selection mode, such as choosing photos from a gallery.*

The Select List pattern is likely to be implemented as a mode of another list. This mode may be entered contextually, due to a request from another application, or explicitly via user action within the list view. Very often, you will be reusing a list that already exists for another purpose. Any type of list may be used, including highly graphical ones such as the Carousel, as shown in Figure 2-35.

Most single-select lists perform the selection and commitment of the selection as one action. Once the item is selected, you should take the user to the next step in the process without forcing him to select a Next button.

For multiselect lists, when the user selects an item, the indicator will change immediately. When the selection method is made again, the indicator will change to the deselected state. If details are available for an item, such as viewing an image or selecting it, the checkbox should have to be hit with some precision, and all other areas may result in viewing details.

Depending on the function, a "select all" may be necessary. Even if functions against all items can already be performed with a dedicated function (e.g., a "Delete All" option menu item, separate from a Select to Delete function), it can be useful for alternative selection modes. The user may want to select most items, and shorten the time required to do so by selecting all of them and then deselecting the few items he wants to keep.

For all multiselect methods, you will have to provide a "Done" function to determine when to move to the next step. This should be contextual, and should carry a label associated with the larger task—"Delete selected items," "Send message to selected recipients," and so on.

Especially for large data sets, the selection process may be involved, so you should "consider" it as a set of user-entered data. If the list is exited before commitment—such as if the "Back" button is selected by accident—you should preserve the selection information. The exit may have been accidental, and when next carrying out the same task (e.g., adding recipients to that same draft message), presenting the last selections instead of an empty selection state may be perceived as helpful. This concept is discussed further in the Cancel Protection pattern.

Since some uses of the "Back" button are deliberate, do not force the user to accept previous selections. See the Clear Entry pattern for some methods of removing user-entered data without placing undue burden on the interface.

Presentation details

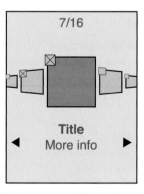

Figure 2-35. *Selection types where there is a strong difference between the in-focus and out-of-focus states, such as the Carousel shown here, should only allow selection when in focus. The selection condition should be displayed for all visible items.*

Selector items should be overlaid on the list items. Within a grid view, for example, the checkbox should be over the image itself, or otherwise not take space from the interface. You will also enjoy doing this, as it reduces the changes needed to make a Select List variant of your existing lists or views.

A variation of the Location Within widget may be employed to indicate the number of selections. This is especially valuable when there is an easily reached limit on the number of selections. You may also choose to use this conditionally; when the selection limit is high, the indicator can appear only as the limit is approached.

To prevent errors, commitment of the selections should be prevented until at least one selection is made. You can do this by hiding or by graying out the relevant functions. You should always try to reduce explicit errors by making them impossible to reach.

Antipatterns

Do not overthink the selector or indicator. Checkboxes work well. Radio buttons may seem to be the best choice for single-select items, but the web form paradigm is not well understood by many users. Novices are very common on mobile devices. Use checkboxes or similar-looking interactive elements for all selections.

Don't let checkbox styling get out of hand. The selected and unselected states should be not just differentiable, but identifiable without comparison, to all users regardless of their vision, and in all lighting conditions.

Do not impose burdensome errors. If the user must be forbidden from making another selection, simply disallow the user from doing so. If a message is required, place it prominently, but away from the focus, and do not make it a dismissal-required modal dialog box.

Be sure the selection method is well associated with the item being selected. Do not place checkboxes by floating metadata, for example, but on the image itself.

Always show selection state, even when the item cannot be selected. If a Carousel is used as a Select List (and it generally should not be), only the item in focus can be checked. However, make sure to show the checkbox for all items in view.

Chapter 3

Control and Confirmation

Quiet, Please

The lights in the theater dim. Voices die down. All eyes stare at the giant illuminated screen and silence overtakes the room. Projecting now, the movie begins. Starting from high above a city, the audience's view mimics the flight of a bird. Slowly, as the view trickles down below the clouds, a row of houses appears. Dropping lower, the view focuses to one house in particular. We enter the house; it's dark, old, and abandoned. Slower now, the camera leads us down the stairs to the basement.

The audience is coaxed into believing something isn't quite right. Attention is focused on the closed closet door, now bringing an increase in fear and tension. Something terrible is about to happen; the audience waits. The camera leads the audience closer to the door. Closer. Closer. Closer. Not a sound to be heard now. Then it happens! The sound of Lady Gaga's "Bad Romance" chimes loudly, breaking everyone's concentration. Heads turn and eyes seek to identify the location of the sound and culprit. The patron embarrassingly finds her mobile phone and switches it off.

Yes, it's annoying when someone forgets to turn off her phone. But hey, people make mistakes. Mistakes happen every day in our lives. Some mistakes go unnoticed, while others can be quite catastrophic. Some mistakes are caused by us, others by objects in our environment. But many mistakes can be prevented!

That Was Easy

How could the mistake of the lady not turning off her phone have been prevented? If only she had an "Easy" button to push that would have allowed her to start over and turn off her phone prior to the movie beginning. Maybe through a distributed cognitive network, her friends could have reminded her. Or maybe the situation could have been resolved if her mobile device detected that she was in a movie theater, and incorporated constraint controls, such as automatically defaulting her device to vibrate.

Unfortunately, life doesn't provide us with a personal "Easy" button to turn a complex situation filled with chaos and mistakes into one of error-free simplicity. Relying on friends can be, well, unreliable. Having our device provide controls and constraints to limit our careless mistakes seems plausible and doable. Therefore, as designers we need to equip ourselves with a general understanding of why we make mistakes to begin with.

Understanding Our Users

Figure 3-1. *Pop ups and similar controls are key to Confirmation, Sign On, and many other types of interactions due to their modality. The user is required to make a choice, and this both focuses the user's attention and prevents her from accidentally performing errors.*

We must accept that we are human and we make mistakes because our body has unique limits. We are limited in our cognitive processing abilities, which are constrained by capacity and duration. We have physical limits in our endurance and strength. We have ergonomic limits in our reach and rotation. We have perceptive limits in what certain electromagnetic and mechanical wavelengths we can detect and filter.

Mixed together with our limitations, we expend a lot of cognitive energy to process and interact with the enormous amounts of stimuli in our environment. Our attention on the task at hand will affect which environmental stimulus needs to be filtered, focused on, and stored. Think of the human mind as a leaky bucket that is constantly being filled. As more and more stimuli are collected through sensory memory, most will be lost due to filtering. Important stimuli will be processed and stored using our working and long-term memories.

Humans have also developed ways to reduce the mental load required to process information. According to Payette, this is possible because cognition is embodied, situated in an environment and distributed among agents, artifacts, and external structures (Payette 2008). We do not solely rely on our individual human limits to process information all the time. We can embed our knowledge of the world in objects that serve as episodic reminders to help us recall. We can distribute our cognitive load to multiple agents or devices. Consider a grocery list. We could try to store all that information in our heads and hope to recall it by the time we get to the store. But it is more effective to situate our cognition in a notebook where we can write down the entire list.

We then no longer need to recall each item. We just need to recall that the notebook contains the list. Further, we can distribute our cognitive load among others. Let's say we are in a baking class that teaches only how to make cakes. Rather than have everyone remember to buy all the ingredients, we can assign each person a specific ingredient to purchase. Can we reduce cognitive load and error even more? Yes. Let's distribute all cognitive load onto technology. What if your refrigerator monitors your shopping habits and cooking behaviors, and can automatically sense which ingredients you need? Then it sends a grocery list order via SMS to your local supermarket. You mobile device can confirm your order was placed, the amount can be charged to your bank account, and you can be notified when your order is ready to be picked up!

Control and Confirmation

Because not all devices are designed today to absorb all our cognitive load, as in the preceding example, this chapter will provide methods that help distribute that load to create enriching situations while preventing user error. Throughout this chapter, cognitive frameworks are presented to help us understand how people process information. These frameworks apply when interactions require control and confirmation:

Control
> This refers to respecting user data and input while protecting against human error, data loss, and unnecessary decision points. This is a key principle of mobile design.

Confirmation
> When a necessary decision point is needed, an actionable choice is modally presented to the user to prevent human error. Before adding modal confirmations, consider the following:
>
> – Is a decision point required here that needs confirmation from the user?
>
> – Will having this confirmation eliminate risk of human error and loss of input data?

Do not use confirmations arbitrarily or excessively. They will increase user frustration by:

- Stopping the user's goal from automatically happening by presenting a confirmation

- Forcing the user to read, understand, decide, and then act on the confirmation

- Increasing unnecessary mental load through the entire process

For example, let's say a user's goal is to enlarge an image on his mobile device by touching it. Would this particular situation require control and confirmation to prevent human error from occurring? And if so, can human error lead to a drastic consequence? In this example, the user's risk for error is low. Security is not compromised, and loss of data is not imminent and likely to follow. Therefore, having a control or pop up modally appear with a confirmation message asking, "Would you like to enlarge this image for viewing?" is not necessary. This will create user frustration.

Let's examine another example. This time a user is at an ATM with a goal to withdraw money. The user's risk for error is moderately high. Security can be compromised and money loss can occur. In this case, having a modal message appear to confirm the amount of money withdrawn may be viable and necessary.

Patterns for Control and Confirmation

The patterns detailed in this chapter are concerned with specialized methods of preventing and protecting loss of input data. In some cases, patterns listed in other chapters will offer alternative methods, and these will usually be cross-referenced within the pattern. For example, Pop-Up may be referenced when creating patterns found in this chapter. See Figure 3-1 for some real-world examples.

Confirmation
> When a decision point is reached within a process where the user must confirm an action, or choose between a small number of disparate (and usually exclusive) choices.

Sign On
> This pattern is used to confirm that only authorized individuals have access to a device, site, service, or application on the device. Theory and principles of privacy and security will only be alluded to, and are not discussed. Please find appropriate references for these.

Exit Guard
> This pattern is used when exiting a screen, process, or application could cause a catastrophic loss of data, or a break in the session.

Cancel Protection
> This pattern is used when entered data or subsidiary processes would be time-consuming, difficult, or frustrating to reproduce if lost due to accidental user-selected destruction.

Timeout
> High-security systems or those which are publicly accessed and are likely to be heavily shared (such as kiosks) must have a timer to exit the session and/or lock the system after a period of inactivity.

Confirmation

Problem

You have created a process that includes points where the user must confirm an action, or pick from among a small number of disparate (and usually exclusive) choices.

Confirmation steps are simple, logical parts of many processes, and can be easily implemented in any number of ways. This pattern describes the best ways to do so on mobile devices.

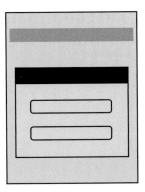

Figure 3-2. For Confirmation dialogs, the process is interrupted to present a series of choices. For touch and pen devices, arrangements like this, with stacked buttons, are common and easy to understand.

Whenever possible, you should use information from current and previous user behavior, sensors, and any other sources to try to present the correct option to the user. When that is not possible or likely, you should present the likeliest choice, and give options to switch to others.

For example, when opening a Messaging compose screen, instead of presenting a Confirmation dialog asking whether the user intends to compose an MMS or SMS message, just open a Compose screen with attachment options. If the user chooses an attachment, the message becomes an MMS message; otherwise, it's an SMS message. The user is still the one making the choice, but he is making it implicitly by actions and choices (selection of the attachments) he needed to make anyway.

Except for error or guard conditions such as Exit Guard, systems can often be designed to avoid explicit choices.

When the step cannot be avoided and user input cannot be divined contextually, you must present the choices contextually, as choices related to the task or screen that preceded the Confirmation screen. Always communicate the consequences of the choices simply and clearly.

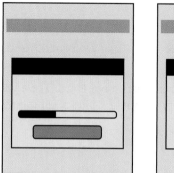

Figure 3-3. *Single-choice Confirmation dialogs are important variations. When the user must be notified of something he cannot alter, but that will affect him, such as a dropped call or frozen system, informing the user even when he has no choice is better than not telling him at all. Processes that require a Wait Indicator widget will also often be in Confirmation dialogs, either with a Cancel button, as shown here, or without one.*

Though many display methods are possible, all will be modal dialogs of some sort, as shown in Figure 3-2. For more details on these, see variations under Pop-Up.

Three key variations exist. A *single choice* (Figure 3-3) is only used to inform the user. For example, let's say a fatal error has occurred and the handset must be restarted. Instead of just restarting the handset for the user without warning, it may be helpful to tell the user that the condition has been reached, in which case you should present a Confirmation dialog with a single Restart button. In cases where the condition is not catastrophic (such as a dropped call), you may choose to have the dialog disappear after a brief time. Including a button then simply allows the user to dismiss the dialog that much faster.

When you automatically dismiss messages, it is usually best to have some other method display the same message. Load the call log, with the item visible, or add an item to the Notifications area. You can determine the best method by examining the OS or application standards.

The *cancel* variant of the single choice includes a Wait Indicator to denote that a process—usually a user-initiated one—has begun and the user must wait to perform other tasks. When possible, include a Cancel button. If you do not, this may become a sort of "no choice" confirmation box, with nothing but messaging and the Wait Indicator widget in a Pop-Up.

By far the most common is the *two or three choices* variant. This is the maximum number of choices that a user can readily comprehend at a glance, while also presenting enough options for most required decisions. Here are some examples of three choices when exiting with unsaved documents:

- Keep working on the document.

- Save the document and exit.

- Exit without saving the document.

Note that these are not ideal labels. See the Exit Guard pattern for more discussion of these sorts of conditions.

In some cases, a larger number of choices must be presented. These will generally be offered as a single Select List in the dialog. You should avoid scrolling, but note that scrolling may be unavoidable.

Interaction details

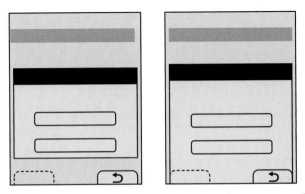

Figure 3-4. *For devices with soft keys or with soft-key-like buttons, there are two methods of presenting the Pop-Up. It may be floating, and the lack of graying out of the buttons and Pop-Up relate them, or it may be anchored to the same edge as the buttons, to tie them together more strongly.*

The Confirmation pattern involves an error or choice that stops the process. Such choices should not be avoidable, but should be contextual to make it clear what application or process they will affect. These should therefore almost always be modal Pop-Up dialogs. See that pattern for additional details.

Whether two or three choices is convenient or a nonstandard UI is required depends on the OS paradigms. If the device OS you are designing for uses soft keys or soft-key-like on-screen buttons, two choices with the dialog docked to the edge with the buttons is the simplest way to present a small number of options, as shown in Figure 3-4. However, this (usually) limits the options to two.

If additional selections are required, they may be loaded in any way needed, such as buttons in the Pop-Up, as pick lists, and so on. Use any soft keys or on-screen buttons for primary supporting actions, such as Cancel. In the earlier example of exiting with unsaved documents, the choices to exit without saving and to save the documents would be in the dialog; the option to return to the application would be labeled "Cancel" and would be the soft key conventionally used for this. See Figure 3-4 for two examples of this.

When working with such devices, avoid disregarding soft keys entirely, for consistency of the interface.

When designing for other interfaces, especially for touch and pen input, selections should usually be in the form of buttons.

Presentation details

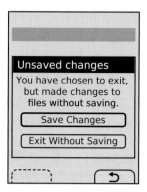

Figure 3-5. *The title, description, and button labels must work independently and in concert to make the Confirmation dialog clear and immediately obvious to the user.*

When the Pop-Up appears, use other methods to notify the user of the stop in the process, such as audio and/or vibration. Follow current system volume settings for the notification. Mobiles are used in quick glances, so the user may not notice such a stop in process for minutes or hours unless he is informed via a nonvisual channel.

If the Confirmation action occurs while an application is running in the background, use any available notification method installed with the OS to bring this to the front. If the only method is a Pop-Up dialog interrupting the in-focus process, use this carefully. Trivial processes, such as your game crashing, do not need to interrupt phone calls.

The Title of the dialog must clearly denote the choice to be made. Do not label it as "Error," with codes or jargon, or with the name of the parent application. For the example used earlier of exiting with unsaved work, "Unsaved work" would be a suitable title.

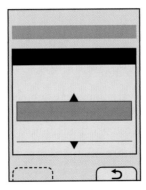

Figure 3-6. *In rare cases, more options than can fit on the screen must be presented. In this case, a scrolling selection box may be offered instead. This is not something a user can understand at a glance, so it should be avoided.*

A description should usually be provided. If the choices are clear or simple enough, you can sometimes leave this out. For the earlier example, a suitable description would be, "You have chosen to exit but have made changes to files without saving." The description and title may or may not be in the form of questions, as the choices will often serve this role.

Button labels must not depend on the description or title. Never present selections such as "Yes" and "No." Do not try to fix these by adding a question such as, "Save these changes?" Users will not read the whole thing, but just glance at the options, and will become confused by your vague buttons.

At the same time, do not let options be too long, or they may not be readable or may not look like selections. Options such as "Delete" and "Save" are much more suitable. For the relatively complex example used previously, options might be:

- "Save changes"
- "Exit without save"
- "Cancel"

"Cancel" may not seem to be the clearest label. It is instead a placeholder for the common term or symbol used in the OS to denote canceling or going back from the current action. In this case, the action is the presence of the Confirmation dialog, so the application will be returned to. See Figure 3-6 for an example of all these messages in one design.

When a choice is clearly the preferred one, or is the safest (e.g., it preserves user data), it should be indicated as such. The OS standards should include a method of indicating button preference such as color, icon use, or border treatment.

Reduce clicks whenever possible. When a selection is made, it should immediately commit the change and proceed to the next step. Never force the user to make the choice, then press some sort of "Submit" button.

Do not overuse the single-choice variation. For example, if an application must quit and a Confirmation dialog is useful, a dual choice would usually be better. Offer options to just quit or to quit and restart the same application immediately.

Carefully consider whether a Confirmation dialog is the right pattern whenever more than three choices are offered. It may be a natural part of the process. The use of the modal dialog implies exceptional conditions or errors and so should only be used when needed.

Unless it is the standard and exclusive way of canceling a process, do not provide a Close button for the Pop-Up dialog. Even if the behavior can be mapped to both the Close and Cancel functions, the Confirmation dialog presents options as a set of required choices. Closing the dialog implies it is optional and can be avoided.

Do not develop new terminology, or use new or nonstandard icons for simple behaviors such as Cancel or Back. Use the OS standard labels or iconic representations.

Sign On

Problem

You must provide a method to allow only authorized individuals access to a device, or a site, service, or application on the device.

Authentication is built into the security model of most devices, but you may want or need to provide additional security for any application, service, or site accessed from the device.

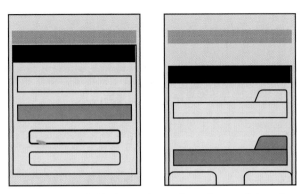

Figure 3-7. *Username and password fields do not change, but Submit and Cancel can change depending on the interaction methods available, and the OS standards. Touch and pen devices are very different from scroll-and-select.*

Security is very often overused. Do not automatically implement it, and consider whether your specific situation requires explicit authentication. Mobiles are personal, and so should only require authentication for first-time entry, or for very high-security situations. Mobile-like multiuser devices such as kiosks will also require authentication.

Personal mobile devices also have built-in security methods, so users can lock them if they are security-conscious. During setup, you may remind users of these features instead of adding additional security to your application or website.

User identity, whenever known, should be used and as much information in the system displayed as possible before authentication gateways are passed. Authentication may then only require the password, as in Figure 3-8.

Authentication must be presented in context, so it is clear what system the user is gaining access to and why the level of security is needed. Entry methods must be designed to encourage ease of access and to prevent miskeying. Users justifiably do not trust vague, general, or inappropriate requests for authentication.

This is an excellent example where common practices are often not just difficult to use, but actually insecure. Use confirmed best practices, and follow the intent and letter of all security laws and regulations that apply to you. Also be aware that some regulations, such as current US CPNI requirements, are not fixed in time, and presume you will improve security as threats change over time.

Figure 3-8. *Anonymous users, such as users signing into an account for the first time, must provide a username and password. When just checking credentials on a second visit, display the user's information and only ask for the password.*

The most common method is to extend desktop paradigms and accept input via form fields, as text or numeric entry using the conventional input methods provided with the device and/or OS: hardware or virtual keyboards and keypads. This works on all devices and platforms, but is not ideal. The difficulty of entering text is exacerbated with the non-word nature of passwords.

For on-screen, virtual Keyboards & Keypads, you may find it helpful to develop a custom keypad, which the user may tap or drag across in a pattern, for increased security or additional ease of entry.

Additional entry methods, such as biometrics, are not yet established enough to have best practices in the general mobile space, and therefore no design patterns exist. Fingerprint readers are the likeliest to become common, and generally work as a single factor (identity), providing most access but requiring that a short password be entered for higher-security actions.

Remember that entry of credentials is not enough to count as a full security methodology. Recovery, reset, and out-of-channel notifications are also required, but are systematic approaches are outside the scope of this book. For very high-security applications, multifactor authentication, biometrics, and token-based security systems (such as RSA SecurID) must also be considered.

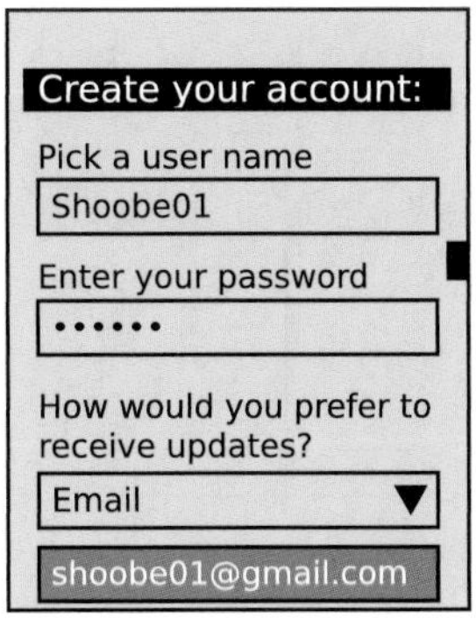

Figure 3-9. *Creating credentials is very similar to entering them for Sign On. When the field is in focus, make it entirely visible. If it will be on the screen for a while, it can be obscured when out of focus.*

When you present a Sign On dialog, it is usually best to make it the only accessible item to avoid confusing entry with other behaviors. This makes it either a complete page or state, or a modal Pop-Up.

When using a form field to enter passwords, you must restrict entry methods to those allowed. If the password is only numeric, disregard alpha or symbol entry on hardware keyboards, and only display the virtual numeric keypad. See the example entry methods in Figure 3-7.

For small, personal-size devices such as phones, MIDs, and tablets, you should generally not obscure or "mask" passwords. This is the common practice of replacing the characters with dots or asterisks. "Shoulder surfing" on mobiles is extremely difficult due to the scale and viewing angles. In addition, the user may be relied on to move to a more private area or simply turn around, behaviors that desktop users cannot exercise. Displaying the entire field will greatly speed entry and reduce errors on entry, all of which not only reduce user frustration, but also increase security by reducing time on entry and reducing use of recovery methods.

When the password field may be on the page for a while—such as during account creation—the password may be obscured when the entire field is not in focus, or after a brief time, as in Figure 3-9. You can also give the user a selection, such as a checkbox, to hide the password. You may appreciate this during demonstrations, so you may hide it while projecting your cool new product at a conference.

An excellent method to enter a passcode for touch and pen devices is to provide a custom keypad, as shown in Figure 3-10. This may use conventional characters, such as numbers, or special symbols to add a layer of obscurity. These are mostly used as an increased security method; the order of the items changes each time, preventing observers from determining the passcode string by looking at finger patterns or examining the screen for marks.

Figure 3-10. *Patterns can be a good way to add security for touch and pen devices, and if the grid changes each time, they provide an extra measure of security when entry can be observed or the device itself is available for others to access.*

Another method that is especially useful for securing device lock screens is to provide a pattern entry, as in Figure 3-11. A grid of numbers or symbols is provided, and a series of adjacent items must be selected as a single gesture, such as an L or U. Although fewer combinations are possible, this can be done with minimal visual input, and may also be configured to learn specifics of the approved user's gesture speed and accuracy. Gestures outside established bounds, though good enough for typing, may require entry of conventional passwords as a backup confirmation method.

You should always consider the entire method of access, and provide links to seek help on entry and to get more information on system security as well as recovery and/or reset methods. For many applications and tools, unusual conditions such as these can simply load your website to complete these processes. However, be sure to weigh the extra complexity, delay, risk of the site not loading, and other factors. This might be best as a stopgap for the first release, but you will need an integrated solution for later on.

When creating usernames and passwords for account creation systems, use tools to check that the username is available and the password is sufficiently secure as soon as the user moves to the next field. Avoid displaying errors after submission.

You should always provide a Cancel function in case the user cannot recall her credentials, or must perform some other task instead. Even for device authentication, there must be a way to revert to the sleep state without inducing an error.

Figure 3-11. *Gestures are lower-security because they reduce the number of patterns and cannot randomize, but are so easy to enter they may encourage locking the device more often.*

You must always label password entry screens and Pop-Up dialogs in the context of the process or portion of the process. Explain the reason for the authentication specifically. Do not say "Sign on to our store," but rather "Sign on to complete purchase." If the title is insufficient, add a description, but keep it brief; users are unlikely to read much, and will gravitate to the entry fields.

Label each field, and be sure to use the same terms each time they are presented, throughout the entire experience. Use parallel labels. "Sign on" and "Sign off" go together. "Sign on" and "Log out" do not, and will be confusing.

Indicate the allowable input methods and the current input method for each field. You may have to add hint text in or under the field, or you may be able to simply lock the keypad into the allowable mode. Be sure to indicate locked entry with an Input Method Indicator. Do not needlessly duplicate information. If entry is locked to numeric values, do not also say "Enter numbers only."

If the user has been identified by device ID, cookies, or something similar, provide the known and nonsecret user information, such as the user's avatar photo and name on any Sign On screens. Although all you know is the password, do not just ask for it with no supporting information. Of course, avoid demanding that the user enter her username when this is already known, or has actually been printed on previous screens (or in the header of the Sign On screen).

Buttons to submit Sign On forms must, whenever possible, be contextually labeled. Use labels such as "Continue" to remind the user she is in a process, and avoid "Sign On" and the even more generic and technical "Submit" whenever possible. If completing the Sign On process will submit another action, make that clear. Use "Sign on and submit order," for example.

When two buttons of equivalent visual weight are presented, the "Submit" choice should be designated as the preferred selection; the OS standards should include a method of indicating button preference such as color, icon use, or border treatment.

The label for the "Cancel" function should always use the common term or symbol designated by the OS to denote canceling or going back from the current action. For devices with soft keys, or OSes with soft-key-like buttons, the "Cancel" function at least should be displayed as a soft-key button. When the Submit button is presented as a more prominent on-screen button, this can entirely alleviate any concerns over prominence of the preferred method.

Antipatterns

If your use case demands hidden fields, use caution when coding or specifying. Do not assume that the default password method built into the OS is secure. Most still reveal the character being typed, and this is basically required in triple-tap text entry modes. If an obscured entry is required due to use as a kiosk, while attached to a projected device or in other shared situations, be sure to specify how obscured it must be. This may require custom fields to be coded.

Avoid using any desktop paradigms without considering the consequences. For example, do not use dual fields to attempt to solve the entry problems associated with obscured entry. The difficulty of entry on a mobile device will make a large percentage of users simply incapable of creating a password with this method.

Be careful to use terminology absolutely consistently throughout the process. If all your documentation refers to a "passcode" (because it is a number), do not suddenly use the term *password* on the Sign On screen. Also try to use correct terms when possible. Enough users are used to authentication systems, and many of them understand the difference between recovery and reset. Do not conflate or confuse the terms.

Exit Guard

Problem

Exiting a screen, process, or application could cause a catastrophic (unrecoverable) loss of data, or a break in the session.

Exit guards can be built into almost any process, though in some cases, preservation of information can be difficult. Plan carefully to ensure that the goal of keeping user data is well designed.

Solution

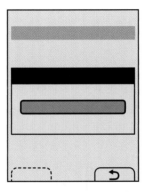

Figure 3-12. *The Exit Guard is a modal dialog that interrupts the process of exiting an application or process. The Cancel soft key will close the dialog and return to the last state without exiting.*

You should simply present a modal dialog that delays the user from exiting immediately. The use of the Pop-Up, as shown in Figure 3-12, keeps the app or function open in the background, and so should preserve any information. The dialog informs the user of the consequences of exiting (loss of data) and requires the user to make choices, at least to confirm exit or return to the session.

Be sure to build systems to prevent data loss even if exiting occurs. For example, auto-save all user entry, save draft messages, and so on. In many cases, even critical entry systems may forgo an Exit Guard as long as recovery is simple and the process may be easily reinitiated. These concepts are discussed further in the Cancel Protection pattern.

For high-security or time-sensitive entry, such as banking, you may find it implausible to auto-save information with sufficient security, or a break in the session may cause too much delay. In these cases, an explicit Exit Guard is useful.

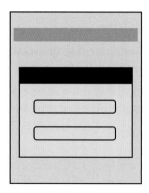

Figure 3-13. *Devices without soft keys will place both buttons within the modal dialog. The top is generally the "Exit" button, and the bottom will "Cancel" the request.*

The most typical variant is a simple modal dialog, informing the user of the risk of exiting and presenting a choice to exit or return to the previous state (Figure 3-13).

Sometimes you may also want to present additional options. These are usually special exit conditions, such as choosing to exit with or without saving, as discussed under the Confirmation pattern.

A variant, shown in Figure 3-15, uses the Wait Indicator and a delay timer to provide one-click exiting, as long as the user is willing to wait a short time. For immediate exit, provide an "Exit now" button for the user to select.

A subvariant of the timer, used for high-criticality applications, reverses the process. The user must select to exit in a short time (or continue pressing the exit button for a designated period), or the process will not terminate and will return to the last state.

Figure 3-14. *Titles, descriptions, and labels must be clear and easy to understand. If additional options are available, such as the example on the right, display those within the same dialog.*

If additional selections are required, you may load them in any way needed, such as buttons in the Pop-Up, as a Select List, and so on. Use any soft keys or soft-key-like buttons for primary supporting actions, such as Cancel. In the example of exiting with unsaved documents, the choices to exit without saving and to save the documents would be in the dialog, and the option to return to the application would be labeled "Cancel" and would be the soft key conventionally used for this.

For devices with soft keys, even if the Pop-Up can be designed to hold all the required buttons, use soft keys for at least some expected actions. If "Back" is always on one of the soft keys, use that function as the "Cancel" function for this screen.

Keyboard and touch input buffers must be cleared the moment the exit selection is made. This will prevent accidental press-and-hold or multiple click-buffered input from skipping past the Exit Guard dialog. If problems continue to occur with this, add a brief delay (one-tenth of a second or less is usually enough) before accepting new input.

You may also wish to have no function selected by default on scroll-and-select devices (requiring a deliberate scroll and then a select to make either choice). Touch and pen devices may wish to consider the location of the dialog buttons relative to the original exit selection mechanism, to prevent accidental or too-easy input; requiring moving to a different part of the screen may provide a sufficiently deliberate action to avoid accidental activation.

Presentation details

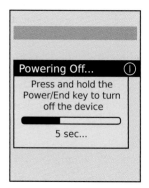

Figure 3-15. *Delays may be used for convenience (left), or to add an additional level of assurance to the guard process (right). When describing press-and-hold or other special key behavior, the key can be iconically represented, and pointed at.*

Timed Exit Guard dialogs must notify the user using tones and/or vibration, to ensure that they are seen. This may also be useful for conventional Exit Guard dialogs, to ensure that the user does not assume the application or process exited, and then does not check back for some time. Use the system-set volume or vibration settings in general, but always at least vibrate, even if set to no output at all. Remember, this only happens on user action, so it is not a notification and should not be detectable to others.

Make sure the Title of your dialog denotes the condition requested, usually something like "Confirm Exit," "Exit?," or another clear and actionable statement. See Figure 3-14 for labeling examples.

If this title is insufficient to explain the situation, because there are special conditions such as unsaved documents, a description should immediately follow the title. To explain a timed exit function, for example, a description might read "Hold the Power button for 5 seconds to turn off the device." Be sure to add functions to this as required to make it reflect the condition accurately. For example, countdown timers should usually change the number in the description as an active countdown so that the user has a sense of how much time remains. If the button is released before the time required to exit, the dialog should remain on the screen for several seconds so that the first-time user, more familiar with short presses and taps, has a chance to learn the process.

If a press-and-hold function can be initiated modelessly (such as powering off the device by holding the Power key, which works from any screen), you should make a dialog appear as soon as the function is selected, to inform the user how to exit, or warn the user of the action.

Button labels must not depend on the description or title. Never present selections such as "Yes" and "No." The best are along the lines of:

- "Exit"
- "Cancel"

where "Cancel" is the OS default label or icon for canceling or going back from the current process, and so should not be ambiguous. In this case, the process is the subset of the previous process, or the exit condition Pop-Up itself. The logic can get confusing if you think about it too much, so don't. If there is no clear default label for cancel, use something unambiguous such as "Continue editing."

When a choice is clearly the preferred one, it should be indicated as such. The OS standards should include a method of indicating button preference such as color, icon use, or border treatment. This may or may not be the safest choice; you will have to determine this during design based on the criticality of the process being canceled and the expected, most common use cases.

This dialog may carry additional cues such as being associated with the location of the button being used, and having a graphic of the button associated with it, perhaps in the title bar. This can even extend to pointing at the location of hardware buttons.

For touch devices, the pop up should spring from the virtual key being pressed, which should not itself move during this action.

Implicit protection methods may use any interactive or systematic methods, or any patterns outlined in this book. To find them, consider how any interactive system can be misused. Often, very small changes, such as clearing the scroll buffer as described earlier, can alleviate them. Larger changes must be considered carefully so that primary use cases guide the design, not rare or edge cases.

When the Clear Field widget is provided, and that method has been used, as long as the session is active, another function will be present that will recover the cleared information and display it in the field. If this is used when there is new content in the field, the user should be asked if she wants to add it to the newly entered information, or replace it with the recovered entry.

Autocomplete & Prediction processes should save user entries as they are being typed, so they are available for auto-complete even when accidentally deleted or an accidental loss of session (such as loss of signal) occurs. When entry into fields may be tedious or repetitive, or recovered user entry is available, display the auto-complete list of options as soon as a match is found with the current entry. To avoid overusing auto-complete, do not attempt to match until a few characters have been typed. The exact number will vary by type of entry and by the processing capacity of the device.

You can leverage the history, such as that used by web browsers, to let users revisit specific locations without recalling and reentering the same address, search, or other information. This should be displayed as a Hierarchical List grouped by session. See Figure 3-17 for a way to combine this concept with auto-complete.

Presentation details

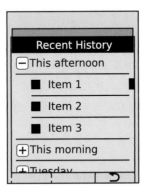

Figure 3-18. *History works best when built as a Hierarchical List. For recovery of accidentally deleted user input, consider contextual use, such as loading the list into a Pop-Up.*

Implicit protection methods are basically invisible to the user, and if you repopulate a form with previously entered data, it usually does not need an explanation.

For details on general Autocomplete & Prediction presentation, see that pattern. For the purposes of recovering user information, an indicator should also differentiate user-entered information from community or spellcheck results, if several types are offered.

Displaying history with a Hierarchical List, as in Figure 3-18, is essentially as described in that pattern. Parent folders or groups should be labeled by the date and time the session occurred, using natural language to account for ranges; "Yesterday afternoon" is clearer than "1:20–3:45 p.m., 7 November."

When you provide an undo process from a clear field, or a history link, make sure there is a clear label or use a well-understood iconic representation. If the OS uses a standard "Back" or "Cancel" icon, this is often a good choice and will fit the existing interactive language even if it is not used in the normal location. Text labels should label either the direct activity, such as "History," or the recovery task, such as "Recent entry."

Antipatterns

Do not preserve secure information such as passwords and financial transaction information without informing the user.

Do not store any information as plain text that can be searched remotely or when stored as backup files, and do not keep all information forever. It is difficult to tell what information is secure to the user, and one person's public knowledge may be another's secrets. Assume everything is worth at least minimal protection.

Timeout

Problem

When designing high-security systems or those that are publicly accessed and are likely to be heavily shared (such as kiosks), you must include a timer to exit the session, or lock the entire system after a period of inactivity.

Timers and sensors are not available for some programming platforms and many web browsers, and so may limit the ability to offer a good experience for some users.

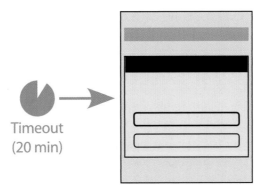

Figure 3-19. *When a period of inactivity is detected, either by a simple timer on-screen interaction, or due to sensors not detecting use of or proximity to the device, a Timeout notification is presented and the session may be terminated.*

Many websites or web services use session timeout as a cost-saving measure to reduce load. Affinity-based load balancing is the default due to simplicity of sharing user data, but it requires that session-startup processes or overhead be left for all predicted users for a particular server.

Mobile interaction, even with websites, may be interrupted more than on desktops, and so should avoid such behaviors. There are other methods as well, so if system and data design is within the scope of your project, consider these as part of the overall design.

If your sessions must expire due to the method by which they have been built, consider making this transparent. Keep a log of the last-used location in the system, and when the user initiates new activity, resume at that state.

Mobiles are personal, so there is generally less worry about security of individual applications than on the desktop web, or for kiosks. Do not use old-school security practices by default, and only use them when necessary.

If session expiry is required for security purposes, try to work around it using sensors to detect when a user leaves or puts the device down for some time. Strict timeout dialogs, as in Figure 3-19, are based on desktop paradigms, where the computers do not have sensors.

When a strict timeout is needed, a modal dialog appears describing the impending loss of the session and presenting methods to preserve the session, as in Figure 3-20. When expired, additional information will be presented, and the user will be required to sign on again.

When you must present an explicit timeout warning or notice to the user, there are several states and variations for doing so.

When the session expires, a Pop-Up will appear that states the session has expired. Depending on the needs of the system (load or security), the session may be restarted from the dialog, or dismissal may send the user to a Sign On screen to reauthenticate.

You may want an alert to appear before the session timeout. This may be very strident, such as a Pop-Up, or may simply be a countdown in the title bar or corner of the screen.

There may be multiple levels of timeout for security. For example, a banking application may require entering the passcode again when the first timeout is passed, but when a second, longer period of time has passed, the complete credentials must be reentered. These are mostly used on shared systems, such as kiosks, to prevent access by others when the original user has abandoned the session.

If secure information is on the screen when a timeout occurs, it may be necessary to obscure the information behind the notification dialog to ensure that no unauthorized users can see the information.

Figure 3-20. *Session and security details should not be routinely shared, mostly to avoid confusing the user. Describe the condition in clear, understandable language, and provide unambiguous, free-standing choices.*

Determining the appropriate timeout requires careful consideration of the user's tasks, their context, and the complexity of their interaction and entry of information for a screen, state, or process. Some guidelines for cognitive load are available, but often the best case is to simply run a test, even with your friends or with other employees, to get a sense of the time required to comprehend and enter information.

Mobile devices should use sensors to replace or supplement the timeout. If you have access to a camera, you can use it to determine whether a user leaves even a fixed device such as a kiosk. Face recognition is now available, and can already just barely be used to determine whether a particular user has likely left the field of view, or more than one person is in view of the screen. Shoulder surfing can be prevented by displaying alerts when more than one user can see the screen. For portable devices such as handsets, timeout can be based on the time the device is left unattended, by accessing the accelerometer information. This timing can be very short, since any application requiring this level of security can generally be assumed to take the user's full attention; if the user sets the device aside, he may understand the need to reauthenticate to continue using the device securely.

When a Pop-Up is presented for timeout, this is a sort of reverse Exit Guard. The user will be presented with options to exit, but the primary condition will be to resume work. This will reset the timer, or load a Sign On screen or dialog as required by security needs.

If the timeout dialog is not used within a certain period of time, it may have a second timer, after which the system returns to its idle state. This is mostly used in shared systems, but may be useful for personal mobile applications where the primary function is consumption. The lost data will be saved as a draft message, as described in Cancel Protection. A notice should appear on this screen somewhere, informing the user of the loss of session and auto-saving of data.

When complete session expiry must occur, be sure the subsequent pages presented are useful and lead the user back into the system. Do not load full-screen errors, or the site home page. When requiring Sign On, use the identified user information when permitted by your security policy.

Continue presenting useful information throughout the expired session process. Once reauthenticated, return to the last location in the system, not to generic start pages or error pages due to loss of session identifiers.

It may be useful to present the Sign On information within the Timeout dialog, as shown in Figure 3-22. Even if this is technically possible, make sure it is not confusing, and does not exceed the space available. This is a case where tasks may need to be broken up to avoid complexity, even though it will slow down experienced users.

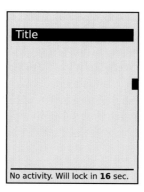

Figure 3-21. *It may be useful to include a countdown timer as the timeout approaches. If the user is otherwise aware of the consequences, or there is room to tell the user what will happen, this may help prevent inactivity before the Timeout Pop-Up is presented.*

Timeout is a very critical condition, so it must be very clear. Anything less intrusive than a modal dialog often will not be seen.

The Pop-Up dialog must be titled and labeled clearly, but with as little technical jargon as possible. Describe the situation not as a "timeout," but by stating the specific reason, such as "Locked due to inactivity" or "Since you walked away…" Emphasize security, and never admit other issues such as load, or discuss a session.

Button labels must not depend on the description or title. Never present selections such as "Yes" and "No." At the same time, do not let options be too long, or they may not be readable or may not look like selections. Typical options are along the lines of:

- "Continue working"
- "Exit application"

However, these need to be contextually relevant—"Continue drawing" and "Stop banking," for example. Cancel and close conditions, as they apply to the Pop-Up dialog, do not really apply here. Though the "Continue" label could be replaced with an action to dismiss the dialog, since it was system-initiated, the explicit label is clearer and more understandable.

If the user must sign on to continue, do not use generic labels such as "Continue working," and state in the description that they must reauthenticate.

If on-screen data is determined to be secret enough that it must be obscured during time-out, you may do this by either making the dialog big enough to obscure all key information, or using a pattern to darken the screen. A simple 1-pixel checkerboard of clear and black (also an old trick before alpha transparency was common) will reduce type legibility as it is laid across. If it is still readable, change the scale of the pattern.

Use nonvisual methods to notify the user of the stop in process, such as audio and/or vibration. Follow current system volume settings for the notification. Mobiles are often used in quick glances, so the user may not notice such a stop in process for minutes or hours unless informed via a nonvisual channel.

If the timeout is very short, or when it begins to become very short (less than 30 seconds), it may be desirable to add a countdown somewhere on the screen, as shown in Figure 3-21. This must be labeled clearly, the timer should count individual seconds, and it should be slightly obtrusive. Make sure it appears from a blank area, instead of replacing another component; this, coupled with the movement of the clock, should make it reasonably visible.

Figure 3-22. *The Timeout notice may also carry other useful information or processes, such as directly allowing reauthentication. Be careful to not provide too much information or too many choices, thereby confusing the primary message.*

Antipatterns

Do not let system or load calculations determine an acceptable timeout. Often, this will be the driving requirement after all, but fight for a user-based method of determining the correct timing.

Do not use desktop security models, or follow the practices of other products, to determine timeout for mobile or kiosk-based systems.

Never permanently expire the session without warning the user at the time the session expires. Otherwise, the user will attempt to continue entering information, and will receive an error message and probably lose the entered information. This can be difficult for web applications on many current mobile browsers due to JavaScript limitations, which is a key reason to consider other methods such as automatic resumption of the session.

Do not overprotect information. You should consider giving different portions of an application different timeout policies based on varying levels of security. Very rarely does information on the screen need to be obscured; transactional protection is much more common.

Chapter 4

Revealing More Information

It's Not Magic!

The audience stares, transfixed, at the man on the stage, hoping to catch a glimpse of his strategy. The man waves a black top hat around, showing its form and lack of contents to the interested spectators. Next, he places a red silk handkerchief in front of the hat. Shouting "Voilà!," the man drops the cloth and reaches into the hat. As the audience "Oohs!" and "Aahs!," a white rabbit hops out of the magician's hat.

Context Is Key

Magic tricks are exciting because we are challenged to figure out what just happened and how it fooled us. We're left to ponder and to discuss with one another the magician's strategy and skill. This curiosity of what and how information is revealed is entertaining to us.

But guessing is not acceptable when designing mobile interfaces. On mobile devices especially, we want to eliminate the confusion. Our users are not stationary, nor are they focused entirely on the screen. They're everywhere, and they want information quickly and to locate, identify, and manipulate it easily.

Understanding Our Users with Norman's Interaction Model

Magic confuses us because it takes advantage of our cognitive processing abilities.

Donald Norman tells us there are two fundamental principles of designing for people (Norman 1988):

- Provide a good conceptual model.
- Make things visible.

A *conceptual model*, more commonly known today as a *mental model*, is a mental representation—built from our prior experiences, interactions, and knowledge—of how something works. It's our representation of how we perceive the world.

The second principle, "make things visible," is based on the idea that after we have collected, filtered, and stored the information, we must be able to retrieve it in order to solve problems and carry out tasks. Norman indicates that this principle is composed of smaller principles such as mapping, affordances, constraints, and feedback.

Figure 4-1. Pop-Up dialogs, regardless of what they look like, are used to present any controls or information the user might need, within the context of the parent page or data object. If you want to move a photo or edit an address, a Pop-Up where the image or contact is visible in the background is often the best way to do it.

Mapping

Mapping describes the relationship between two objects and how well we understand their connection. On mobile devices, we're talking about display-control compatibility. On a mobile device, controls that resemble our cultural standards are going to be well understood. For example, let's relate volume with a control. If we want to increase the volume, we expect to slide the volume button up. If we want to read more information in a paragraph, we can scroll down, click on a link, or tap on an arrow. Problems occur when designs create an unfamiliar relationship between two objects. On the iPhone, in order to take a screenshot, you must press and hold the power button and home button simultaneously. This sort of interaction is very confusing, is impossible to discover unless you read the manual (or otherwise look it up, or are told), and is hard to remember.

When designing mobile interfaces:

- We should use our knowledge of cultural metaphors. We understand that an "X" icon, when clicked, will close the page or window with which it is affiliated.

- We should use proximity. Make sure the display and control you are using have a *proxemic relationship*. In other words, an indicator whose function is to expand or reveal more information must be close enough to the information it will affect.

Affordances

Affordances are used to describe that an object's function can be understood based on its properties. For example, a handle on a door affords gripping and pulling. The properties of the door handle—its relative height to our arm's reach, that the cylindrical shape fits within our closed grasp—make this very clear. If an object is designed well and clearly communicates its affordance, we don't need additional information attached to the design to indicate its use, such as signs and labels.

When designing mobile interfaces:

- We expect all buttons to do something and change a display's state.

- We should consider that physical buttons afford pushing, pulling, or turning.

- We should consider that screen buttons afford touching, selecting, and clicking.

- Depending on context, images, words, and graphics may afford selecting.

- If we cannot recognize that an object is supposed to reveal more information, the user will ignore it, assume it is decoration (and therefore not functional), or not understand how to interact with it. Interfaces that have no affordances, such as interfaces that require gestures but have no indicators at all, are a real concern in this area.

Feedback

Feedback describes the immediate, perceived result of an interaction; it confirms that action took place and presents us with more information. In a car, you step on the accelerator and that action has an immediate result. The feedback is that you experience the car moving faster. On mobile devices, when we click or select an object, we expect an immediate response. Feedback can be experienced in multiple ways: a button may change shape, size, orientation, color, or position, or very often a combination of these. A notification or message may appear, or a new page might open up. Feedback can also appeal to other senses using haptics (vibration) or sounds.

When designing mobile interfaces:

- Be sure to design actions that result in immediate feedback. This will limit confusion and aggravation while making the user's experience more satisfying. Delaying feedback can even result in the user performing other actions and spoiling the process.

- Create a change of state (contrast, color, shape, size, sound) that is measurably different from its initial state.

- Use confirmations when user data could be at risk of being lost. See Chapter 3 for more information.

Constraints

Restrictions on behavior can be both natural and cultural. They can be both positive and negative. You may remember playing with a toy consisting of different colored plastic shapes and a cylinder with those shapes as cutouts on the cylinder's surface; the idea was to fit the yellow cube, for instance, through the square cutout in the cylinder; the red triangle through the triangle cutout; and so on. The cube would fit through the square cutout, but not the triangle cutout, and so forth. The size and shape of the objects are *constraints* in making the correct fit. This is an example of natural constraints (though still learned). Cultural constraints are applied to socially acceptable behaviors. For example, it's not socially acceptable to steal from someone or throw your friend's phone out the window to get her attention.

When designing mobile interfaces:

- Use constraints to reduce or prevent user error. When you accidentally press Delete instead of Save, you should be provided a constraining confirmation message that requires your action.

- Use constraints to fit content and interaction to the size of the viewport, and the device.

- Use constraints so that unimportant buttons become inactive but remain visible.

Norman's Interaction Model is a framework that you should refer to when using patterns to reveal more information. For a better understanding of his model, refer to his book, *The Design of Everyday Things* (Basic Books).

Designing for Information

A good way to start thinking about the topic of interactive design is that it is about display-ing information. A discussion of detailed information architecture is beyond the scope of this book. However, interaction design as it pertains to presenting detailed information and results is well within the scope of our discussion.

Displaying detailed information requires an understanding of the user, his context and goals, and the information available. In many cases, information should not be hidden behind a link or other action, and should be immediately available; how useful would the clock on your mobile be if it was behind a "Current time" menu item?

If this seems extreme, consider many of the systems we encounter every day, and that are regularly griped about. For example, say you're checking on an airplane flight. Once you sign on, most airline websites still require that you click on your itinerary and so forth to simply find out whether the flight is on time—even if you only have one flight stored under your identity.

But much other information is of a second-tier nature, and demands user input to be displayed. You must decide:

How to access the information

Access methods, such as links, are discussed in Part III, in Chapters 5 and 6. The patterns in this chapter are almost entirely used for "drilldown" or getting further information, but explicit decisions must be made about the information architecture and must be understood by the project team if they are to make the right decision.

How to display the information

Display may be of two different types:

Display the full page

Full pages are generally part of a process, and the user should ideally not have to go back to the previous page in order to view other information. Doing this repeatedly can be perceived as bouncing back and forth confusingly (i.e., pogo sticking). However, for processes where large amounts of content will be entered or consumed, it is the correct method.

Reveal in the context

For information that can be understood through just a quick glance, or to help the user decide how to proceed, the information should be revealed quickly and easily within the context. A Pop-Up is a contextually revealed item, because the dialog is visibly a child of the page from which it was spawned.

Sometimes the simple facts of the information available will demand full-page presentation of information where the user's context and tasks may otherwise demand that it be displayed in context. Use access and display widgets carefully to provide access to the information, or reconsider whether the information architecture can be redesigned to make this simpler.

Patterns for Revealing More Information

The most common method of revealing information, that of displaying another page entirely, is not covered in these patterns, simply because there is very little to say about it. Linking widgets and the many display patterns, such as those listed in Chapter 2, cover these functions. In some cases, patterns listed in other places offer alternative methods, and these will usually be cross-referenced within the pattern. For example, Windowshade is very similar to Fisheye List and the two may both be used for the same purpose when implementing the same product on different platforms.

The patterns detailed in this chapter are concerned with specialized methods of presenting more information, which have no other uses (Lynch 1960):

Windowshade

This pattern is used when a displayed element must be able to easily reveal a small or medium amount of additional information, without leaving the current context or page.

Pop-Up

This pattern is used when a displayed element must be able to easily reveal a small or medium amount of additional information, while remaining associated with the current context or page. See Figure 4-1.

Hierarchical List
Use this pattern when a large set of information must be presented. The information is hierarchically ordered, and this structure is relevant to the user.

Returned Results
When users have explicitly requested subsets of information, the narrowed data set that results must be displayed in a meaningful manner.

Windowshade

Problem

You are displaying an element that must be able to easily reveal a small or medium amount of additional information, without leaving the current context or page.

Although the Windowshade pattern is built into many application frameworks, it will require scripting in web browsers, and so will not work in some older or low-end devices.

Solution

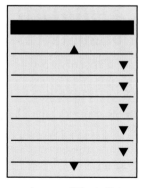 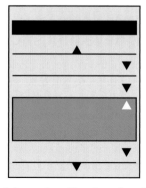

Figure 4-2. *The apparently normal Vertical List on the left reveals itself on the right to be a Windowshade list when an item is selected and the line item expands to reveal additional information. The arrow icons give a hint as to this extra information.*

Items are provided with an indicator to imply that more information is available, generally by defining upper and lower bounds or by enclosing the indicator in a box. When selected, this area expands vertically to display additional information or interactive elements, as shown in Figure 4-2, including additional links and input or form areas.

The term *windowshade* implies that there is a small container that expands downward (similar to a roll-up window shade in a house). Other terms are used to describe these, but this one is unique and is the least ambiguous.

Do not confuse this with the Fisheye List. Windowshade requires an explicit action from the user to open. It may be used as an alternative to the Fisheye List for devices without a hover state, such as those where touch/pen is the primary or only interface.

The basic interaction is very simple, and limited. The available variation is in the selection of the bounding area and indicators of the action itself. Three basic types exist:

- The title of a section may be used as the only visible item. When expanded, this becomes a title bar forming the top of the expanded area.

- For pages or other areas that are relatively strictly vertically oriented, top and bottom bounds can be defined around an area. Any number of summary items may be in this area, and when selected, the bottom bound slides down to reveal even more information.

- Similarly, for design purposes or because the whole width is not occupied by the content, the summary information is in a complete box of some sort. When expanded, the bottom of the box slides open in a downward direction. When open, the box is still complete, and contains the entire expanded content.

You must explicitly and deliberately select a model during design of either *one-open* (all others close when one Windowshade is selected and opened) or *all-open* (the user can open or close as many as she wants). There is no absolute answer; it depends on the type of information and how the user interacts with it. Lists, for example, should usually be one-open, but the visually similar faceted search options list should be all-open.

Horizontally expanding items are plausible, but you should use them only when the overall design of the element—or preferably of the entire OS—is arranged along this axis. Some desktop websites use vertical labels along the sides, and when selected they open to reveal the content of that section. A mobile interface could operate in the same manner.

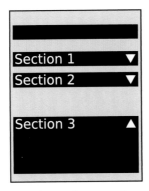 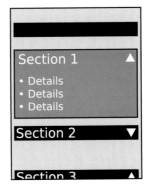

Figure 4-3. *A boxed title element becomes a larger boxed informational area when it is a Windowshade.*

Interacting with a Windowshade should be very simple. Selecting the closed area using whichever method is available and preferred by the user will expand it.

If you have placed several elements in the closed state, the expansion function must be a clearly defined single Icon, Link, or Button. Selection of other items in the closed state will therefore still be allowed, such as going straight to a new page from another link in the closed state. Use this carefully, and leave plenty of room around the controls to avoid accidental activation.

You should always make a control available to close the expanded area. This should almost always be the same control as the open control. It may be an explicit control, or when it is a bounded area such as the title bar, the closed content area at the top of the expanded area. See the two states in Figure 4-3.

No automatic scrolling should take place during opening. The area will expand downward, even if it goes out of the viewport. You should make an exception if almost none of the expanded area is visible. To make it clear that the action has occurred, the page may scroll up enough to reveal that additional information has been made available.

Presentation details

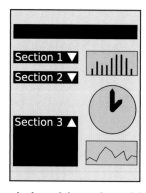 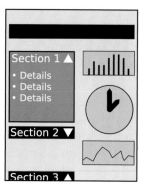

Figure 4-4. *Windowshade modules can be used for narrower layouts, not just full-width.*

You can expand any element using this pattern, even those much smaller than the viewport, as shown in Figure 4-4. Be careful to never expand an item in such a way that it is much larger than the viewport.

Except as described earlier, do not scroll or move the title or other elements that are visible in the closed state when the area expands. This acts as an anchor for the user to understand what has happened to the display. For this reason, most Windowshade areas use a title alone for the closed state label.

An indicator should be adjacent to the title or integrated into the closed state label. Down arrows and "Expand" text are typical. These indicators (whether graphics or text) will change to indicate the state. When they are open, you should change the state, to an Up arrow or "Hide" text, for example.

Design any bounded areas, whether lines or boxes, to be clearly perceived as boundaries, and not just background design elements. They must function this way both when expanded and when collapsed.

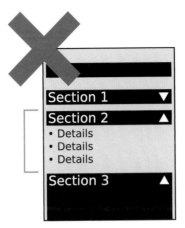

Figure 4-5. *Do not use Windowshade without indicating a bounding area around the expanded area, or the collapsed area, for that matter.*

Avoid use of this pattern without sufficient bounding of the labeled area. Do not simply expand titles, links, images, or other general areas (see Figure 4-5).

Don't break bounds when opening an area. If a box expands, make sure there are sides for the entire vertical space, not just the closed state, then gaps, then a bottom.

Use iconography or other indicators to make it clear the item will expand in place. Generic links or "more info" indicators may cause confusion as to the expected action; as a result, you will reduce the rate of clicking items, or users will not immediately understand they are on the same page.

When possible, use animation of any sort to indicate the opening action. Even if this is a stand-in, and displays only an empty box during the opening (then fills with content when open), this is better than the entire box simply appearing. Users may otherwise believe they are viewing a different page, or a Pop-Up.

Try to avoid areas that expand to be larger than the viewport. If larger expanded areas are used, this may cause confusion regarding what section of the page is revealed, especially if the Windowshaded items are routinely opened and closed.

Pop-Up

You must display a small or medium amount of additional information, while remaining associated with the current context or page.

Pop-Up dialogs are built into pretty much every mobile platform, and for many other methods they can be used to build custom versions. Many of these have enough built-in functionality that they may be misapplied, so use them carefully.

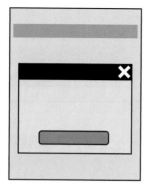

Figure 4-6. *A typical floating Pop-Up, with functions such as buttons inside the window.*

A Pop-Up is a child "page" that is smaller than the viewport and that appears on top of the parent page or displays context that spawned it.

For mobile, these should almost always be "modal," with the Pop-Up having exclusive focus. Elements on the parent cannot be accessed without dismissing the Pop-Up.

The parent can be seen behind and around the Pop-Up, but should be clearly disabled.

Pop-Ups may be free-floating, with space around all sides as in Figure 4-6, or may be anchored to one side of the viewport or parent window, as in Figure 4-7. Anchoring to the top is useful when the Pop-Up is related to a condition in the Annunciator Row or another top-margin item, such as the URL strip of a browser. Anchoring to the bottom is most useful when on a device with soft keys, or soft-key-like on-screen buttons, which will be used in place of button actions within the Pop-Up itself.

The content of a Pop-Up may be absolutely anything, including all the other types of content or interaction described in the rest of this book.

If room allows, and there is value (viewing content on the parent, or providing access to other on-screen elements such as virtual keyboards), the Pop-Up may be moved around the screen.

For larger devices, such as tablets, or in rare cases for handset-size devices, you may find that a nonmodal or "semimodal" Pop-Up is the best choice. This means the parent window can be interacted with, or clicking outside the dialog dismisses it. You have to base this decision on the expected use of the entire system, such as the relationship between the content on the parent window and the Pop-Up. Whatever choice you make, try to enforce it rigidly across the entire application.

Interaction details

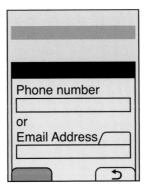

Figure 4-7. *A Pop-Up anchored to the bottom, with soft keys logically and visibly attached to it. Simple forms, such as sign-on, can reside in Pop-Ups.*

Only allow one Pop-Up to be open at a time. This is easily solved by simply not allowing access to the parent, but since this is sometimes possible, it is a good principle to keep in mind.

Opening a Pop-Up can be allowed via a typical Link, or as a result of other user or system actions. There is no universally recognized design method to launch or make the user aware that the action will load a Pop-Up. Within an application or service, however, try to use Pop-Ups consistently so that the user can learn which items load new pages, and which pop up instead.

Always allow your Pop-Up to be dismissed. Even for a step within a required process, the Pop-Up should be able to be dismissed, so the user may view his context and take other actions that may only be allowed within the full page. Dismissal may be via a dedicated function such as a "Close" icon in the corner of the dialog, or as a part of the primary content such as a "Cancel" button adjacent to the submit functions. If a dedicated "Back" function is provided, this should usually dismiss the dialog also.

Actions the user must take within a Pop-Up should be considered "page-level." Whether the user is agreeing that he has been presented with information, is confirming a condition, or is submitting form information, use buttons (or button analogs such as soft keys) to commit the action. Buttons imply page submission or major action, so the dismissal of the Pop-Up will be expected.

Avoid scrolling within a Pop-Up. Long text elements, such as legal agreements, should be in full pages, and not Pop-Ups. Selection lists within Pop-Ups may scroll, however. They may work best with only portions of the Pop-Up scrolling, instead of the entire frame. This will allow any submission buttons to be revealed at all times, making the presence of actions clear, even if the actions are forbidden until scrolling has occurred.

A Select List is often used to select large numbers of items, even in a Pop-Up dialog. The single-select variant is most useful as you can use a single action to make the selection and you do not have to have additional controls fixed in the dialog or elsewhere in the viewport.

When the device OS uses soft keys or soft-key-like on-screen buttons, you can place buttons that control actions in the Pop-Up in there. This is nice because you do not have to fit these two buttons inside the smaller window. If additional actions are required, they may be loaded in any way needed, such as buttons in the Pop-Up, as a single Select List, and so on. Use the soft keys for primary supporting actions, such as "Cancel," as shown in Figure 4-8. Avoid disregarding soft keys entirely, for consistency of the interface.

Presentation details

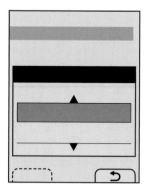

Figure 4-8. *A floating Pop-Up with associated soft keys. A scrolling list of selectable items or actions is within the Pop-Up.*

Make sure the Pop-Up is clearly defined as not being a portion of the page. Use OS-level framing and other treatments to make this clear. Shadows are another valuable way to differentiate the Pop-Up.

Control items especially, including close buttons, and soft keys should appear similar or identical to the OS standard controls. This way, users will not have to learn a new interface language.

Controls should also be as clear as possible so that Cancel, Accept, and Close buttons or functions are immediately obvious to the user and require no interpretation.

The state change of not just loading the Pop-Up but also obscuring the parent is also critical to communicating this. The parent should usually be overlaid with a white tint or black shade that allows the user to view the content, but is different enough that it is clear the content and interactive elements are superseded by another.

When the Pop-Up is nonmodal or the content of the page is key to making choices within it, the background should not be obscured. Be sure there is sufficient border, shadow, or other treatment on the Pop-Up to ensure that it is clearly recognized as a different element, on top of the page.

Antipatterns

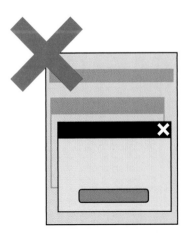

Figure 4-9. *Do not let a Pop-Up spawn another Pop-Up. Generally, Pop-Up windows only work when modal, so anything that can cause more than one to appear is bad.*

For small devices especially, such as handsets, consider ways to avoid using Pop-Ups. Generally, there are other solutions. Many mobile browsers and OSes do not deal with Pop-Ups well, and may display them poorly or confusingly.

For devices with soft keys or soft-key-like on-screen buttons, avoid use of buttons within the Pop-Up. Anchor to the bottom of the screen (or whichever side has the buttons) and use those instead.

Do not allow a Pop-Up to spawn (be the parent for) another Pop-Up (see Figure 4-9).

Do not overuse Pop-Ups by including navigation, tabs, and so on. Present the single information or interactive concept, and then allow the Pop-Up to close when the user dismisses it or the task is completed.

Do not display a Pop-Up exclusively to present advertising. If an Interstitial Screen or Advertising is required, use a full-screen display instead. However, actual content, even if heavily branded or ad-funded, is reasonable as long as it is relevant to the page in some way and was requested by the user.

Hierarchical List

You must present a large set of information. The information is hierarchically ordered, and this structure is relevant to the user.

Arrangement of information is a purely design-oriented task, and except for some interactive methods that may be difficult to accomplish on nonscripting browsers, this may be used anywhere.

Solution

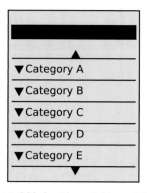

Figure 4-10. *A Hierarchical List that is folded, with no children visible.*

A Hierarchical List can display hierarchically ordered information in a comprehensible and quickly accessible manner. The parent-child relationships are exposed visually (Figure 4-10), and users may fold and unfold the list to view only the parts they need (Figure 4-11).

Displays of this type can be supported by even very simple interactions, with vertical scrolling only, and in surprisingly narrow spaces.

The depth of the list (the number of tiers included) is arbitrary, but it depends heavily on the size of the display area and the complexity the user is willing to endure. Typically, more than three tiers is quite complex, and it may be worth it to consider redesigning such lists just from an architectural point of view.

The user's exploration of this hierarchy, by opening and closing items, is critical to her understanding of the relationships. Do not reveal too much for the user, and allow exploration. When the ability to expand and collapse items is not available for technical reasons, consider other methods to display the information.

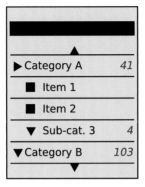

Figure 4-11. *A Hierarchical List opened to one tier. Note the counters on all parent categories, indicating the number of immediate children.*

You may load a Hierarchical List that is either closed or open when first displayed to the user.

Closed lists only display the top tier, as a simple vertical list. The user must select a parent item to reveal any children within.

Open lists reveal a portion of the list when first loaded. This may be the expected location, the current location within a system, the last location used, or other relevant sections. Do not reveal arbitrary sections just to communicate that the list opens.

A key variation is in the methods of revealing (or hiding) child list items. This will have serious impacts on design and the information architecture (and sometimes even the organization of the data repository). These options are discussed in detail in the following sections.

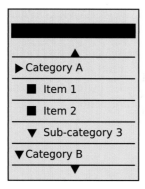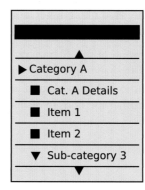

Figure 4-12. A Hierarchical List comparing the "single" (left) and "dual" (right) parent methods. In the latter, the parent is only used to reveal children, and the first item in the child list contains details of the parent level.

Hierarchical Lists are types of lists, and usually are variants of the Vertical List. They may encompass most any other subvariant, and include features from the Select List, Thumbnail List, and so on, as needed.

You should consider ordering the list so that any item which is a parent to others is displayed "twice." An example might help make this clearer:

- **Trees**
    - **All Trees**
    - Deciduous
    - Coniferous
- Shrubs
- Grasses

Here, the boldface items are currently selected items. By selecting "Trees" I have opened the subsidiary categories so that they may be selected as well. However, there is sometimes confusion over which tier is selected when the level below is opened. This solves that by giving a landing page, or default selection of "All Trees."

Indicating a parent category like this is not always required or possible, depending on the data set, and in cases where the second listing of the parent element is not needed. For example, if selecting a US state and the list is by region, there is no value in selecting a region alone. The region parents may, on selection, only open or close to reveal the child states. The two variations are shown in Figure 4-12.

An interaction commonly used when not including parent items at subsidiary levels is to use dedicated reveal functions and allow the user to fold and unfold the list items directly. This decouples the opening and selection, but can induce other issues with comprehension. Consider the data set and users carefully.

In this case, selection of a parent again will collapse or hide the child listing of that parent.

For lists where the "reveal" action is separate from the list item itself, selection of the item title will typically select or load more information (depending on the user task). Selection of the reveal item will immediately display the children of that item.

Selection of the "hide" action (the reveal becomes a hide when open) will collapse or hide the indicated child listing.

Presentation details

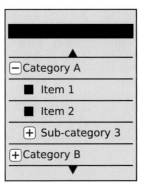

Figure 4-13. *Other common indicators that an item in a Hierarchical List is a parent (and may be opened) are + and − symbols. Note that in most cases, an icon for the leaf node is also used.*

Make sure you have a good understanding of the hierarchical relationship of all items before you begin designing. Use position as the primary indicator of the hierarchical relationship between items. Top-tier items will be left-aligned; the second tier indented slightly; and so on.

Indicators should be present to make clear that the list may be opened. Arrows or plus symbols (Figure 4-13) are most common. In many cases, the indicator will reside in the left column, and indenting will correspond to the width of the indicator. Text indicators such as "More..." can be useful in small quantities, but they take up much more room than graphics for small screens, and for larger quantities the repetitiveness can make them difficult to scan.

Indicators of position and action in the Hierarchical List are basically just a repeated, very well-ordered version of the Indicator pattern. See that pattern for guidelines on position, shape, and interaction details.

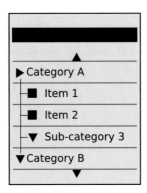

Figure 4-14. *A Hierarchical List with lines depicting the relationships between tiers.*

A count of items within a parent category can be very useful to both indicate that there are additional items, and—for some data sets—to help the user decide which categories to explore. These line-level counters are not quite the same as Pagination or Location Within widgets, and are more like yet another Indicator of the "content beyond" type.

When a set of child pages is revealed, the indicator should change state to indicate that it can be used to close or hide the list.

Additional indicators of hierarchy may be useful, such as type weight (bold for parents, especially when not selectable), color, and size. An optional but well-used method is to provide guidelines among and between list items at different tiers, as shown in Figure 4-14.

Antipatterns

Use caution with excessively complex interaction methods. Simple devices that require option menu selection, or actions that are difficult to discover, such as unique gestures, may impede use of the list.

Likewise, avoid trying to express overly complex list hierarchies of information with simple lists like this. Generally, after more than about three tiers of information, the relationships between the levels of information become too complex to be immediately understood.

Avoid opening too much of the list on entry. On smaller screens especially, it may be unclear how the list is ordered, or the parent-child relationships may be unclear at first glance.

Consider the task at hand, and the user's need to understand the system. Many data repositories are hierarchical, but the architecture may be unimportant to the user. Do not reveal internal process, organization, or jargon unless needed by the end user.

Do not force an informational hierarchy on a data set just to use this display method. The user will adopt a mental model of the system or information that is incorrect, or will never understand the information at any level.

Avoid using too many interaction methods alongside the Hierarchical List. For example, a Fisheye List may seem like a good way to provide some extra information before opening, but will likely add confusion. Items such as this that act in another axis are especially prone to adding confusion.

Returned Results

When users have explicitly requested subsets of information, you must display the narrowed data set that results in a meaningful manner.

Returned results may be suitably displayed with page refresh, and with any method you wish, and so can work just fine on any platform.

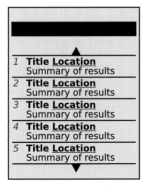 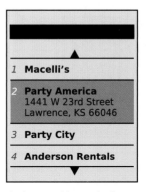

Figure 4-15. *A Returned Results list may be simple, include several lines of information (left), or use display methods such as the Fisheye List to solve display density problems.*

You will show the Returned Results as an ordered list or other array. This may be a page, or as a Pop-Up displayed over other contextually relevant information (such as a map or graph).

In some cases, part or all of the information subset may be implicit or automatic, such as repeated searches and the use of sensors such as GPS as some of the input.

The most common type of Returned Results is a simple Vertical List. Additional information, such as the order of results, relevance, and domain, must be presented as well. Add-ons may also be used when relevant, such as a Fisheye List or Thumbnail List, as shown in Figure 4-15.

You may also wish to present the information contextually, as an overlay or add-on to other information visualizations. This may include very small amounts of information, suitable for placing in an Annotation. These results will themselves be points, and may appear over maps, graphs, charts, and similar visualizations.

In certain cases, you might want to present prioritized information within the same display type. This may be customized information (e.g., stock information when searching for a company), or paid placement results. See Figure 4-16.

Even if multiple information types are offered, the most relevant display type should be presented by default.

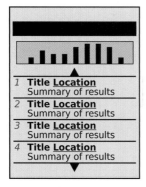
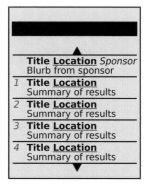

Figure 4-16. *Returned Results may be accompanied by sponsored results (right) or other targeted information such as a graph of stock prices when searching for a company (left). Note that sponsored results are not included in the numbered listing.*

Typically, items in the Returned Results list will need to be selected, either as choices for another process in order to view more details, or to start a process (such as navigating to the selection).

The selection mechanism may be the entire item, or a specific portion of the item display area. Be sure to use the correct type of selection and selection indicator for the process; choice selection will be different from viewing details.

Pagination or Location Within controls and indicators should always be used for list-type displays, and may be relevant for contextual display.

You can display multiple types of results on one page, as overlays, adjacent to one another, or as easily switchable alternative displays. For example, if a visualization overlay is used, the results may be more useful for some users as a list. Subsequent or related displays may be shown as adjacent panels of a Slideshow, as Tabs, or in other ways that allow the user to quickly flip between data sets.

Modifications to the parameters that resulted in the display of results (such as a text search string) should be easily accessible. They may be on the screen, available as options, or available for direct modification by going "back" to the previous screen. All parameters used, even those that were not directly selected by the user on the initial display of results, should be available for modification; see the Sort & Filter pattern. Less often, you may wish to use the Search Within widget to further modify the result set.

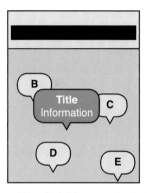
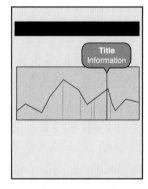

Figure 4-17. *Returned Results laid over contextually relevant information: for the map (left) a point in two-dimensional space, for the graph (right) a vertical line representing values on a single axis, such as time.*

Display the reason the results have been shown, generally as a title. If search results are being displayed, the search terms must be printed on the screen.

Number the results so that an order is clear. For devices with support for Accesskeys, make sure the numbers cannot be confused with the accesskey labels. Color, alignment, and treatment (e.g., light blue and italics) are generally sufficient ways to differentiate.

When space is available, and the relevance parameters can be clearly communicated, you can display the relevance as additional information with each result. This may be a relevance factor (a percentage is typical), by listing terms that are relevant, or by listing a summary of the information that is most relevant to the result. In addition to this, you can also include the implicit relevance of display methods such as position on a map. Distance from your location, or other physical measures, are similar, and you should display them in the same manner.

If the domain is not entirely restrictive (searching only internal documents) or entirely nonrestrictive (the whole Internet is available), an indicator of some sort must be placed by each result. Types of documents, when not all the same, should also be displayed, so the user is aware before he attempts to view a movie or download a file.

When information is displayed contextually, it should appear as an Annotation, as in Figure 4-17, with a pointer indicating the precise location in the contextual system, and a head that may be selected for information or contain a label of the item.

Paid placement results must be differentiated from the other results. They should not, for example, be numbered with the other results, and you should not count them in the total results in the Pagination or Location Within information.

Availability of the item for selection must be indicated. Whether the entire displayed area for the item is selectable, or only certain portions are selectable, an indication of selectability should be present. Each type must be differentiated so that links to view further details appear as links, and selection for a process appears as a choice.

Antipatterns

Avoid error conditions for valid entry of information. Displaying errors for no results will dissatisfy users, so solve the problem for them. Options include:

- Expanding the search parameters; for local search, change the distance until results appear.
- Correcting misspellings, but not unless there are few or no results.
- Using common, similar searches to replace results.
- When no results must be displayed, making it clear that none was returned. Do not excessively obscure this message with paid placement, or other helpful information.

In all cases, inform the user of the results display of the changes, and offer an easy method to see the original results.

Use paid placement very carefully. Advertising revenue is a way of life for many products, but make sure items within the results list are at least slightly related to the original intent of the returned results list. Do not use paid placement "results" as a replacement for no returned results. When users figure this out, they will have less faith in all the results.

Avoid making parameters too difficult to discover, understand, or modify. Try to explicitly state all parameters, even if they are available via other interfaces; for example, results on a map may have the radius constrained by the zoom level of the map, but this may not be clear unless it is also available as a search term, or explicitly stated on the results display.

Summary

Wrap-Up

As you learned, components are sections of a page that may take up the entire viewport, occupy a smaller section, or even appear modally in front of other displayed elements. In this part of the book, you became familiar with specific component patterns and how to use these patterns appropriately to display and organize information. When used effectively, these patterns will allow mobile users to have an interactive experience that is enriching and satisfying. When incorporating these component patterns into your design, consider:

- The user's needs and his task-specific goals

- How the user's mental model and cognitive processing abilities will influence the design of the mobile interface

- The fact that the design must be visible, provide appropriate feedback, and use constraints to prevent and/or minimize human error

- That the context of use will dictate how, why, and what information is going to be accessed and interacted with

Pattern Reference Charts

The following pattern reference charts list all the patterns found within each chapter described in this part of the book. Each pattern has a general description of how it can apply to a design problem while offering a broad solution.

Cross-referencing patterns are common throughout this book. Design patterns often have variations in which other patterns can be used due to the common principles and guidelines they share. These cross-referenced patterns are listed in the following charts.

Chapter 2, "Display of Information"

This chapter described the importance of displaying information based on the user's mental model and how we organize this information cognitively. Failure to abide by these principles will most likely cause the user to become lost, confused, frustrated, and unwilling to continue. To prevent this, this chapter explained research-based frameworks, tactical examples, and descriptive mobile patterns to use.

Pattern	Design problem	Solution	Other patterns to reference
Vertical List	Display a set of information, usually text or a text representation, that uses horizontal space inefficiently.	Rather than using horizontal space inefficiently, display a set of information vertically using an entire allocated space.	Infinite List Thumbnail List Select List Location Jump Search Within Sort & Filter Titles
Infinite List	A Vertical List is called for, but the information set is very large and is not locally stored, so retrieval time is inconveniently long.	Reveal small amounts of vertical information at a time because the information set is very large, and not locally stored.	Vertical List Thumbnail List Select List Location Jump Search Within Sort & Filter Wait Indicator
Thumbnail List	A Vertical List is called for, but additional graphical information will assist in the user's understanding of items within the data set.	Use a Vertical List with additional graphical information to assist in the user's understanding of items within the data set.	Vertical List Select List Infinite List Carousel Grid Titles
Fisheye List	A scroll-and-select device is targeted, and a Vertical List is called for, but small amounts of additional information can be displayed that would assist in the user's task.	When a scroll-and-select device is targeted and a Vertical List is called for, this can be used to reveal small amounts of additional information that can assist in the user's task.	Vertical List Windowshade Pop-Up Pagination Location Within Titles
Carousel	Present a set of information, most or all of which consists of unique images, for selection.	Display a set of selectable images, not all of which can fit in the available space, but that can be scrolled through using many methods.	Pagination Location Jump Grid
Grid	Present a set of information, most or all of which consists of unique images, for selection.	Display a continuous array of selectable images, only some of which can be seen at once due to the limited size of the device viewport.	Simulated 3D Effects Pop-Up Slideshow Location Jump Pagination Position Within Carousel Stack of Items

Pattern	Design problem	Solution	Other patterns to reference
Film Strip	Present a set of information that is a series of screen-size items or can be grouped into screens for viewing and selection.	A series of screens are displayed as a continuous strip, with small spaces or markers between them. When a screen is centered, it fills the entire viewport. When scrolled, part of two screens and the divider can be seen at the same time.	Carousel Fixed Menu Revealable Menu Pagination Location Within
Slideshow	Present a set of images or similar pieces of information for viewing and selection.	Each image is presented full-screen, with a function to transition to the previous or next image in the series.	Pagination Location Within Film Strip Infinite Area
Infinite Area	Complex and/or interactive visual information must be presented to the user. The information can be presented as a single image that is so large it must be routinely zoomed in to so that only a portion is visible in the viewport.	The \entire data set is considered to be a large, two-dimensional graphic, and smaller subsets can be viewed as though "zoomed in" to portions of the larger image.	Simulated 3D Effects On-screen Gestures Wait Indicator
Select List	Selections, either individual or multiple, must be made from a large, ordered data set.	A method of selection, of the line item, or by adding checkboxes to the elements displayed, can be added to almost any display method, such as the Vertical List, Grid, or Carousel.	Vertical List Grid Confirmation Wait Indicator Titles

Chapter 3, "Control and Confirmation"

This chapter described appropriate control and confirmation patterns that can be used on mobile devices to prevent costly human error resulting in loss of input data. When human error may occur, you can incorporate modal constraints and decision points as preventive measures. When considering confirmation controls, consider the user's context of use, because an overuse of these constraints and decision points during low-risk situations will cause user frustration by increasing processing time and mental load, and delaying or stopping the task.

Pattern	Design problem	Solution	Other patterns to reference
Confirmation	A decision point is reached within a process where the user must confirm an action, or pick from among a small number of disparate (and usually exclusive) choices.	Present choices contextually—usually as a modal dialog—and simply and clearly communicate the consequences of the choices.	Exit Guard Pop-Up Wait Indicator Titles
Sign On	A method must be provided to confirm that only authorized individuals have access to a device, or a site, service, or application on the device.	Consider whether your specific situation requires explicit authentication. Mobiles should only require authentication for first-time entry, or for very high-security situations. Mobile-like multiuser devices such as kiosks will also require authentication.	Pop-Up On-screen Gestures Titles
Exit Guard	Exiting a screen, process, or application could cause a catastrophic (unrecoverable) loss of data, or a break in the session.	Present a modal dialog that delays the user from exiting immediately (the app or function is kept open in the background), informs the user of the consequences of exiting (loss of data), and requires the user to make choices, at least confirming the exit or returning to the session.	Wait Indicator Pop-Up Titles
Cancel Protection	User-entered data or subsidiary processes would be time-consuming, difficult, or frustrating to reproduce if lost due to accidental user-selected destruction.	Processes must be designed to protect user input. Methods must be provided to recover previous and historical entry.	Clear Entry Autocomplete & Prediction Hierarchical List Input Areas
Timeout	High-security systems or those that are publicly accessed and are likely to be heavily shared (such as kiosks) must have a timer to exit the session and/or lock the system after a period of inactivity.	Try to avoid the use of Timeout as a solution to load and for security. If sessions must expire due to the method by which they have been built, consider making this transparent.	Pop-Up Sign On Exit Guard Titles

Chapter 4, "Revealing More Information"

This chapter described how to appropriately design to reveal more information. As a designer, you need to become aware that your users, devices, and networks all have limits. Screen size will limit the amount of information that can be displayed at a time. A device's OS will limit processing and loading times. Our memory limits cause us to filter, store, and process only relevant information over a limited period of time. If we disregard these important principles, we can expect the mobile user to encounter performance errors, dissatisfaction, and frustration.

Pattern	Design problem	Solution	Other patterns to reference
Windowshade	A displayed element must be able to easily reveal a small or medium amount of additional information, without leaving the current context or page.	Items are provided with an indicator that more information is available, generally by defining upper and lower bounds, or by enclosing the indicator in a box. When selected, this area expands vertically to display additional information or interactive elements.	Fisheye List Pop-Up Titles
Pop-Up	A displayed element must be able to easily reveal a small or medium amount of additional information, while remaining associated with the current context or page.	A Pop-Up is a child "page" that is smaller than the viewport and that appears modally on top of the parent page or display context that spawned it.	Annunciator Row Titles Exit Guard Link Advertising
Hierarchical List	A large set of information must be presented. The information is hierarchically ordered, and this structure is relevant to the user.	Display hierarchically ordered information so that it is comprehensible and quickly accessible. The parent-child relationships are exposed visually, and users may fold and unfold the list to view only the parts they need.	Fisheye List Vertical List Select List Titles
Returned Results	When users have explicitly requested subsets of information, the narrowed data set that results must be displayed in a meaningful manner.	The displayed page will show Returned Results in an ordered list, or displayed over other contextually relevant information (such as a map or graph).	Vertical List Thumbnail List Fisheye List Select List Pagination Location Within Accesskeys Titles

Additional Reading Material

If you would like to further explore the topics discussed in this part of the book, check out the following appendix:

Appendix D, "Human Factors"
 This appendix provides additional information on our human sensation, visual perception, and information-processing abilities.

PART III

Widgets

The word *widget* can mean a number of things, even within related Internet technologies. Even the savvy user may be confused by the lack of common terminology and the lack of any inherent meaning. The term may apply to bits of code, applets, engines, and GUI elements.

However, the scope of this book, and of this part of the book, is solely concerned with mobile GUI widgets. These widgets are display elements such as buttons, links, icons, indicators, tabs, and tooltips. Numerous additional elements (sometimes called GUI widgets), such as scroll bars, are discussed as components and functions in Part I.

The functionalities of the widgets discussed in this part of the book are to:

- Display a small amount of directly related information
- Provide an alternative view of the same information, in an organic manner
- Provide access to related controls or settings
- Display information about the current state of the device
- Provide quick access to indexed information

The widgets that we will discuss here are subdivided into the following chapters:

- Chapter 5, "Lateral Access"
- Chapter 6, "Drilldown"
- Chapter 7, "Labels and Indicators"
- Chapter 8, "Information Controls"

Types of Widgets

Widgets for Lateral Access

Whether your information architecture is organized hierarchically or laterally, its presentation and access are affected by the potentially small mobile display. One option to consider is to use lateral access widgets such as Tabs, Peel Away, Pagination, and Location Within to assist the user in quickly navigating through and selecting this content.

Widgets for Drilldown

Using an information architecture that is structured hierarchically allows content to be laid out from general to specific while depending on parent-child relationships. This drilldown, top-down approach is effective in providing users with additional related content and commands within multiple information tiers. Patterns, such as Link, Button, and Icon can be used to access these child content types quickly.

Widgets for Labels and Indicators

In some situations, it may be necessary to use small labels, indicators, and other pieces of information, such as a Tooltip, Wait Indicator, and Avatar, to describe content. Mobile users each have unique goals. Some require instant additional information without clicking. Others may need additional visual cues to assist them while quickly locating information. In any case, you must present the information labels appropriately while considering valuable screen real estate, cultural norms, and standards.

Widgets for Information Controls

Finding specific items within a long list or other large page or data array can be challenging. With appropriate controls, such as Zoom & Scale and Location Jump, to locate specific information quickly, the user can instead quickly locate and reveal information on mobile devices.

Helpful Knowledge for This Section

Before you dive right into each pattern chapter, we would like to provide you with some extra knowledge in the chapter introductions. This extra knowledge is in multidisciplinary areas of human factors, engineering, psychology, art, or whatever else we feel is relevant. Due to the broad characteristics of widgets, we find it helpful for you to become knowledgeable in the following relevant area: wayfinding.

Wayfinding Across Content

Whether interacting on a PC, kiosk, or mobile device, your users can easily get lost when navigating content. To reduce their frustration of being lost, you can use visual, haptic, and even auditory cues to help guide users in getting to the place they need to be. When designing a navigation system, you must provide those cues to answer the following user questions:

- Where is my current state or position within the environment? Where am I on this page?

- Where is my destination? Where do I have to go to achieve my end goal?

- How do I get to my destination? How am I going to navigate across content to achieve my end goal?

- How do I know when I have arrived?

- How do I plan my way back? Are there alternate routes I can take?

Kevin Lynch, an environmental psychologist and author of the book *The Image of the City* (MIT Press), determined that we rely on certain objects to help us identify our position within an environment. Let's examine how these objects act as cues and can be used to improve navigation:

Paths
> These are the channels along which a person moves. Examples are streets, walkways, transit lines, and canals. On mobile devices, paths are the routes users take to access their desired content. These paths can follow both lateral and hierarchically organized structures. Help the user define routes by clearly labeling, color-coding, and grouping related content. Use location within widgets to define the user's current position along the path. Provide alternate paths to access the same information.

Edges
> These are linear elements that define boundaries between two phases, such as walls, buildings, and shorelines. On mobile devices, edges can include the perimeter of the viewport, or of fixed menus, scroll bars, and annunciator rows. Use edges to appropriately contain navigation.

Nodes
> These are focal points, like distinct street intersections. On mobile devices, these may serve as graphics, labels, and indicators to describe small pieces of content.

Districts
> These are areas within boundaries that share common features, such as neighborhoods, downtowns, and parks.

Landmarks
> These are highly noticeable objects that serve as reference points.

Getting Started

You now have a general sense of what widgets are as well as a physiological visual perception framework to reference. The component chapters will provide specific information on theory and tactics, and will illustrate examples of appropriate design patterns. And always remember to read the antipatterns, to make sure you don't misuse or overuse a pattern.

Chapter 5

Lateral Access

What a Mess!

Whether you're a college student, a design professional, or a book author, you have experienced the clutter of notes, reminders, memos, drawings, and documents scattered across the surface of your desk. There comes a point in this chaotic, unorganized display when your "tidy instinct" begs for some order.

If you're lucky, you quickly find materials you can use: a binder, file folders with colored tabs, paper clips, even a stapler. You grab the content, and sort and filter it as a means for organizing and creating order. As you organize, you may classify the data by such lateral relationships as:

Nominal
    Using labels and names to categorize data

Ordinal
    Using numbers to order things in sequence

Alphabetical
    Using the order of the alphabet to organize nominal data

Geographical
    Using location, such as city, state, and country, to organize data

Topical
    Organizing data by topic or subject

Task
    Organizing data based on processes, tasks, functions, and goals

Having now integrated your organizational skills with those office supplies, you can marvel at your clean desk. On its surface lay a faceted arrangement of folders. Each folder, containing related content, is clearly labeled with colored tabs to allow for quick and easy access.

Figure 5-1. *Tabs can be used explicitly, styled to fit the space and serve more as indicators, or presented more as options as in the icon strip. Either one follows the principles of wayfinding to help the user know her location and decide where to go next.*

Navigation Structure

As discussed in Chapter 2, we understand the importance of organizing an information structure across a single page, or an entire OS. To recap, we know that there are two main types of organization with information architecture:

Hierarchy
Organizes content based on top-down, parent-child relationships.

Faceting
Organizes based on information attributes without imposing any parent-child relationships. The structure is based on heterogeneous content, with each item sharing the same level—being just as "important"—within the information architecture.

Lateral Access and the Mobile Space

Content across a device OS must be organized and designed to follow a consistent information architecture to ensure a positive user experience. However, when considering mobile devices, the potentially smaller displays play a significant role in determining how this information architecture is designed for interaction. Smaller screen sizes affect the amount and type of content presented, and the user's ability to successfully search, select, and read this information. To account for this, consider presenting the information laterally, at the same tier level in the information architecture.

Use the appropriate widget to provide access to the information, and to make the relationship between items clear.

There are several reasons to use lateral access widgets across the mobile space:

- Mobile displays can be much smaller than a PC, reducing the amount of information shown on a viewport's current state. Unlike PCs, where information is designed for 1024 × 768 resolution, common smartphones still (in 2011) have screen resolutions at 320 × 480 pixels.

- Even for larger screen devices, pointing is often very coarse, with the common use of capacitive touch. Many interactive items cannot be as small as they would be on a similar-size computer with a mouse or similar interface (trackpad, trackball, pen, etc.). In addition, the user's finger and hand may occlude information during interactions, which may be much of the time the device is in use.

- Globally, feature phones still make up about 80% of the mobile market. These devices have a common resolution of 240 × 320 pixels, though some can have much smaller screen resolutions, down to 128 × 176 pixels.

- Because the screen real estate is so valuable, it's important to prioritize features and content related to the user's high-level goals.

- A mobile user's attention is competing with all the stimuli in the surrounding environment. Therefore, the access to the content must be detectable, even at a glance, and afford its function appropriately.

- According to Fitts's Law, there is a direct correlation between the size of the target, its distance from the user, and the amount of time it takes to select it. Therefore, a smaller display requires larger buttons for gestural interaction—which, in effect, will reduce the amount of content that can be placed on the screen.

Use of lateral access affords the following benefits:

- Categories of information do not have to be labeled within a long list, or as flags on arbitrarily listed items.

- Lateral access limits the number of levels of information a user has to drill through to access priority information.

- Lateral access can reduce "pogo sticking," or constantly returning to a main page to drill into subsidiary pages, searching for the right section.

Follow the Principles of Wayfinding and Norman's Interaction Model

As you are now aware, the size of mobile displays can greatly influence how you should organize and design your content. To make sure your users can navigate across this content, consider applying research-based frameworks tied to wayfinding and Donald Norman's Interaction Model.

Wayfinding

In the introduction to Part III, the principles of wayfinding are discussed in detail. You can apply certain wayfinding principles to lateral access widgets to ensure that your user can navigate across content with ease.

Districts

Within the viewport, sections of information that are grouped or bounded by visual elements that follow the laws of proximity will be perceived as being related and separate from other components.

Tabs

Place related tabs adjacent to one another to define a perceived group. These tabs may share common visual design elements, such as size, shape, color, and form. If you are using multiple groups of tabs, make sure these are clearly distinguishable from other tab groups. Selected tabs must be bound to the related area of content.

Pagination controls and indicators

Most often these widgets should be placed immediately above and below the content unique to the page. Another option is to anchor the widget (in the title bar, for example), so it is available at all points on the page. If the page does not scroll, a single fixed-location item logically becomes an anchored widget.

Landmarks

Distinct visual objects that are easily detected and identified from other elements will likely be perceived as reference points. In the mobile space, landmarks can aid the user in quickly finding key navigation features, as well as providing a reference to their position within the browsable content. When designing lateral access widgets as landmarks, consider shapes that are well understood by the user and afford the correct action.

Norman's Interaction Model

As Donald Norman said, "Make things visible!" (Norman 1988). Lateral access elements must follow the principles of legibility, readability, and conspicuity discussed throughout this book. Consider the principles discussed in the following subsections.

Feedback

Feedback is a noticeable and immediate response to a direct interaction. Using feedback appropriately confirms to the user that his action took place. Feedback can be visual, haptic, and audible. If feedback is too delayed or does not exist, the user will become frustrated or lost after the action is completed.

Even if there is essentially no delay in the actual loading of the new content, explicit feedback may be needed to make it clear, if the change is not clear at a glance.

Consider how feedback can be applied to:

Tabs

An immediate response must occur after a user selects tabs. The selected tab may change color and size and will display its related content in the viewport. Adjacent tabs that are not selected must remain unaffected. Be sure to follow conspicuity guidelines when selecting appropriate colors to contrast with the background.

Simulated 3D effects

Whether you are peeling a back corner to reveal the page underneath, or rotating an object, feedback must occur to indicate the action was successful. As discussed in detail in Part I, images are transformed incrementally within our visual perception process. When we see an object move and transform its shape in incremental steps, we have an easier time understanding that the two objects are related or identical. Therefore, use appropriate transition speeds throughout the entire action.

Pagination

Feedback can be used to display indications of page jump. Those pages not accessible due to current context can be grayed out. For example, "Back" or "First page" links should be visible but inaccessible when the user is already on the first page.

Mapping

Mapping describes the relationship between two objects and how well we understand their connection. We're able to create this relationship when we combine the use of our prior knowledge with our current behaviors. To quickly recall these relationships, we develop cognitive heuristics, or cultural metaphors. These metaphors reinforce our understanding of the relationship of the object and its function.

Some common metaphors you can use to access information on mobile devices are:

Tabs

Tabs are understood in principle to correlate to the file folder tab. We expect this tab to be selectable and an identifier of a folder that contains related content. If we see one tab, we will expect to see adjacent tabs that are separating additional content. Too many grouped tabs, however, will cause our users to become cognitively overloaded during the decision making process, thus resulting in slower performance and user frustration.

Cubes

Cubes show content on one or two faces, and will be expected to have content on all other sides that share the same axis. Users will expect faces to rotate. Constrain the rotation to one axis; otherwise, users will become frustrated with their inability to quickly navigate and correctly select the content. Animation of the rotation from one face to the other must be provided. Otherwise, the user may not know her current selection's position on the cube.

Peeled corners

Pages that are slightly peeled back are understood in principle to correlate to pages in a book. We expect access to the pages underneath the top layer. A gestural interaction that allows for turning the page will be expected. Animate the transition from one page to the next.

Other 3D effects

Aside from multisided shapes, effects such as shadows and transparency can be used to simulate layers and depth. The user will understand that information can occupy virtual space, appearing modally in front of other layers, while hiding others.

Constraints

Restrictions on behavior should be appropriately implemented to eliminate or reduce performance error while laterally accessing content. Consider using the boundaries of the display (edges and corners) during gestural and scroll-and-select navigation. The edges and corners provide an infinite area over which to move our fingers or cursor. This can significantly reduce the amount of time it takes to navigate across pages. Consider this effect when using a circular or closed navigation structure.

Circular navigation

A key tactical consideration, mentioned several times in the patterns, is whether browsable data sets are "circular" or "dead-end". Circular lists simply go around and around. When the user is at the "last" item in a list, continuing to view the next item will display the first one in the list. This is useful for faceted views or other cases where the ordering is unimportant. Consider integrating the Location Within widget to indicate current position.

Closed navigation

For dead-end lists, there is a definitive start and stop, and (aside from links such as "Back to top") there is no way to go to the other end. These are useful for information with a very ordered, single-axis display. Most hierarchies and much date-sorted information should display like this. An example is a message list that starts with the most recent message. Consider integrating the Pagination widget to indicate current position.

The severity of your constraints will vary widely depending on the criticality of the feature. An entertainment application may be designed for delight, and be expected to draw attention and be used while still; therefore, it has very few constraints. An SMS application could be used by anyone, anywhere, under any conditions, so it must be easy to use above all other considerations.

Medical and public-safety applications may have additional constraints. Even consumer device features may have legislative or regulatory constraints that must also be addressed.

Patterns for Lateral Access

Using appropriate and consistent lateral access widgets will provide an alternative way to present and manipulate content serially. Within this chapter, the following patterns will be discussed, based on how the human mind organizes and navigates information:

Tabs
>Based on the concept of file folder tabs, and used to separate and clearly communicate sets of pages or features at the same level in the information architecture. See Figure 5-1.

Peel Away
>An organic and animated representation of a page being flipped over to reveal a second page behind it.

Simulated 3D Effects
>Display an alternate view to the content on the page using 3D graphics. When device gestures or viewer movements are used, the items affected will follow the presumed physics or correctly represent the space they occupy.

Pagination
>Serially displays a location within a set of pages, and offers the ability and function to navigate between pages easily and quickly.

Location Within
>Uses an indicator to show the current page location within a series of several screens of similar or continuous information. This is presented with an organic access method.

Tabs

Problem

You must provide access to a small number of items at the same level in the information architecture, while also clearly communicating this hierarchy of information.

Tabs are a very common interactive method and they can be implemented on almost any device or platform. Some will have built-in tab implementations, which may restrict the methods you can most easily choose from, but also help to enforce consistency across the OS.

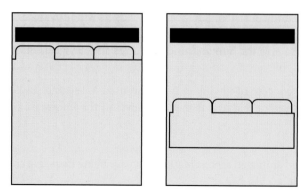

Figure 5-2. *Tabs may be used across the application or service, as shown on the left, or to control access to a portion of the page. The tab may serve as the page label, or a separate title may be needed.*

Tabs are based on real-life objects: the tab labels that stick up from file folders. The folders contain information that takes up too much room to just leave on the desk. To make sure they work, follow the same principles as the paper and file cabinet:

- Clearly label what is inside the item associated with each tab.

- Indicate when you have the one you want selected, or open.

- Make sure all tabs and labels are visible at once, or it is clear there are more to be seen if you scroll or otherwise interact with the tab bar.

Tabs do not actually have to look like file folder tabs, with rounded corners, or even with any container at all. But the principles still apply. For interactive applications, one (and only one) tab is always opened, the contents are visible, and the tab indicates which item is currently open. See Figure 5-2.

Tabs are generally only used for sets of pages or features between three and eight items. When you use tabs with only two items, the design may fall apart, and you must carefully redesign to make it clear which is selected. Larger sets, especially when they do not all fit on the page, may cause the user to become lost. If very large lists of items within the information architecture must be displayed, consider a more list-like pattern instead, or reconsider whether the information architecture is correctly built.

A tab can sometimes overlap with a Fixed Menu. If you provide access to a number of other screens as part of a menu, and the menu items are visible on more or all of the pages, you may be confusing the two patterns. Whenever the selection also indicates current position, it is a tab. Be clear about which one you are using.

Variations

Aside from graphic design distinctions, there are basically two axes upon which tabs may vary.

Tabs may be used across the site, application, or process, as an indicator and access point to the highest level of navigation available without exiting. Or they may be used within a single page or page template to provide access between broadly equivalent types of information without logically leaving the page.

All the available tabs should be visible at all times. However, this is often not possible due to labels being too large for the screen. Tabs that do not fit may fall off the sides of the screen, as in Figure 5-3. And those that do fit will appear as focus shifts. In the most extreme case, only a single tab will fit the screen, and sideways scrolling is required to see any other tabs.

Interaction details

Figure 5-3. *When not all tabs can fit on the screen, make it clear with truncation, arrows, and similar design tactics that more are available. Note how this example does not use a container for the tabs, except to highlight the current one.*

For scroll-and-select devices, tabs are most useful when the remaining content only requires vertical scrolling. In this case, pressing the Left or Right scroll key anywhere on the page will switch focus to the current item in the tab bar, and subsequent presses will switch focus to the next item in line. Pressing "OK/Enter" will select the tab and switch to that page, section, or function.

When using a tab bar to switch out content for a subsection of a page, sideways scrolling among those tabs will become available when vertical scrolling has brought the tab bar into focus. This works very much like it does for subscrolling of a text area in a form. The current item will be in focus, and others may be focused in the usual manner. While in this tab row, any other tab rows may not be accessed.

For scroll-and-select devices, tab rows should always be circular lists. When the user is at the end of the list, another directional press in the end-of-list direction will jump the user to the front of the list. This is especially important if your tabs do not all fit in the viewport.

For touch and pen devices, always allow the user to directly interact with the tabs. However, this can be a problem as the tabs must be correspondingly larger to allow for the touch target. Again, unless the labels can be small enough or the device is large, tabs may not be the best option. For cases where not all tabs fit on the screen, scrolling (by dragging in the tab bar) may allow browsing of the whole tab row.

You may combine other interactive mechanisms such as the Film Strip with tabs, to allow multiple interactive methods to access the other screens with On-screen Gestures while still making the location in the system clear.

For any selection mechanism, selecting a tab when the tab row does not entirely fit on the screen will cause the current tab to become centered immediately. If this is not technically possible, it will assume the centered position when the new content loads.

Presentation details

Figure 5-4. *Sometimes only a single tab label can or should fit on the screen. Be especially careful to indicate that more is available, with prominent arrows or, even better, by indicating relative position in the total list of tabs, as shown on the right.*

Tabs really only work well when they are arrayed horizontally, or when all of them are on one row. Vertical tabs of any sort are rarely understood by the user to be tabs, and so will not be discussed here.

You may save some vertical space in some cases by using the highlighted tab as a Title for the page or component. Ensure that it is clear, readable, and sufficiently prominent to serve both roles.

Tab labels are generally text. Icons may be used to support them, but tabs will rarely be understood sufficiently without text. Use labels in a consistent manner, and ensure that they are clear and jargon-free. See the Titles pattern for more discussion of labeling. You may sometimes get away with truncating the text, or allowing it to marquee (scroll) when the tab is in focus. This can be helpful to regain space and to make each tab a consistent size, but make sure they are still readable.

Tab rows should be presented as though they are dead-end lists, with fixed endpoints. Only when an overflow action is selected will the user jump to the extreme other tab. Animation may be used for tab rows that do not fit on the screen—rapidly scrolling back to the other end of the list—to more clearly communicate the position that has been selected.

When not all tabs fit on the screen, you must provide some indication that there are more tabs. There are three basic methods for doing this:

- Fit whatever is possible, so parts of tab containers and labels are on the screen. This is the clearest method, but some interface designers can perceive it as "messy."

- Only fit complete tab labels on the screen, including as few as one tab. The presence of additional items is only communicated with an indicator (such as an arrow) at the end of the list with additional items. This can be more difficult to discover, so you should design it carefully to ensure that the indicators are clear and visible.

- Use "magnification" to indicate the position in the complete list in a tab-like manner, while providing a much smaller number of labels at a readable size.

Any indicators that there are more tabs off the screen must reflect actual content with the dead-end list paradigm. That means when the end of the list is reached, you remove all stubbed tab containers, arrows, and other indicators. See Figure 5-4 for some indicator options.

When scrolling through the tab list, make sure any visible but nonselected tabs look different from the selected one, even without comparison, so that users do not mistake it for the page title.

Antipatterns

When only one or a few tabs are visible, ensure that the tab paradigm is clear and that it is obvious the tab is not just a page title, but is one option of many from which to choose.

Clever solutions for space rarely work, so follow best practices and existing working methods before attempting to develop your own solutions. A second row of tabs is always perceived as subsidiary to the top row, and is not read as a second row of text would be.

Avoid using tabs for both high-level and in-page navigation, as two different rows. If needed, differentiate the two in some key way to express the hierarchical difference.

When tabs are used within a page (such as providing a switch between overview, detailed specifications, and reviews), avoid refreshing the entire page. Logically, and by convention that users are familiar with now, this should only load the requested information and leave the rest in place, without even flickering.

When scroll-and-select is the only selection mechanism, and page interaction uses a virtual cursor, tabs are not suggested as they may be slow and difficult to select.

Do not use conventional circular list presentation, where the last tab is adjacent to the first tab. Users may lose track of their position in the list.

Avoid using Pagination ("Page 1 of 6") or other generalized Location Within widgets to indicate how many tabs are available. These patterns are associated with other access methods, and so may disrupt building an accurate mental model of the information architecture.

Peel Away

Problem

You must display a small amount of deeper information, or provide access to related controls or settings, in an organic manner.

Peel Away requires notable graphics processing, full-screen display, and gestural interaction to work well. Many modern OSes have such features built in or can support them. Websites generally should not use these features, even if they support the underlying technology.

Solution

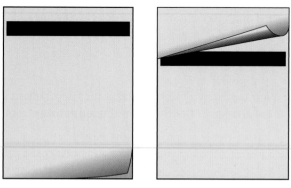

Figure 5-5. A closed Peel Away on the left, with the front page and a hint in the corner, and opened on the right to reveal almost a whole screen for the back page.

The Peel Away pattern is like the Tabs pattern, and many others in this chapter, in that it simulates a real-world interaction. Here, you use simulated 3D effects to pretend the page is a piece of paper, and the user can peel or flip it back to view a second page behind it (see Figure 5-5). Compare this to the Pagination pattern, especially the simulated paper effect versions.

This pattern only allows for a single page of additional information, often much less. You can think of it as comparable to presenting the same information in a Pop-Up. The value of this pattern is in the more direct relationship between the primary content on the front, and the revealed content on the back, or the page behind. A distinct hierarchy of information or control is established, and the user feels she is in control of the experience more than she would be if she were clicking a Link to reveal a dialog.

This pattern is not particularly suitable for scroll-and-select devices. Even if a virtual cursor is available, the action is not as natural. It is not worth using this except on touch and pen devices.

Variations

This is a fairly singular pattern, with no real variations presenting themselves aside from details of the graphic design.

Refer to other patterns such as Simulated 3D Effects for similar methods.

Interaction details

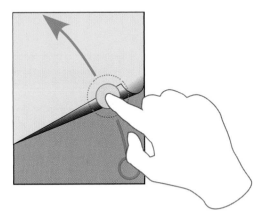

Figure 5-6. *The Peel Away pattern only really works for touch and pen interfaces. It can be tapped or, as shown here, pulled open by a drag gesture.*

You can make a Peel Away operate in one of three ways. These are not precise variations, as they should be largely driven by the interaction standards upon which the OS operates. For example, if drag and hold is not used otherwise in the OS, it should not be used in an individual application.

- Tap the peeled-up corner (or an icon or other control) to reveal the back page.

- Drag back the peeled-up corner, as in Figure 5-6, to reveal the back page. The drag action will, much like the Film Strip, have to move a certain distance before it commits; releasing before this occurs will snap back to the closed position. The dragging is a direct interaction, so it must follow the user's drag speed, including inertia.

- If the back page contains only content, and there is less information than a full screen, a drag-and-hold action may be used instead. The front page can be pulled away, but never "flipped over" so that it sticks. As soon as the drag is released, the front page will fall back into place.

These variations also allow you to use the pattern for any touch or pen device. If your platform does not support dragging well, use the tapping interaction.

These may sometimes be combined. For example, both dragging and tapping may be supported. Closing the panel reverses the actions in all cases. Use the same method or methods to both open and close.

Except when impossible, because you are using the drag-and-hold method to reveal it, you may use any interactive methods on the back page. This allows you to use form controls or links, and reveal settings panels or help documentation with links to other sections.

Presentation details

 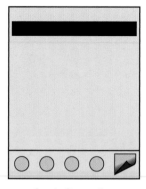

Figure 5-7. *The Peel Away hint, when closed, may bear a mark to indicate what content is on the second page. When a Fixed Menu is on the front page, this will interfere with the hint, and an icon may be needed instead.*

There is no point to using a Peel Away function if no one knows it's there. You must design an affordance, usually by showing the page as being peeled back slightly from the lower-right corner. If space is at a premium on the front page, this can be minimized, but you should never eliminate it, or the function will be difficult to discover. To encourage use, the effect can be exaggerated at page load, with the peeled-back edge settling (animating) slowly in a second or two.

You can provide other hints, such as an icon as a translucent element over the folded-back page, or on the part of the back page that is visible. This serves to indicate to the user what information is there. This is most useful when the icon is easily understood, such as if the back page provides help information. Figure 5-7 shows two options for hinting.

When a Fixed Menu is used, and therefore blocks the bottom edge of the viewport, it may be necessary to use an icon in the lower-left corner of the menu to activate the Peel Away function instead of allowing direct selection of the peeled-back front page. The Peel Away does not work well when it is not up against the edge of the viewport, so you should avoid moving the "peeled part" up.

Always animate the opening, so the front page peels back and progressively reveals the back page, though very rapidly. Without the transition, the relationship may not be clear to users. Use physics models, or set timing to simulate the actual behavior of paper.

When open, you will probably want to place an image of a page folded back, as though caught by a staple, in the upper-right corner. This function is essentially the reverse of the closed state. The user can tap or drag on this to easily return to the front page.

The margins, type, and general look of the back page should be similar to the front page so that there is a clear relationship between the two. A title should be presented on the back page, but may be in much smaller type than normal, as a note that the information consists of details or disclaimers, or is continued from the front page. Naturally, you may need to either place the title lower or allow some of it to be obscured by the peeled-back top page.

Antipatterns

Do not use the Peel Away pattern to display a second page of information that just continues directly from the front page. Simply Scroll the page or use the conventional Pagination methods instead.

Unless it is explicitly an "Easter egg" and intended to be hidden, always show the peeled-back edge. Do not make users guess or ask friends how to get additional useful information. Delight only comes when self-discovery is easy.

Do not mix interaction paradigms, only allowing dragging to open and tapping to close, for example.

Do not use this pattern if the entire front or back page scrolls. Though subsidiary areas may scroll, the pattern relies on simulating a piece of paper, and this does not work well if the edge of the paper can scroll off the viewport. Additionally, if scrolling is suitable, typically the additional information can simply be farther down the page, and the Peel Away is not needed at all.

If the animations required for this cannot be supported, use a different pattern that does not require animation or 3D effects.

Simulated 3D Effects

Problem

You must display a small amount of directly related information, provide access to related controls or settings, or provide an alternative view of the same information, in an organic manner.

Simulated 3D Effects do not always require a 3D display, but do generally require significant graphics processing power, and may require access to sensors (which browsers often cannot get).

Solution

Figure 5-8. *Individual elements on a screen, or the whole screen, can reveal related information—or settings, as shown here—as a part of the element, instead of using conventional but unrelated items like a Pop-Up.*

Extending the physical-simulation concepts of the Tabs and Peel Away patterns, other types of Simulated 3D Effects can be used to pretend the screen, or items on it, are dimensional, physical objects. The user can see the sides, rotate or flip them (as in Figure 5-8), move them aside, or look around, all by changing the point of view or interacting with the objects directly.

Note that this pattern is labeled as "Simulated," as it is displayed on a flat screen. It also does not require actual 3D display technology, and surprisingly dimensional results can be achieved with conventional 2D displays and very careful design. Some display methods can even be simulated without particularly involved graphics processors. If the pattern seems useful for your design, investigate ways to accommodate it with the technology available.

This pattern works best when direct manipulation of the screen or the device is used. See the On-screen Gestures, Kinesthetic Gestures, and Remote Gestures patterns for more details on direct interactions. This pattern will rarely work well on scroll-and-select devices without accelerometers.

Variations

To use Simulated 3D Effects, you will pick one of three methods of interaction, and the associated methods:

Screen gesture
> Items, or the whole screen, will react to drag actions.
>
> - The selected item may be flipped to another side. Usually, these are represented as rectangular prisms, flipping 90 degrees at a time.
>
> - The items may be moved around the screen. This should not be confused with Stack of Items, which allows actually manipulating the items selected, but may look similar.

Device gesture
> The device senses position relative to the viewer, using accelerometers, machine vision (cameras), and other sensors. The screen simulates an environment in which items live and are affected by movements of the device.

Viewing point
> Using machine vision or a 3D display, when the user moves relative to the screen, items declared to be above others may simulate parallax and allow the user to look under them to other items or to the background.

In the event the device has two screens, one on each side, a device gesture that rotates from one side to the other may display related information on each screen, without using simulated 3D effects at all.

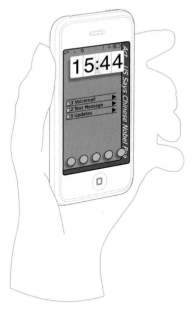

Figure 5-9. *Simulated 3D Effects use space that would otherwise be unavailable, allowing simple gestures to instantly reveal information. Here, a news ticker is on the right frame, and simply tilting the phone reveals it. Another item, such as a stock ticker, could be on the left frame edge.*

When the user selects and directly manipulates items, the items will be rotated to view information or controls on another face of the element, or they will slide away so that the user can see past the item or to the top of another item or the background. You should restrict actions to only one of these types of movement for each object. Generally, it's a good idea to use one type of movement for any class of item, across the whole OS or application.

Make sure all interactions on a movable element do not interfere with any other gestures. For example, only allow clicking on items within the element, and ensure that any drag actions will rotate or move it. This means you cannot have things such as Scroll within the 3D element, or even allow the use of Mechanical Style Controls such as the spinning date selector.

When device gestures or viewer movements are used, any items affected will follow the presumed physics or correctly represent the space they occupy. Common uses are:

- Seeing around or behind items in the front
- Observing the side of an item
- Seeing "inside" the device, such as under the edge of the viewport. (see Figure 5-9)

When first used, and after major changes, the frame of reference is typically reset to assume the backdrop is level; subtle changes will then allow items to either react to the movement (fall to the "downhill" side) or stay in place and allow the user's head movement to view "past" the item.

Presentation details

Details of the presentation are almost unlimited, offering you more scope to explore than any other interactive pattern. From geometric shapes to natural scenes with leaves and grass, anything is possible. Make sure the style used matches the OS, or the brand of the product or service, and that it is still understandable and usable.

These standards may settle down once the pattern is used more widely, user expectations become more set, and best practices are established. But for now, you have room to develop those standards.

Any shapes you display must "look 3D," so the user expects to find additional information with an interaction, even when closed. This may be as subtle as providing a shadow to indicate the item has height. Since many items have these visual effects, try to add additional details, such as movement, or build imperfections into any machine-vision or accelerometer-based systems so that minor actions will cause the other faces or items behind to peek out.

Any items you place on the screen that are not directly related to the interaction reveal method should also generally exist within the simulated 3D environment. Lighting changes will fall on them; items that move must pass over or under them, or bump into or bounce off them, and so forth.

Antipatterns

Three-dimensional interfaces require commitment, and cannot be well implemented with half measures. Do not implement some 3D features if conventional layout elements will interfere with the illusion. For example, if items move about the screen, the typical grid of icons must also exist within the simulated space and interact with the simulated physical world. The icons must not simply be painted onto the backdrop, act static, and be ignored by other objects.

Avoid using secondary interactive methods such as menus to change the on-page elements. Selecting a menu to "Rotate" does not work the same as a gesture. If support for scroll-and-select devices is required, include the basic functions under options or via accesskeys instead.

Avoid use of multiple types of interaction methods. Movement via On-screen Gestures will tend to change the user's frame of reference enough that he may tilt the device. If Kinesthetic Gestures are also employed, the user will rapidly lose control.

Pagination

The user's location within a series of screens—which continue to display a single set of content—should be clearly communicated. You must provide easy access to other pages in the stack.

Pagination control can be implemented very simply in any type of interface. This does not mean that quite complex, graphical, or interactive methods may not be used instead.

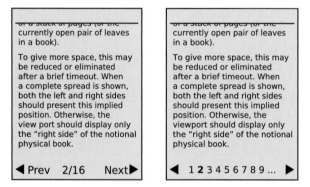

Figure 5-10. *A variety of types of information may be used for Pagination. The key features of the location and moving forward and backward are always present.*

The many screens displayed serially for large amounts of content may be considered pages, as though they are part of a bound paper item.

You will then display page numbers and a sense of the relative position within the total set. Integrated with the display is a method to move between pages easily and quickly; see Figure 5-10. You may also want to offer methods to jump farther than the immediately adjacent pages.

When building a system with large amounts of information, consider if a multipage view is even correct. In many cases, Infinite List can solve the same display problem without building a pagination widget at all. Look at the OS standards as well; you can use methods such as Film Strip to present a series of pages of content of any sort, and often this should be the default method. Consider other methods of discovering information in large sets, such as the Search Within pattern, and do not just require users to scroll to find information.

Note that widgets that only present the location, but do not have any role in the interaction of moving between pages—even when such interaction exists on the page—are considered Location Within indicators, and not Pagination widgets. Please see that pattern for comparison.

Location Jump is a very similar type of interaction, but uses index points related to the content (such as dates or the first letter of a name) instead of the display methods (page numbers) used for Pagination.

Variations

Three basic variations exist:

Widget-based
A section of the page is dedicated to the display and control of items concerning pagination.

Organic
Natural displays are becoming more common now, especially on touch and pen devices. These simulate "machine-era" presentation methods to imply the same information as the widget view. The display methods and interaction are integral to the design of the entire page, and are not in a specific location.

Hardware
On devices with hardware keys, you may map some to key page functions when inside an application. Certain devices designed around page content, such as eReaders, may have dedicated page control buttons.

Interaction details

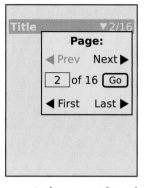

Figure 5-11. *If too many functions are required, or cannot fit on the page, a Revealable Menu can be used instead. This may also be a suitable way to present options when gestures or dedicated keys are used for basic functions.*

You should always present pagination controls as an integral part of the OS, application, or site. Make all controls visible or immediately accessible whenever possible. When it is needed, you may offer a larger "control panel" style of interface, as in Figure 5-11, which users can open into a Pop-Up dialog by accessing a related component such as the current page.

A number of controls are available. Many are paired sets, and will be considered as a single entity. A large number of these controls must be prioritized for display and access; you would almost certainly not want to use all of them. The following are listed in priority order for most uses, though yours may vary:

- Move "forward" and "back" to the immediately preceding and following pages.

- Jump to a specific nearby page. As many of the closest pages as can fit conveniently in the available space are listed. Selecting a page number (or analog such as an icon) will move to that page immediately.

- Jump to the next displayed set of nearby pages. Although the key goal is to display a new set of pages from which to choose, this should also change the page view to the first or center page in the listed set.

- Jump a small number of pages forward and back. This will usually be either 3 pages (just enough to be inconvenient to use "next" several times), or 10 pages, as it is a larger, even value. Which one you use will depend on the type of information-browsing problem being solved.

- Load the first or last page in the list. In some cases the list is of arbitrary length, and there is no value in the last page (or it is technically infeasible to load), but the first page link may still be useful.

- Jump to an arbitrary page, by selection from a list or by directly typing.

For each of these, you should only display pages that actually exist; gray out pages that are not accessible due to current context. For example, "Back" or "First page" links should be visible but inaccessible when the user is already on the first page.

The forward and back controls may be easily activated on touch and pen devices, either by tapping an indicated corner of the page (right is next, left is previous) or by gesturally "flipping" the pages with a drag action. Details on a related drag action are available in the Peel Away pattern, but that is for one page only, and so should not otherwise be confused with this pattern.

Some of the more involved jump methods may also be initiated or carried out with gestures, as well as actions such as press-and-hold on the page flip section. However, these are not yet consistent and well defined, so users will not understand or discover them without training.

Hardware key mapping may be similar to Accesskeys but is usually best mapped to the Directional Entry keys. When a five-way pad is available and is not used for other purposes such as a cursor, Left and Right should map to "Previous" and "Next." Volume keys or other key pairs can also be used, but may be difficult to discover.

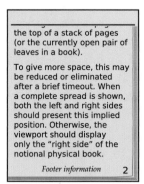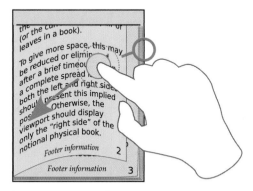

Figure 5-12. *Page flipping may be displayed as though the interface is composed of a stack of actual pieces of paper. Represent the relative number of remaining sheets, and display gesture-initiated page transitions as though the pages are flipping.*

The Pagination widget should be available wherever the user is most likely to need to change pages, or be aware of his location. Most often, the solution is to place the widget immediately above and below the content unique to the page. Another option is to anchor the widget (e.g., in the title bar), so it is available at all points on the page. If the page does not scroll, a single fixed-location item logically becomes an anchored widget.

Labels should be clear, but they must be as brief as possible to allow more items to be placed on the page. Usually, labels themselves can be eliminated; "1 of 16" is generally just as clear as "Page 1 of 16 pages," and is often much easier to scan and read.

Use caution with nomenclature. Due to the lack of an actual physical object and the hypermedia nature of dropping into the middle of a stack, paired terms such as "Previous" and "Next" can be confusing for some users.

You can use arrows and page icons to good effect. Right arrows mean "forward" and left arrows mean "back." Chapter jump icons from video players (an arrow pointing to a vertical line) are often encountered, but so far have not been consistently employed, and so will require discovery and training.

Organic display of pagination can imply position by graphically showing the current page as the top of a stack of pages (or the currently open pair of leaves in a book), as in Figure 5-12. To give more space, this may be reduced or eliminated after a brief timeout. When a complete spread is shown, both the left and right sides should present this implied position. Otherwise, the viewport should display only the "right side" of the notional physical book.

Pagination controls can easily get out of hand. In any reasonably complex system, it is easy to find justification for every method of page control. Avoid this, and attempt to only include the minimum set needed, so they can be placed on the page and be easy to find and use.

Realistic representations of pages must accurately reflect the information. For example, if a stack of pages is shown to the side, to indicate that more are available, the relative number shown must be of a plausible size to represent the number of pages and must change the indicated size as pages are flipped. Do not always show 10 pages, whether 320 or zero remain.

Avoid cumbersome entry or methods. For example, do not present pull-down lists of pages that are longer than about a dozen items, even on large-screen devices. Offer typed entry or some other solution instead. See the Mechanical Style Controls for some useful widgets.

Avoid overdoing mapping of shortcuts to keys. For example, mapping "Beginning of document" to the Up key induces a risk that the user will accidentally perform this action, and cannot return to her previous location. There is no clear "undo" for navigation in a document like this, so it is not worth you trying to solve this by implementing such a control.

Location Within

You must describe the location within a series of screens that contain alternate views, or that continue the display of a set of content.

Whether an interface is composed of text or graphics, the Location Within pattern is easy to implement. Some OSes will have a built-in style for at least some cases, which should be used.

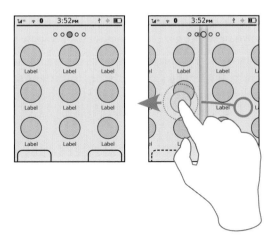

Figure 5-13. *Graphical indicators may be of any shape. Dots are common, but bars or other shapes can be used and integrated with other design elements, instead of floating. Note how during transition the indicator also transitions, instead of just jumping from one to the other.*

When several screens of similar or continuous information are presented with an organic access method, an indicator is usually required so that the user understands his position within the system. Two very common uses for this widget are to show position within a Slideshow or a Film Strip. Each has the interframe interaction integrated with the basic pattern.

This overlaps heavily with the Pagination widget, to the degree where it may be unclear which is being applied. If the indicator is integral to the method of changing pages, it is a Pagination widget, and if it is a static component, it is a Location Within indicator.

See the Location Jump pattern for another variation of providing access using a similar presentation method.

Location Within widgets are used to convey the specific context of a page within a series. The method used depends on the amount and type of information being presented. There are two basic types.

Text widgets list the current page of a set of pages—"1 of 16," for example. This is often encountered in slideshows or other contexts where much other metadata is often displayed. These are suitable for any number of screens; numbers are very space-efficient ways of displaying the information up to sets of hundreds or thousands of items.

Graphic widgets indicate the position as a single item among a field. This may overlap with certain implementations of the Tabs pattern, but the entire set of pages must be visible on the page and may not overflow. The dot and bar indicators to show position within the Home screens, shown in Figure 5-13, are a typical use for these indicators.

You can also combine these two variations, such as when a graphic indicator cannot fit all the graphics on the screen; the graphics are used to indicate transitions between individual pages, and a text indicator may be used to provide overall context. Graphic indicators may also have numeric values attached to them.

Interaction details

There are no direct interaction details with the indicate-only Location Within pattern.

Presentation details

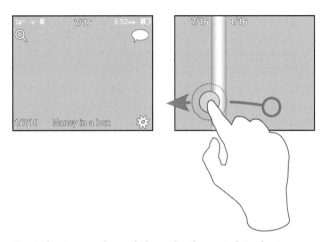

Figure 5-14. *Location indicators may be used alongside other metadata about a page, as when viewing galleries of images. To make location clear when scrolling through a series, the indicator may remain and may change to note the centered item or, as shown here, may attach a page number to each item.*

For text items, labels should be clear but as brief as possible to not make the element too prominent, and to allow more items to be placed on the page. Usually, labels themselves can be eliminated; "1 of 16" (or even "1/16") is generally just as clear as "Page 1 of 16 pages," and is often much easier to scan and read. Remember, your users are probably moving between the items, so the change will make things clear. See Figure 5-14 for a typical implementation.

Graphic indicators may be any shape, but they should not imply action or activity. Triangles, for example, may imply direction or action, which is not necessarily true. The indicators must be in the orientation of the page transitions. For a horizontal sliding Slideshow, for example, the indicator must be horizontally aligned also.

The current position must be clearly indicated. A color, size, and/or contrast change must be clearly different from the other page indicators, and must be prominent enough to stand out from the background or other elements on the page. Use shadows or other effects to make the indicators stand out, or to make them appear to be on a layer over the page content.

Any indicators may disappear when the screen is not transitioning. After a few seconds, fade the indicators. Use this carefully, and allow a method to easily reveal the indicators so that a sense of location is available at all times.

Make sure any graphic indicators are visible when transitions between pages occur. The indicator must transition, by sliding between positions instead of simply changing the focus item. Alternatively, text indicators may change to label each visible item during the transition.

You may find yourself integrating other components, such as previous/next arrows, and then thinking they should also be selectable, and then realizing you are making a Pagination widget. This is not always true. It may be instead that you have just decided to place the indicator and control items for the underlying Slideshow or Film Strip adjacent to the Location Within. Such design license is always acceptable, and will happen all the time. Just be sure you recognize it so that you are always using the correct components in the correct manner.

Antipatterns

The key risk with this pattern is not using it at all. Do not avoid using indicator patterns in an attempt to prevent page clutter. Many implementations are very clean and simple and do not interfere with use of the screen otherwise.

Chapter 6

Drilldown

Get Ready to Push!

Driving cross-country in your car can be quite exciting—whether you're stopping in small, quaint towns that are hardly noticeable on a map, flying down coastal highways with perilous views below, or enjoying the endless horizons of the plains. However, that state of happiness usually breaks immediately when you notice the low-fuel status icon has now appeared in your gas gauge.

This icon's design creates more user mental load than necessary. If you're in the middle of nowhere, and you haven't seen a gas station or fuel information sign in miles, and now, on top of your uneasiness, you must calculate and predict how far you can go without running out of gas, you're wondering:

- Did this status light just come on?

- Or has it been on for several miles?

- How many miles am I going to have to walk to find a gas station after I run out of gas?

- Do I have the appropriate walking shoes?

What you really need is the ability to quickly access additional related information.

Figure 6-1. *Iconic labeling allows you to add information and selection methods directly to graphical or visualized data elements. The user knows what the interesting stuff is on the page, can tell the difference between the various types of information, and has an accurate expectation of what will happen if he selects an item. Even if the element opens additional details, hints of the data are presented contextually and invite users to find other features without exploration.*

Maybe We Won't Have to Push

Maybe the fear of running out of gas will come to an end; we're not referring to alternate energy sources here, but rather to an improved display design on our dashboard.

Here are some suggestions:

- Provide more surface-level information. For example, imagine if this status icon was interactive. Pushing it may reveal numerical information about how many miles you have left before you run out of gas. Also, make the initial warning light blink on and off, and then pause in a calculated repetition.

- Because our short-term memory can only hold about three chunks of information at a time, we usually store chunks that are important to the task at hand. If a warning status light appears without undergoing periodic state changes, our current attention may not focus on it and we'll assume it's not even there.

- Provide drilldown access to additional information. For example, provide a button that is linked to your GPS navigation map. This could reveal your current location and the manageable route to the nearest gas station that you could make before you run out of gas, put on those comfortable walking shoes, and push.

Drilldown and the Mobile Space

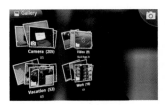  

Figure 6-2. *Improved screens, processors, and input methods increasingly allow the use of natural-looking objects. These communicate their content and interaction organically, and so hold the promise of innate, learning-free use. They are not quite there, so design items like the Stack of Items carefully, and don't be afraid to provide labels.*

Throughout this book, we mention the importance of following a consistent information architecture to ensure a positive user experience across a device's OS. In Chapter 5, you learned that it is important to design the interface to ensure that the user can effectively access this content because:

- Mobile displays are smaller than PCs, reducing the amount of information shown on a viewport's current state. Common smartphones still (in 2011) have screen resolutions of 320 × 480 pixels. Many feature phones still have resolutions of 240 × 320.

- Smaller displays require larger buttons for gestural interaction, which, in effect, will reduce the amount of content that can be placed on the screen.

- On larger mobile displays that require the use of capacitive touch, interactive items cannot be as small as they would be on a similar-size computer with a mouse or comparable interface.

- The user's finger and hand placement may occlude information during interactions, which may be much of the time the device is in use.

- Mobile screen real estate is so valuable that content and related features must be prioritized based on user goals and needs.

- A mobile user's attention is competing with all the stimuli in the surrounding environment. Therefore, the access to the content must be detectable, even at a glance, and afford its function appropriately.

- There is a direct correlation between the size of the target, its distance from the user, and the amount of time it takes to select it.

Drilldown access also requires these considerations. But unlike lateral access whose information architecture is based on the same tier levels, drilldown is concerned with accessing related content based on hierarchical parent-child relationships, hence the name "drilldown." You are accessing additional detailed information relating to the current content or the state of the device.

The fact that drilldown content is being accessed may appear directly on the state currently being displayed, or it may cause the user to jump around, whether within the current page or by opening a new one. If the user is required to access content on a new page, you must consider use of wayfinding principles to give the user an understanding of her current location as well as a clear path to follow back.

There are several reasons to use drilldown widgets across the mobile space:

- They allow the user to access more specific related content.
- They provide the user with visual cues that more related content is available.
- They allow the user to immediately notice the hierarchical relationship of the content.
- They allow the user to gain additional information, or change the state of the device, without removing himself from the original context.

Use of drilldown widgets affords the following benefits:

- They present immediate relevant information with a portal to access more specific content.
- They efficiently use valuable mobile screen real estate.
- They can allow for quick jumping to a specific section within the current page.
- They can provide additional surface information without removing the user from the context.

Patterns for Drilldown

Using appropriate and consistent drilldown access widgets will provide an alternative way to present and manipulate content hierarchically. In this chapter we will discuss the following patterns based on how the human mind organizes and navigates information:

Link
> A selectable, content-only item that provides access to additional pages or content.

Button
> A distinct, visual element, within any context, that enables the user to initiate an action, submit information, or carry out a state change.

Indicator
> A visual representation—dual-coded with an image and text—that initiates an action and submits information similar to the actions of a link and button. See Figure 6-1.

Icon
> A clear and understood visual representation that easily provides the user access to a target destination or direct function.

Stack of Items
> A stacked set of related, selectable items that can be dispersed to reveal the contents, which provide further selection or access to each item. See Figure 6-2.

Annotation

Reveals additional content details without entirely leaving the original display context. This additional content may also carry functionality.

When to Use Links, Buttons, and Icons

Knowing when to use these types of drilldown widgets can be challenging to understand. Use this chart as a reference to guide you in that process.

Pattern	When to use it
Link	Use a link when a new page of related content must be loaded.
	Use a link to jump to additional content within the current page.
	Use a link to open a Pop-Up dialog containing relevant content.
Button	Use a button to initiate an immediate action.
Standalone	Use a standalone button to initiate an immediate action without additional user input.
In-conjunction	Use in-conjunction buttons with other user inputs or controls (radio buttons, spinners, checkboxes, etc.) to commit these user selections.
Delayed input	Use a delayed input button to interrupt the submission to request additional user data. A modal Pop-Up dialog will likely be used to retrieve this information.
Indicator	Use an indicator to initiate actions of linking, commit actions, and state changes.
	Use an indicator to visually describe the type of activity that will occur when initiated.
Content beyond	Use a content beyond indicator to visually explain what type of content will be loaded if the link is followed. This is typically an icon in front of the text label.
Type of action	Use a type of action indicator to describe the type of activity that will occur when the link is selected. For example, a "Refresh" label can be accompanied by a revolving refresh icon.
Manner of action	Use a manner of action indicator to describe the way the action will be carried out. The icon should indicate that the action may go forward or backward in the process, opens a pop up, or performs some other type of action.
Icon	Use an icon to provide access to disparate items or functions, in a glanceable manner.
Fixed	Use a fixed icon to clearly explain, within the image, its function or target destination.
Status	Use a status icon to indicate a change with the current condition. This may be an external change such as the current weather, a system change such as inbound messages, or a user-initiated state change such as switching from scroll to select mode.
Interactive	Use an interactive icon to carry out a behavior directly, such as enabling WiFi. This icon does not provide immediate access to any target application, site, or information.
Stack of Items	Use a stack of items when information can be represented as thumbnail graphics, and all items in the group appear in a virtual stack which can be shuffled or expanded.
Annotation	Use annotation when more information should be presented for an item in focus, such as a pinpoint on a map or chart. An annotation is smarter than a tooltip, and may offer links or actions.

Link

You must provide a function to allow access to related content, from arbitrary locations within a page.

The Link is a very common action element, available in one way or another on every platform.

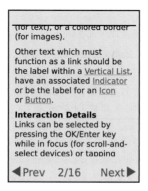

Figure 6-3. *Links are clearly differentiated from the text in which they appear, and cannot be confused with any other text styling.*

Links are simple, textual interactive elements which provide access to additional information, generally by:

- Loading a new page of content
- Jumping to the portion of the current page which carries the additional content
- Loading a Pop-Up dialog with the relevant content
- Revealing an adjacent page using the Film Strip pattern

Links are almost purely hypermedia elements. Although links were not invented by the Web and are not exclusive to it, the ubiquity of websites and links can serve as a good guideline for display standards and use of the pattern; your users can have a single understanding of the OS, applications, and the Web if consistent methods are used.

Links are generally found in content-rich areas, as shown in Figure 6-3, when used in application or OS contexts. Links to additional information within explanatory text, or for cross-referencing within help systems, are common uses outside the Web. Most other text found within an application or OS will be labels for lists, icons, buttons, and so forth.

Links alone are rarely suitable for these situations, so this chapter discusses numerous interactive methods you can choose from for your specific needs.

There are no true variations of the link. Links may be found anywhere on the page, and are differentiated from nonlinked content with a change in color and underline (for text), or a colored border (for images).

Other text which must function as a link should be the label within a Vertical List, have an associated Indicator, or be the label for an Icon or Button.

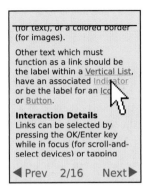

Figure 6-4. *Hover states must be indicated by a change in the link, usually a color change. The change is an important indicator that the correct item will be selected, as well as that it is indeed an active item. Visited items may also indicate this condition with a change in color, exemplified here with the word "Button."*

Links can be selected by pressing the "OK/Enter" key while in focus (for scroll-and-select devices), or by tapping directly on the link text for pen and touch devices.

Be sure to support focus states, or hover, by changing the link color as shown in Figure 6-4, or possibly the text style. Of course, this will only work for devices which support hover, or focus-without-selection. Pen and touch devices generally do not have this capability.

Figure 6-5. Additional information about a link should be presented immediately after the link text, and not as a part of the link. File types and size may simply be text, as shown here. Use an icon or the name of the provider (if short) for content that will load a new application or leave the site.

Inline links should always be a different color from any styled text. Whenever possible, they should also be underlined, but be sure to follow the OS standards. Simply using color can be confused with style and not perceived as a link. If needed, dotted or dashed underlines, or slightly desaturated color, can be used to reduce the visual prominence of the underline.

Some systems will use highlighting (a field of color behind the whole link text) instead of an underline. This is easy to see, and you should follow it if it is the OS standard, but some users may not immediately understand the meaning.

Underlines should never obscure descenders. Underlines must break, or otherwise integrate cleanly and clearly. If they don't, pick another typeface.

For most inline-style links, the color of the link should reflect the current condition:

- Nonvisited

- Visited

- Hover (when available)

- Active (mouse down)

In certain cases, the visited state is of no significance. If the information is dynamic, or the entire experience is used so that all links would be "visited" on a typical viewing, that state may be disregarded.

Links in lists use their position in a list, as a series of repeated items, to indicate that they are selectable. To avoid excessive horizontal rules (the separators for the individual line items), you usually should not underline links in lists.

The title for each link should be unique, clearly communicate the information on the target, and be easily scannable.

Additional information, in text or as a graphic, should be included after the link to communicate when an atypical target is presented, as shown in Figure 6-5. When a link leaves a site, loads a new application, or will initiate a file download, indicate this. Since normal links do not have icons, any reasonable icon after a link will usually be interpreted correctly. Text following the link should display the file size of downloadable items, and the type of file, such as "(384 kb PDF)".

Linked images should be bordered with the same colors as text links, when no other communication is available. When possible, add a description of what will happen to the image, or add a descriptive text link immediately adjacent to the image. Avoid use of simulated text or buttons in the image; the lack of interactivity (such as hover or active states) may confuse users.

Antipatterns

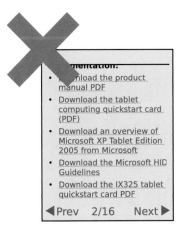

Figure 6-6. Do not use repeated or unnecessary prefixes, such as "Click here" or "Download," as shown. Additional information, such as file types, should not be part of the link text. Compare the scannability and readability of this with the example in Figure 6-5.

Do not use links to activate functions. Even within a web page, use buttons and other form elements for action behaviors.

Avoid using generalized text such as "Click here." Such text is not meaningful, and makes the content hard to scan for the relevant information, as shown in Figure 6-6. Especially in lists, try to use unique but descriptive phrases. Rather than "Click to view the manual," simply use "View the manual."

Do not use underlines for any other elements on the page, such as for emphasis. User-defined styles may still be allowed when editing text in word processing programs, of course.

The Link pattern defines behaviors for the linked text only. Do not change the background for an entire row or other container in which a link occurs or this will become another type of pattern.

Avoid changing the type style for link condition changes, such as hover. Switching to bold, except for a small number of typefaces, will make the text wider, causing strange jumps, confusing the user, and perhaps causing the page to reflow. Other changes may be suitable and appropriate, such as switching from a dotted underline to a solid.

Button

Problem

You must allow the user to initiate actions, submit information, or force a state change, from within any context.

The Button is probably the most common element across all mobile devices, and is built into every platform.

Solution

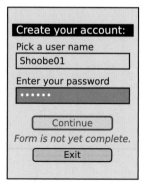

Figure 6-7. *Buttons are used in conjunction with input fields and selectors. The button submits or commits the selections. When fields or selections are required, a good way to avoid error messages is to disable the submit button. Be sure to label this condition so that users are not confused by it.*

Use a Button to initiate an immediate action. A button is not simply a "more important" Link, and should not act as such. Button-like action behaviors may be initiated with an Icon and label pair, if space or other layout or style considerations require it.

Variations

Standalone buttons will perform an action, or change modes immediately. They will also do the following, with no additional user input:

- Switch from email list view to composing a new message.

- Begin to synch email with a remote server. Click again to stop the process.

In conjunction with radio buttons, checkboxes, or any other user inputs or controls, a Button will commit these user selections; see Figure 6-7. Within websites, buttons are used to submit (or cancel submission of) forms to remote servers:

- With a text field, the Button submits a search.

- With radio buttons, the Button sets the USB connection mode.

- With a set of Mechanical Style Controls, the Button saves changes to the time of an alarm.

Delayed input interrupts the submission to request additional user data. Usually, a modal Pop-Up dialog will be used to retrieve the information. Ellipses follow the text label of the Button to indicate the process will continue, such as:

Share via:

- [Bluetooth…]

- [Email…]

- [SMS…]

- See the Exit Guard pattern for some related examples.

Interaction details

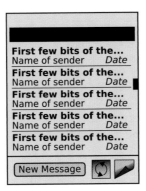

Figure 6-8. *Buttons may be combined or grouped with icons, which should then be of similar vertical height, as though they are related buttons. Consider the Page design carefully when placing buttons. For certain types of applications, you should place buttons at a regular location on the page or even in an area such as a Fixed Menu.*

Depending on the action, after the user selects a Button, the page may or may not change. If there is no change (such as with the earlier synch example), you must make the button change state to indicate the process is ongoing, or change the label to the reverse action (e.g., "Cancel Synch").

Users may select buttons by pressing the "OK/Enter" key (on a five-way pad) when the Button is in focus or by tapping directly on the Button for touch or pen devices. In many cases, especially in Pop-Up dialogs and full screens where an action is key, a button should be in focus by default so that no scrolling is required.

When user entry is required (for in-conjunction buttons), until sufficient entry has not yet been made, the submit action should be made inactive. This should be communicated with text adjacent to the button.

Presentation details

Figure 6-9. *Graphical icons can be used to support the messaging of the button.*

Buttons must be very easy to see and activate (especially for touch/pen devices). Background color and contrast must make the button stand out from the page background appropriately. Remember color-deficit users and glare, and make sure the text on the button has sufficient contrast to be readable.

Buttons should usually be about twice the height of the default page text, to provide room around the label. Smaller buttons can be used for less-important or less-used items, but they should never be smaller than about 120% of the vertical size of the smallest text. Consider click target sizes for touch and pen devices, and do not make selection areas smaller than 10 mm.

The width may be fixed, or it may vary based on the size of the text. Try not to have too much empty space. Vertical size must be the same for equivalent actions, as in Figure 6-8. An "OK" and "Cancel" pair, for example, must be the same vertical size even if you change the width to save space.

Buttons should generally appear to be raised above the page slightly. When a button is selected, it should appear to have been pressed and be level with or below page level slightly. Buttons are momentary-contact items, and this activation display should cease when the user click action ceases.

For scroll-and-select devices, the button in focus will have a thicker border or other indicator. Inactive buttons must be indicated by being grayed out. Be sure to select a sufficiently saturated background and high enough label-to-background contrast that graying out is clear.

Use the writing style already defined for the rest of the OS, your application, or your site. Be consistent with capitalization in all buttons. Be clear about terminology, and avoid labels that require reading. Rarely should buttons say "Submit" or "OK." Descriptive, freestanding labels such as "Connect" are always better.

For additional clarity, consider using icons inside the button, as in Figure 6-9. Use the most obvious graphic possible, with multiple encoding (e.g., a green check for submit and a red X for cancel). Generally, graphical icons should be to the left of the text, but if direction indicators are used, place the icon to the side indicated.

If graphics are not supported, or cannot be relied on (as with low-end web applications) for your platform, you can use text entities instead. The following table shows a selection of reliable ones.

Icon	Typical use	Entity	Decimal	Hex
⊕	Add	`&oplus;`	`&#8853;`	`&#x2295;`
⊗	Cancel/remove	`&otimes;`	`&#8855;`	`&#x2297;`
√	OK	`&radic;`	`&#8730;`	`&#x221A;`
⇐	Back/previous	`&lArr;`	`&#8656;`	`&#x21D0;`
⇒	Continue	`&rArr;`	`&#8656;`	`&#x21D0;`

Naturally, more common items such as "<" or even normal characters such as "+" can be used as well.

Antipatterns

Carefully consider whether a series of buttons should be used to make a choice. A more suitable solution is often another type of selection (such as radio buttons) with the Button used to commit the selection. This also allows adding other behaviors, such as a checkbox to indicate that the selection should be the default from then on.

Avoid making the default button perform an unrecoverable, destructive action. Users must make deliberate choices instead. See the Exit Guard pattern for more on this behavior.

Avoid color schemes that can be misconstrued as being grayed out and inaccessible. Likewise, avoid grayed-out states that are so well designed they look like they are just attractively gray buttons.

Indicator

You must allow the user to initiate actions, submit information, or force a state change, from within any context.

An Indicator is simply a link with an adjacent graphic (or text icon). It can be easily implemented on any platform.

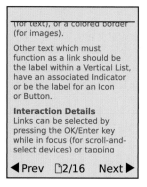

Figure 6-10. *Indicators are text labels with graphics used to indicate that the item is selectable, and what it will do. This pagination example uses arrows to emphasize the "previous" and "next" actions. The icon on the page location indicates that it is also a link, allowing direct access to page controls.*

The Indicator pattern, as shown in Figure 6-10, is a type of action initiator between a Link and a Button. You always use indicators with text labels, and they may perform any action: linking, state changes, and commit actions.

There is significant overlap between these three patterns, and the Icon pattern; in some cases deciding which to use depends on consistency and style.

You can use this pattern to express a hierarchical relationship between items. The Indicator would be considered more important than a conventional underlined Link and less important than a typical Button. Use caution with this, and try to use parallel controls for similar types of actions as much as possible. See the Button pattern for some additional discussion of this.

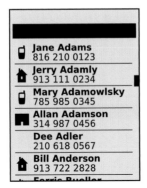

Figure 6-11. *Icons in the call history list on the left are used to denote which phone type was used; selecting the item will dial the indicated number. The same list on the right has no direct actions, but will load additional details for each number as indicated by the right arrows to the right of each line item.*

Indicators are expressed in a limited number of ways, but can indicate three different types of meaning:

Content beyond
> Explains what type of content will be loaded if the link is followed. This is typically an icon in front of the text label. Examples are the file type (e.g., Acrobat, Word) or types of objects to be viewed (e.g., photos, videos).

Type of action
> Describes the type of activity that will occur when the link is selected. This will usually reinforce the wording, instead of adding additional information, to assist with glanceability. For example, a "Refresh" label can be accompanied with a revolving icon indicating either a reload or a refresh.

Manner of action
> Describes the way the action will be carried out. This is typically in addition to the label, and so adds to the description. The icon should indicate that the action goes forward or backward in the process, opens a Pop-Up, or performs some other type of specific action.

All of these are shown in the examples in Figure 6-11.

There are few special or innovative interaction methods available for this pattern. The text will be selectable, usually in the exact same manner as the Link.

Whenever it is technically feasible, the indicator icon should also be selectable. As long as the text label indicates focus, on hover-state interfaces, there is no need for you to make the icon change also, although you may if it would assist in clarity or improve the interactivity of the product.

Presentation details

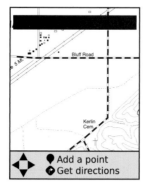

Figure 6-12. *Indicators can be mixed with pure icons, links, and buttons. Use labeled indicators when an icon may not be clear enough, or a text link would not be immediately scannable.*

Indicators are almost entirely associated with text; even when the text is positioned alone, or as a part of a list, the graphic should be inline with the text, and therefore immediately to the left or right of the text.

Which position is used will often carry meaning, and so should be carefully considered. If the function is "forward" or "next," for example, the Indicator icon will be facing to the right. It may be a good idea to place the indicator to the right of the text in this case. The "next" and "back" indicators used alongside text labels in the Pagination pattern are a good example.

Graphical icons are the most common type of Indicator used. When used without a text label, or when the icon becomes the most prominent item (with text supporting it), this becomes an Icon pattern instead. For the Indicator pattern, the supporting icon should not be much larger than the vertical height of the text. See Figure 6-12 and the associated patterns to combine the interactions in one space.

For use of the Indicator in lists, also compare it to the Thumbnail List pattern.

Icons may be of any style, but like the text, they must be consistent and must match the brand and other design guidelines of the site, application, or OS.

Indicator icons may be special text characters, when working in constrained environments. See the list in the Button pattern for some useful, universally available text items. If special typefaces can be loaded, you can use more interesting text icons, often with great efficiency in space, speed, and complexity of the code.

Avoid placing indicators above or below the text. Centered below the text, for example, may appear to be the most space-efficient location, but in fact is likely to be perceived as on another line, and either nonsensical or associated with other items on the second line. If this layout is needed, use an Icon instead.

Do not use an Icon or Indicator just to be consistent, or to add visual flair. Ensure that all indicators are accurate, are truthful, and clearly explain their purpose.

Do not use indicators with a clear meaning that they do not actually carry out. A common error is the use of the right arrow to mean "more" when the specific item loads as a Pop-Up, or loads a new application entirely. This should only be used when a "next" page is loaded, preferably by visibly sliding in as a Film Strip item.

Icon

Problem

You must provide access to disparate items or functions, in a glanceable and easy-to-select manner.

Icons for selection and display are common, and will be easy to implement on all mobile platforms.

Solution

Figure 6-13. *Icons are used as shortcuts to highly used items, even within interfaces that are otherwise not icon-centric. This is a typical idle screen for previous-generation smartphones and many current feature phones.*

When using an interactive icon, you must immediately display a change and provide any other relevant feedback. If the actual state change will take time, such as enabling WiFi, and taking time to power up and find a network, provide some sort of interim condition before the "on" condition is switched to. These will usually take the form of a small Wait Indicator.

If additional information or settings are available for an interactive icon, double-tap or Press-and-Hold behaviors may be used to provide access to these screens.

Some types of status icons will also not load a new page, but will simply change the state and indicate this immediately by the Icon changing state. Any other state changes, such as a mouse pointer or cursor changing state, must occur at the same moment as the Icon.

Status icons may also behave as normal, fixed icons, displaying current conditions but loading the target page when selected. For example, a calendar icon with the current date and an overlay with the count of remaining appointments is very much a status icon, but selecting it simply loads the calendar application.

Presentation details

Figure 6-16. *Interactive icons must immediately appear to be interactive, with clear controls and an obvious indication of the state change.*

The design of icons is an entire topic, with entire books dedicated to these principles. There is very little difference between the use of icons on desktops, kiosks, and workstations over the past several decades, and their use on mobile devices. Do keep in mind typical mobile requirements such as visibility in all lighting conditions, or under movement.

All your icons should be very similar in size to one another. You may have to follow existing standards, but if you are developing the framework, try to restrict what sizes can be used. When you are making icons for your application, follow best practices for the platform, even if it is not technically restricted.

Icons should generally carry a text label. This will usually appear centered, below the icon, and should not wrap to a second line. Labels to the right or left of the icon will often fall into the Indicator pattern instead. In certain cases, it may work to have the labels truncate instead of taking up additional room. When the icon is in focus, the label may "marquee," or scroll, to reveal the entire label.

Functions should imply the functionality. Switches, buttons, and other actions contained within the icon should have 3D effects or look like standard on-screen controls. Use form-like controls (e.g., radio buttons) or simulate Mechanical Style Controls, such as switches and buttons. Indicators often work best with shadows and simulated bleed and glow effects as well.

Antipatterns

Avoid automatic thumbnail or Avatar images for icons. One web page or person will look very much like another at thumbnail size, and a key part of icon design is the design of the outer bounding shape. These may be used in a Thumbnail List and other places where the image supports a text label, but they are not generally suitable for icons.

Information in an icon that looks like data must be real data. Do not, for example, make an icon for a calendar or clock with fixed values for the date or time. Use the status icon style and indicate actual data, or use another icon style.

Try to label each Icon, in all conditions; even if you use them in a carousel, where additional information may be presented with the in-focus item, use short labels for each visible icon whenever possible. Not all icons are clear on first use, so users will have to take time to bring each item into focus before selecting anything.

Stack of Items

Problem

You must display a set of closely related items, which can be represented as icons or thumbnails, in a manner implying the hierarchy and providing easy display of the contents.

Rotating and moving images, especially smoothly, requires relatively powerful or specific graphics processing, and will not be available on all platforms. Ensure that you have the technical capability before designing such an interactive method.

Figure 6-17. The "Photos" stack in the left frame is tapped, and the thumbnails move into their final Grid positions as the page behind changes. Note the title changes and the processing icon in the title bar during the transition, as well as the importance of the labels below each stack in this example.

A Stack of Items is a design pattern that is exactly as it sounds. A set of thumbnails are arranged so that they appear to be actual items, such as a pile of photos stacked on top of one another. Only the top one is completely visible, but to make the stack clear, others stick out from the sides. Selecting the top item opens the stack, revealing all the thumbnails and providing further selection or access to each item. See Figure 6-17.

The concept of drilldown for the Stack of Items inherently encompasses two tiers of the hierarchy, instead of only the usual one tier. Opening the stack to reveal the images within is one, and clicking the linked icon to view the image details (or otherwise load the target state) is the other.

At first glance, this pattern appears to be a graphical variation of the various expanding folder views such as the Hierarchical List, and should therefore not be used in favor of the simpler, more common display methods when they will do fine.

The difference is the content; for list displays, the primary displayed item is text, and any graphics are supporting only. With a large collection of images (or similar, graphically representable content), if drilldown is required, use a Grid display using Icon thumbnails of each image to open folders, or use the Stack of Items if the platform will support it.

Two basic interaction and display paradigms exist for all interfaces like this:

- The transition from stacked to open takes place on a static screen, without a transition to another space. This requires space for the stack to open into, or that other items on the page move out of the way. When folded, these items may or may not move back to their original position. This is the least common method used.

- The transition from stacked to open takes place at the same time as the underlying page changes. When folded, the stack may be in a relatively constrained area. The unfolded state may then occupy space as it needs, as the available space is dedicated to it. The page transition is obscured behind the stack that is expanding; the user simply will not notice it happening.

The Stack of Items pattern is still relatively new in production devices, so it has not established many common variations. It has been well used in many prototypes, and in large-scale touch devices (such as the Microsoft Surface). If your application is capable of supporting such a pattern, try to take time to explore alternative methods of using this pattern that may tie to others. For example, if sensor-driven tilt is used in the Grid display of items when open, it may be useful to enable tilt when stacked, either for viewing or as a method to open the stack. For other ideas along these lines, see Simulated 3D Effects.

Interaction details

This pattern represents a simulated physical set of objects, so even with simple tap-only behaviors, it is really most suitable for touch and pen devices. Consider the complete experience carefully before including it in a purely scroll-and-select interface.

Simple selection of the stack (such as a tap for touch and pen devices) will open it and initiate the transition to the expanded state. When you use it with scroll-and-select devices—or the scroll-and-select interface of a mostly touch device—you must include in the highlighted state the entire stack, including the thumbnails that are "sticking out."

Other interactions—such as drag to fan-open or certain kinesthetic gestures—should only be used if similar gestures are commonly employed in the rest of the OS.

While the stack is unfolding, do not allow selection of individual cards. This will prevent accidental input from the user trying to chase moving items. Do not disable any other functions. Selecting the "back" or "cancel" button (or gesture) will immediately stop the unfold action and reverse the animation to the closed state.

If room allows, place a Wait Indicator in the title bar or other location to indicate this condition; the system may become occupied, or lock up, and this will help to explain the temporary loss of input.

There is no organic provision for collapsing an unfolded set of thumbnails. You must use the device's default "back" function, or place a dedicated on-screen control, on the screen or as a part of a Fixed Menu or Revealable Menu.

Presentation details

Show each item that unfolds as a thumbnail. When collapsed, the stack will look like a physical stack of cards, occupying a space only slightly larger than the top thumbnail. Shadows and other effects (as long as they do not interfere with viewing the thumbnails) should be used to emphasize this. The edges of individual items should be clear in all cases, to assist with differentiating the images from the background. This helps to imply that the items are selectable and not just decoration or parts of the background.

When stacked, the cards should look like a stack, though a messy one, with some cards sticking out from the edge in some manner. Never fold them up so far that only the top thumbnail is visible. The top item in the folder (using whatever method is relevant for the content type) is used as the label for the stack. Sometimes you may want to allow the user to select which item serves as the label or badge for the stack, but generally it will have to be automatically generated, and so may not represent the contents. Be aware of this as a possible pitfall.

Put a text label under the folded stack, as described in the Icon pattern. Individual thumbnails may or may not be labeled, depending on their relevance. Critical differentiators (e.g., video or still photo) may be indicated with icons laid on top of the thumbnail.

Antipatterns

Do not use this pattern if it cannot be built to operate smoothly and responsively on all the targeted devices.

As with all such simulations and effects, avoid overusing or misapplying this pattern. This may result in an unusual interaction, which could be perceived as nonstandard. Therefore, ensure that it adds value when used.

Do not use a generic icon/graphic to describe the stack (e.g., iPhone grouping); be sure to indicate it's a folder very clearly.

Do not present a Stack of Items that simply reloads the page as a Grid. The unfolding action should always be animated to show the thumbnails moving from the stack to the final display location.

Annotation

Problem

You must be able to attach additional information to a data point within a dense array of information, without leaving the original display context.

Any interactive infographic that demands such additional information can generally support an Annotation. Layered display is easy to add to almost any platform, but some attention must be paid to precise location ability when used in Infinite Area and similar displays.

Figure 6-18. *A typical example of a reveal label. When it is originally selected, a small pointer indicator is used alone. When it is selected further, a label with details is revealed, as shown to the right.*

High-density information displays, such as maps, charts, and graphs, always carry additional layers of information, which may be best understood by allowing the user to view certain details about a specific point. Conventional methods of drilling in deeper, such as the Link or Button, are unsuitable for finding out about a specific data point, partly because they remove the user from the original context.

Instead, you can employ a special type of widget, exemplified by the *pinpoint*. An iconic element points to the information selected, and presents (sometimes only after further selection, or in another area of the screen) a label and additional options.

Do not confuse this with a Tooltip. A deliberate action is required to reveal the information. Tooltips, instead, are transient and are initiated by hover or are automatically presented when the system determines the user needs help.

Content within an Annotation is instead discrete information, or may even carry functionality, and is not simply help or explanatory text as in a Tooltip.

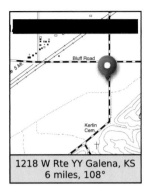

Figure 6-19. *A banner attached to the edge of the viewport can also be used as a label. This is most useful for large amounts of information, which would otherwise clutter the display to the point where it would not be seen clearly.*

There are numerous ways in which you can make the selection of an item reveal additional details, without entirely disregarding the original context. Many of these, such as split-screen variations, are not yet common enough to be considered patterns, but these three are very common:

Fixed label

When an area is highlighted by the system, or selected by the user, a label bearing text and/or icons is displayed as a part of the graphic pinpoint indicator.

Reveal label

When originally selected, either by the user or by the system, a small pinpoint indicator is placed, as in Figure 6-18. This is usually generic, and often has no label, or a very simple one such as a shape, color, or single letter. When the user selects the pinpoint again, it will expand to reveal a larger label. In this mode, the label is functionally identical to the fixed label.

This is especially useful when multiple pinpoints may be needed, such as graphically represented search results. The small label can carry an identifier such as a letter, and the user may then subsequently select details on any individual result, without as much clutter.

Banner

A small pinpoint indicator is used to mark the location on the data array, as in the reveal label. At another location on the page, usually a strip anchored to the top or bottom of the viewport, is the label text, or graphics, as in Figure 6-19. This is most suitable when a large amount of data must be presented, or a number of selectable options should be selected. In this case, the style and functionality of any Fixed Menu or Revealable Menu used within the OS or application should be carefully considered, and may be reused without modification.

You can use any of these three options in conjunction with one another, within a single interface. For example, when single selections are made, the fixed label is used, and for search results, the reveal label.

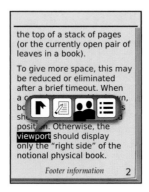

Figure 6-20. *Annotations can be used in nongraphical data sets, such as charts or, as shown here, to provide options for selected text. This also exemplifies how multiple selections may be made within a single pinpoint label.*

Selection details are simple and mostly outlined in this pattern's "Variations" section. Note that the selected information set can be of any type, including maps, graphs, charts, tables, or text, as shown in Figure 6-20.

In some cases, reveal labels may convert from the small to large display when hovered or when the item is in focus, instead of by explicit selection. This is, of course, only suitable for interfaces with a focus separate from selection, such as scroll-and-select and certain pen devices.

When more than one pinpoint is visible, use the reveal label or banner style, and only display the label for the item in focus. This also serves to simplify selection of options from any other menu systems that may be present in the application.

When a reveal label is expanded, make sure it gets out of the way of other actions the user might request. If a selection is made somewhere else on the page, especially somewhere adjacent to the label, it should disappear and revert to a pinpoint only.

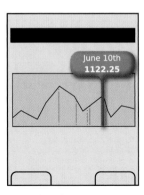

Figure 6-21. The type of pointer used must be selected to comport with the data being indicted. The expanded label can be quite far from the content being pointed out if the indictor is extended properly. Here, a one-dimensional scrolling graph has a pinpoint that indicates a position on the horizontal axis. The intersection point on the graph could have been selected instead, if this is the only data shown on this date.

Pinpoints must be easily selectable. For touch or pen devices, follow best practices for the size of the pinpoint.

For scroll-and-select devices, the pinpoint currently in focus must indicate it is selected, and must clearly differentiate this from any nonselected pinpoints.

The pinpoint must clearly indicate the location in the data set referenced. Use appropriate selection methods:

- Pointers for two-dimensional coordinates

- Lines for positions on one-dimensional graphs, as in Figure 6-21

- Highlights for text, or fields within charts

Pinpoints and labels must clearly indicate that they are not part of the page context. Use borders, shadows, and transitions to make this clear. It is often best to make it appear as though the pinpoint is floating above the image or protruding from the page. You can use certain practices described in the Simulated 3D Effects pattern to emphasize this.

When only one item may be selected, be sure to not imply that there are multiple options, by offering a Button or underlined Link separate from other selectable items. If needed, to clarify the click action, a false button (or link) can be added to the area, as it might be used in Advertising.

Labels should never exceed the space available. Do not let them simply float off the page. Label text should not end in ellipses or wrap to a second line. Multiple lines of information may be displayed, but each line should carry its own information.

Chapter 7

Labels and Indicators

Down Under and Backward

Many of you have had experiences traveling or moving to another country. The opportunity to become immersed into another culture can be quite exciting while also a bit intimidating. Having recently moved to Australia from the United States, I've been encountering many cultural differences that I have been forced to quickly adapt to. Not all have been so easy to grasp. Here are just a few.

Phone Numbers

The Australian Communications and Media Authority, which maintains and administers the telephone numbering plan, established a Full National Number (FNN) that is composed of 10 numbers: 0x xxxx-xxxx. The first two digits are the area code; the next four generally make up the Call Collection Area and Exchange. The last four numbers define the line number at the exchange.

Mobile numbers also have 10 digits but follow a different structure: 04yy yxx-xxx. Originally, the y digits indicated the network carrier. But now that Australia allows for Wireless Number Portability (WLNP), like the United States, there is not a fixed relationship between these numbers and the mobile carrier.

So, giving my new mobile number out to people was a bit confusing. I was always following the landline format of 0x xxxx-xxxx and getting a few confused looks.

Gas Types and Pricing

So much to choose from, and so little knowledge! There's E85, ULP (Unleaded), E10 (ULP + 10% ethanol), PULP (Premium), UPULP (Ultra Premium), Diesel, and LPG (Liquefied Petroleum Gas). Unlike the United States, where gas is sold by the gallon, Australia sells it by the liter and it is priced in cents. A typical petrol price of ULP may be 135.9. Having a US pricing format embedded in my head, I was shocked at first to think that gasoline was $135 per liter, though my sense quickly rationalized this was a wrong deduction.

Date Format

My first experience with this confusion occurred when I was applying for both my visa and my overseas health insurance and I had to enter my date of birth. The visa application was clear in the format I had to enter: day/month/year. The insurance form was not. It was a paper form with blank squares for each separate character.

The empty boxes had no label under them, just ☐☐-☐☐-☐☐☐.

In Australia, this format is culturally understood. However, for me it's quite unclear. Do I enter my month or day first? Each of those fits within the constraints of the provided format. Yet each clearly yields an entirely different result.

Understanding Our Users

As designers, we can limit these confusions by designing better UIs that meet the needs of our users. To do this, we must consider who are users are, what type of prior knowledge they have, and the context in which the device will be used.

Users and Their Prior Knowledge

Mobile users come from a variety of cultures with a multitude of experiences, skills, and expectations. If the devices' UIs cannot meet these user requirements, their experiences will quickly turn into frustrations while their actions result in performance errors.

To prevent this, start by identifying your users early in the design process. Use methods of observation, interviews, personas, and storyboards to gain insight into their needs, motivations, and experiences. Refer to this collected data to validate and guide your design decisions throughout the design process.

Context of Use

Unlike desktop users, mobile users will use their devices anywhere and anytime. These users may be checking an email while walking down the street, snapping photos during a traffic jam, or lying in bed playing a game.

In all of these situations, external stimuli are present that affect the amount of attention the user can use to focus on the task at hand: lighting conditions, external noise, and body movements are all stimuli that affect one's attention. Consider the effects of how bright sunlight on a glossy screen limits our ability to distinguish details, colors, and character legibility. Similarly, when the device is shaking up and down in the hands of the user as he walks down the street, selecting an element on the screen becomes prone to error.

These external stimuli are not always controllable, but the device's UI can at least be designed to limit the negative effects on the user experience. Using labels and indicators can redirect the user's attention away from the external stimuli and back to the task at hand.

Labels and Indicators in the Mobile Space

Labels are either text or images that provide clear and accurate information to support an element's function.

Indicators are graphical elements supported by text to provide cues and/or user control on the status or changes.

As you are now aware, displayed information is not always easily understood or easily detectable. The use of labels and indicator widgets can provide assistance with these issues through surface-level visual cues and functions. These visual cues and functions can help our users better understand the current context within the device. Labels and indicators can be used in the mobile space to:

- Indicate the current status or progress of the state of the device (see Figure 7-1)

- Present additional information, tips, or advice to clarify the current context

- Present information, especially text and numerical data, in the most appropriate and recognizable format for the context and viewer

- Provide user controls for operations involving access to remote servers

Visual Guidelines for Labels and Indicators in the Mobile Space

Figure 7-1. *As we get better and better hardware and networks, we ask more of them. Delays appear to be inevitable. Do what you can to label, and communicate within context, so that users have some information to occupy and guide them.*

The following are some visual guidelines for using labels and indicators on mobile devices:

- When using small text for a label, there must be a luminance color contrast from the background. The ISO recommends at least a 3:1 ratio between the luminance of text to the luminance of background color.

- When using color to show detail, luminance contrast is the most important principle. Using black and white together has the strongest contrast. Other effective combinations are yellow on black and dark blue on white (Ware 2008).

- Use symbols that are easily understood and represent the intended function. Use symbols familiar to your users, not to the designers and developers.

- When using animation, objects that have a constant motion may become habituated to the user, and may end up being ignored.

- If an object must grab a user's attention, consider having it emerge, disappear, and re-emerge every few seconds or minutes into the visual field. Be mindful of the overuse of multiple objects moving on a screen. The user will become easily distracted.

- Objects used to signal the user should emerge and disappear every few seconds or minutes to reduce habituation.

- Use motion, such as rotation or fill, to indicate an ongoing status change. Static objects provide a lack of feedback when it comes to progress or status. Users might feel such an indicator has "frozen up" if it lacks motion, especially if it has arrows or otherwise evokes movement.

- When possible, use images that evoke meaning, emotion, or desire to your users. These images will increase the user's motivation to attend to these objects. If objects lack these emotional values, users will overlook and avoid them. One example is to use recognizable images as avatars that represent people we know. This works effectively well in contact lists. Better to avoid "default" objects that are repeated (Saffer 2005).

Patterns for Labels and Indicators

Using appropriate and consistent label and indicator widgets provides visual cues and functions to communicate the current context of the device. In this chapter, we will discuss the following patterns based on how the human mind organizes and navigates information:

Ordered Data
  Presents textual information in the most appropriate and recognizable format for the context and viewer, while considering user expectations, norms, and display size constraints

Tooltip
  A small label, descriptor, or additional piece of information that is displayed to further explain a piece of content, a component, or a control

Avatar
  An iconic image or profile used to represent or support the label for an individual

Wait Indicator
  An indicator used to inform the delay status of a loading component

Reload, Synch, Stop
  User controls for operations involving access to remote servers, including override

Ordered Data

You must display information, especially text and numerical data, in the most appropriate and recognizable format for the context and the viewer.

The presentation of Ordered Data is largely a matter of design, and can be implemented with any technical means. For many interfaces, such as all websites, it is important that only the proper data be sent, so this is implemented at the server side. Any level of client software will work equally well.

Content types that are routinely displayed have become regularized in their display. You have to present them in specific formats so that they are easily recognizable to the users.

The types of data being discussed are, for example:

Names
> Should the first name appear first or last? Should it include the middle name, initial, or neither?

Times
> Should you use 12 or 24 hours? Are seconds needed?

Dates
> The order of items may vary, and abbreviations are rampant. In some cases, preceding zeros may be relevant.

Days of the week
> This includes languages as well as abbreviations, all the way to single letters in some cases.

Locations
> This includes city, state, and then zip code. But is the state abbreviated conventionally, is it abbreviated with two letters, or is it spelled out? And this is in the US only.

Units of measure
> This includes °F or °Fahrenheit, miles or mi., kilometers or km, and so on.

The list goes on. Some of these may be specific to a population, and may be technical terms or jargon only understood by the user community.

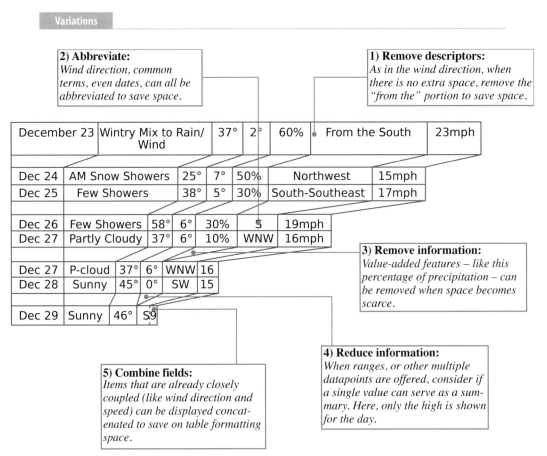

Figure 7-2. *A table showing how tabular data (for The Weather Channel mobile web) can be compressed or expanded as space allows. This shows many methods of displaying several types of ordered data.*

The two key challenges you will discover in presenting ordered data are true for all systems, not just mobiles:

User expectations

In many cases, this can go all the way to universally accepted standards for notation. Units of measure, for example, have standards for long and abbreviated display that should always be followed. Many others have cultural norms instead. Dates are displayed in several formats, commonly flipping month and date, without labels to say which is which.

If your product will be employed by users from different backgrounds, try to use unambiguous display methods whenever possible. For dates, avoid abbreviating months as numerals; 4/11 can be April 11 or November 4, so the only slightly longer "Apr 11" is entirely clear.

Available space

For columnar display in any interface, but a special challenge on many mobile devices, ordered data types such as dates may have design pressure to be compressed to the point they can become unreadable. Do not just rely on common sense, but refer to universally understood display standards to ensure that the compressed format will be readable. Also consider use of variable formatting based on the available size; if a larger device is used, or the screen is rotated, display larger amounts of information. See Figure 7-2 for an example.

Interaction details

This is a presentation pattern, and it generally supports no interaction at all.

You may wish to add additional details to explain any ordered data that may be confusing. For example, if a date is shortened to a large extent, a tooltip or some interactive method may present further details. Information can also be optionally presented in multiple formats; for example, a date-time code of "Yesterday" can present a tooltip that says "19 March 2011 at 11:21 a.m."

All normal interactions are available for any of the information presented. A common method is to provide additional information on each line, accessed by the use of a Link, Button, or other common method.

Presentation details

There are far too many of these formats to discuss in detail. We already provided some examples earlier in this pattern. Instead, we will use this space to discuss another method of displaying some content types: relative display. This is most useful for dates and times, but as these are very commonly encountered by many types of systems, they are worth detailing.

Relative dates and times use the conventional format users would when in conversation. People don't say "I called him at 2:17 p.m." when they can say "I called him about five minutes ago." Communicating in natural manners like this reduces cognitive load, making interfaces more usable and speedier.

Timings can change depending on the context, but one example of how to do this is:

- Within the current hour, show the time as minutes ago; over 20 minutes, round to the nearest 5 minutes.

- Within the current day, or within eight hours, show the time as the number of hours ago.

- Within the past two days, show the day of the week, and the time range—morning, afternoon, evening, night.

- Within the past week, show the day of the week.

- Older than this, but within 12 months, show as Mmm/DD.

- Older than 12 months, show as the year alone.

Examples:

- 20 minutes ago

- 8 hours ago

- Tues. morning

- Tuesday

- Nov 20

- 2008

Antipatterns

Do not rely on user settings to solve cultural norm issues. The vast majority of users will not perform such a setup. Even if forced as part of a first-run process, defaults will be used to speed it up. Customization may be allowed, but ensure that default conditions are clear and unambiguous.

Do not make up abbreviations. If one does not come to mind, it doesn't mean it doesn't exist, just that you don't know what it is. Use reference works to find the correct abbreviation. Always find good references, preferably by the governing body of any technical organization, and do not trust hearsay. Street labels in the United States, for example, are being rewritten somewhat randomly by local departments, and the abbreviation for Lane is now often "La" instead of "Ln." This sort of change can be confusing.

Do not use the default label values without understanding whether they have meaning. For another roadway example, many digital mapping services use "Street" as the default type of roadway. However, in many cities, saying "Street" means it runs the opposite way of an "Avenue." This carries as much meaning as assuming that unspecified times of day are "AM."

Don't go overboard with interactive display methods. Avoid interfering with the primary interaction of the display, such as selection of a line item. Although tooltips can be helpful, too many can make it difficult to read the rest of the data set.

Tooltip

You need to add a small label, descriptor, or additional piece of information in order to explain a piece of page content, a component, or a control.

The Tooltip is a common interactive element, included in desktop OSes and some mobile platforms. For others, you may need to design and manually build the display method.

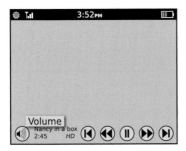 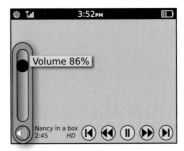

Figure 7-3. *Tooltips are most often used to explain controls which do not carry a label, such as these video playback controls. Due to familiarity and use of standard symbols, the labels have been eliminated; a Tooltip helps anyone who may be confused by appearing after a delay which may indicate indecisiveness. Tooltips may also show the current state of a control setting, such as the volume level on the right.*

A Tooltip is a transient, usually contextual, informational assistance widget. Tooltips are initiated by the user hovering over a potentially interesting target, or they are automatically presented when the system determines that the user needs help—such as during her first visit, or when there is a change in the system since her last visit.

The information presented in a Tooltip will usually be a helpful label, or an addition to the content. Use this information to clarify short labels or icons or to explain jargon, requirements, or systematic needs that may not be clear.

When a function such as a Tooltip is called for, but it may be deliberately displayed by the user or has interactivity in the information label itself, this is instead an Annotation. See that pattern for more details and comparisons.

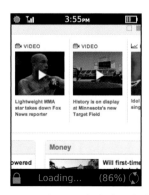 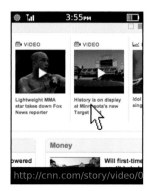

Figure 7-4. *Banner-type tooltips are locked to the bottom of the window or viewport. They can display status information, as shown on the left, or they can display hover-state information. The target URL as shown on the right is a perfect example of information that cannot easily fit into the page, but that some users will want to be able to see before selecting the item.*

Floating tooltips are the traditional Tooltip encountered in modern desktop windowing OSes. A small box appears adjacent to the mouse pointer, as in Figure 7-3, above every other item on the page. A small amount of content (almost always just text) populates the box.

Banner tooltips occupy a strip, generally anchored to the top or bottom of the viewport or window. These are most suitable for larger amounts of text, or when a message can appear on the page most of the time. This may carry labels for hover states as the floating labels, or information on the current state or mode of the application, such as a browser status bar. See Figure 7-4.

Banner tooltips that may use the style and functionality of any Fixed Menu or Revealable Menu used within the OS or application should be carefully considered, and may be reused without modification.

For mobile devices, you will often find no explicit, built-in method to present tooltips, and you will have to develop the behaviors and presentation yourself. For any systems that support hover states (mostly scroll-and-select), any item in focus may display a Tooltip label. For any device, including touch and pen devices without a hover paradigm, tooltips can be used by the system to point out unused tools or new features. The lack of context makes this less immediately useful, and the items to be highlighted must be carefully selected to avoid overselling a single new feature.

Certain desktop systems allow entering a "help mode" where tooltips appear instantly (though still on hover), for all page items, but this has not yet become common in mobile and so cannot be considered a pattern yet.

The information labels, whether floating or in a banner, may not be selected in any way. They should be built so that they do not exist as far as selection mechanisms go; clicking an item behind the Tooltip is not impeded in any way. The Tooltip is simply disregarded, except for its visibility.

Ideally, the Tooltip will not even visually obscure other items. Any label being displayed must get out of the way of other actions the user might request. If any selection is made elsewhere on the page, including entering text input fields, it should disappear immediately.

Presentation details

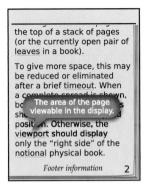

Figure 7-5. *Tooltips can be used to present all sorts of contextually relevant information that cannot, or should not, fit in the page, and not just for items that are already interactive—for example, definitions of jargon in technical descriptions. Any number of methods can be used to communicate that the item will reveal information, but it should not be entirely hidden.*

The banner Tooltip may operate in one of two modes:

Always present
> The Tooltip appears in a permanently allocated space, which is empty when no messages are present. Labels for hover states appear and disappear without delay.

As needed
> Labels and the surrounding box will appear with the timing of the floating Tooltip, but the Tooltip always appears in the same place on the screen instead of adjacent to the relevant section.

You should present any floating Tooltip after a brief pause, to avoid interfering with use of the interactive elements on the page. The delay is intended to reflect hesitation on the part of the user as well.

Only one Tooltip may be present at a time. Moving to another element will overwrite the previous Tooltip or label with the new one.

A floating Tooltip should generally disappear after a few seconds. This timing is infinitely variable depending on the user context expected for the application. If the user is likely to not be paying attention to the screen when the Tooltip appears, the timing may need to be very long, or a banner style should be used.

Any floating Tooltip must be clearly not part of the page context. Use borders, shadows, and transitions to make this clear. It is often best to make it appear as though the Tooltip label is floating above the image, as in Figure 7-5, or protruding from the page. Certain practices described in the Simulated 3D Effects pattern may be used to emphasize this. Base the color, size, and other styles on the desktop standards, whenever possible.

The text inside any Tooltip should only be one line long whenever possible. If it must wrap, never exceed two lines. Avoid truncation, although some labels (such as URLs) make this unavoidable.

Antipatterns

Do not rely on a Tooltip to solve interaction and interface design problems. If icons are unclear, make better ones, or apply fixed labels, for example.

Labels should never exceed the space available. Do not let them simply float off the page. Label text should not end in ellipses or wrap to a second line. Multiple lines of information may be displayed, but each line should carry its own information.

Avoid overburdening an already complex, interactive interface with tooltips. Though their noninteractive nature may seem to indicate that they can be added freely, they are still observed and must be understood by the user. This may interfere with speed of comprehending the entire interface. Additionally, when multiple small floating items are on a page, it may not be clear what the layer relationship is between them.

Avatar

Problem

You must provide a glanceable representation of a person for use in various contact-listing contexts.

Visual representations of any sort can technically be displayed by any platform. Providing the images can be a notable challenge if you have to make the images, or make arbitrary images fit in the space.

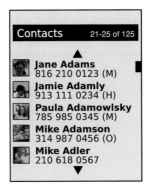

Figure 7-6. *Avatars add quick scanability to lists; instead of making the user read each item, the user may recognize regularly used images for faster access.*

Text is accurate and precise for listing or otherwise describing people. However, text labels can be slow to read, and this is often undesirable in a mobile context. Whenever possible, make the interfaces glanceable and scannable.

Much like the Indicator pattern, an Avatar is an iconic image used to represent or support the label for an individual, such as a contact in an address book.

This pattern is addressed separately from the Icon or Indicator pattern because the Avatar image represents a real person, as in Figure 7-6, not a concept or action, and because in many cases the image itself cannot be directly controlled by the designer or the user of the device.

Figure 7-7. *Avatars are a good way to label an individual in a very small space. Selecting these should reveal details of the contact, or options relating to the contact.*

The two basic types of Avatar are different from other variations in the book in that they vary by individual use. Each one can, and very often will, appear in the same interface.

The first, and most obvious, is any *custom representation*, as in Figure 7-7. This may be a photo, or a personalized icon or other graphic. It may be loaded by the user of the device (as when a photo is taken of a contact for use on that device alone) or it may have been created by the owner of the avatar (as when the profile image from a social network is automatically loaded).

When no custom image is available, as will occur very often with first-time use, a *generic icon* or set of icons is often used as a placeholder. Whenever possible, avoid truly generic icons.

As described in the Thumbnail List pattern, irrelevant images may be simply left out. The specifics of your interface will determine whether this is a workable solution. During design, consider the likelihood of most or all items being populated with image content, and make decisions about use of the Avatar pattern accordingly. See Figure 7-8.

Interaction details

 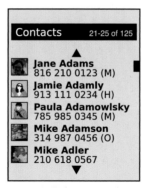

Figure 7-8. *When custom Avatar images are not available, either don't use an image at all (as on the left) or use a generic placeholder that is appropriate. On the right, one of several different stand-in images is used for each unassigned Avatar.*

Avatar images may be selectable or not. Be sure to follow the general paradigm of the context in which they are used. If they are used in a line item, such as an address book, you may make them selectable as an extension of the name, as a part of the whole line, as individual items with different behaviors from the other labels, or not at all.

Always consider how clearly the interaction which will follow selection of an Avatar can be communicated. If it is not perfectly clear, add additional information, such as an overlay icon as described in the Icon pattern. Otherwise, consider simply making the Avatar image display only.

When you display an Avatar without a label—such as in threaded messaging or an SMS conversation—the Avatar should always be selectable, and generally either display the user information or offer options for interaction with the individual as a Pop-Up or Annotation.

Presentation details

Since the image itself often cannot be created, the designer must be careful when selecting borders, background, sizes, and aspect ratios.

Avatar images are almost always square, to account for both landscape and portrait orientation source photos, without adding masking bars to part of the image or causing inconsistencies in the display and alignment of page items.

Generally, to fit to a square, the long sides are cropped off, and the short side is fit to the image dimensions. Occasionally, it may be useful to crop the outer 10% or 20% of the whole image, to ensure that a recognizable face is in the image area. Base all such decisions on the final rendered size and aspect ratio; aside from source imagery varying widely, even very poor images will generally be much larger than the Avatar image.

Though some Avatar selection systems allow manual image cropping, this—as well as photographing within the mobile device to generate images—is not within the scope of discussion here, but should be designed to generate the best possible images and adhere to the design principles of the application or process. Feel free to explore alternative methods; facial recognition software, for example, is now reliable enough that it can be used to automatically crop Avatar images.

When practical, use avatars alongside text labels. When selectable, use the existing paradigms, discussed under the Link pattern, to denote that an image is selectable.

When no supporting label text is associated with an Avatar image, consider use of the Tooltip, when practical and supported by the device's interaction model. If an Annotation or other interaction may conflict with the Tooltip, make sure label information such as the user's name is in the interactive selection mechanism as well.

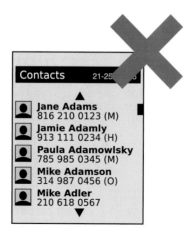

Figure 7-9. *Avoid using the same stand-in image for each Avatar (or so few variations that they repeat excessively). The image should add meaning to each individual use, not just fill a gap in the design.*

Avoid using too many generic icons, or only having one style of graphic that serves as a placeholder for every empty slot. See Figure 7-9.

Always keep in mind the value of the Avatar, and do not fill it with arbitrary content to populate a flawed or boring design.

Wait Indicator

Problem

You must clearly communicate processing, loading, remote network submissions, and other delays.

The Wait Indicator is common, and is built into many platforms. However, many default messages violate key principles of usability and user experience, not to mention mobile design principles. Even when they are provided, you likely will need to build a custom interface to provide a good experience.

Figure 7-10. *A short Wait Indicator is just a moving graphic that communicates a delay in loading. Explanatory text may or may not accompany the indicator.*

A variety of wait indicators are used to inform users of delays that are imposed by technical constraints.

This widget may be used in many ways, as it best serves the needs of the user and the constraints of the design.

Some of the indicators described here are used almost exclusively on an Interstitial Screen. Read that pattern for implementation details, such as cancel functions, that are not considered here. Others may appear as portions of an interface, to communicate loading of modules or individual images.

Another common implementation is the loading bar, often used for web browsers, which may not disable the existing page until some or all of the new page has been retrieved. The Wait Indicator occupies a space along the edge of the viewport, and does not interfere with use of the old (or new) page.

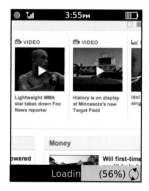 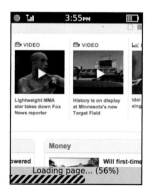

Figure 7-11. *Longer delays, when the percentage or time of completion is known, are typically displayed as bars, filling in from left to right. Shown here are two status bars, indicating loading of a web page. On the left, a spinning indicator (to the right) shows continuous progress even if the loading is very slow; also note the reversed text to ensure visibility. On the right, the bar itself has an animation, with the stripes constantly moving to the right to imply movement.*

There are a number of variations on this basic theme, all of which will be considered here, as they address the same basic issue:

Short wait

A one-part indicator used when the wait is very short or when the length of the delay is unknown. This consists of an animated, endless graphic. See Figure 7-10.

Long wait

A two-part indicator, usually using a horizontal bar to indicate percentage complete and a text counter of size, percent, and/or time until completion. The long wait is used only for processes that take a reasonably measurable amount of time. This cut-off is almost infinitely variable, and depends entirely on the user's perception of the speed of the system. Sometimes even five seconds, when the user is highly focused on the system, will seem too long for a unitless short wait indicator. See Figure 7-11.

Ghost

Part of the lazy load concept, heavily used with images, where the content itself acts as the indicator. Think of an image loading slowly, going from fuzzy to sharp, or loading top to bottom.

The entire practice of loading web pages as the content is received is a sort of lazy loading implementation, as shown in Figure 7-12. Consider how slow loading of your page (or of many pages if you are designing a browser-based system) can influence the user, and whether the design may need to be optimized to communicate the delay better.

These variations may sometimes be used in conjunction. If, for example, ghost loading of an image is not entirely clear (and it often is not), you may add an additional short wait indicator on top of the image until it has completed loading, as in Figure 7-13.

Figure 7-12. *Wait indicators may take any form that suits the needs of the design and still communicates the delay and progress. Here, images not yet loaded have an animated progress indicator taking up the entire image space. These could also add a percent complete, if needed, and become long indicators.*

Interaction details

The Wait Indicator is the widget that informs the user of status only. It has no direct interactivity itself, though it may be used with buttons, soft keys, or other interactive elements.

See the Interstitial Screen, Infinite Area, and Reload, Synch, Stop patterns for the most common interactions to be used with the Wait Indicator.

Presentation details

Figure 7-13. *Ghost progress indicators use the lack of completeness of the actual content to indicate loading. A typical case is shown here, where images are loaded progressively. To add clarity, or for thumbnails not loaded at all, a short wait indicator is overlaid on top of the image. This is not always needed.*

All indicators must remain moving or impart some other sense of movement if the load speed is too slow. Bars with internal animation are a very common solution to this.

Display the units in which any measurements are displayed. Do not just display "67" but "67%," for example. You may display times in a unitless manner if the format is clear, and individual seconds are counted. For example, "5:15 remaining" is very clearly five minutes and 15 seconds, as long as the seconds change every second. There is no need to add additional labels.

Always make sure any visuals and units accurately map to the percent of task complete. Ideally, timing benchmarks should be made and the indicator updated with percentages in relation to time remaining for the task, rather than tasks pending in the execution queue.

You may save space by having labels appear on top of the indicator bar. However, ensure that labels are visible at all times. Consider reversing the text color pixel by pixel as the bar progresses, to preserve maximum contrast.

When a Wait Indicator is used alongside a content area or in conjunction with the content (overlaid), the indicator will disappear when loading is completed.

Do not use an indicator that is too seamless or unobtrusive, or the user may not notice that it has completed. One minor mitigation is to allow the "fully loaded" state to persist for a few more seconds and give the user more of a chance to see it.

In some cases, another indicator may be needed to indicate that there is no loading. This will usually appear in a permanent status bar, which is where the loading bar would appear. Typical messages are "Idle," "Completed," "100%," and the like. Make sure the message is clear, is contextually relevant, and does not rely on jargon or excess familiarity with the system. This status area may constitute one state of a banner-style Tooltip.

Long delays, especially on mobile devices, may cause the user to not pay attention to the screen when the task is completed. For very long or important tasks, you may wish to announce the end of the delay with:

- A blinking LED
- A backlight that is turned on
- Vibration
- A tone

or a combination of these. If these are used, generally a "completion" message must be made available, to make it clear what has happened and what the annunciator is indicating.

Do not use the unitless short indicator just because the delay is unknown. Do whatever is possible to get an estimated time to completion. If that is impossible, be very clear in text descriptions that it will take approximately a certain time (e.g., two minutes) and alleviate the wait in some other way, such as with a tone on completion.

Do not mix paradigms. Develop a system of wait indicators, and use the same type of indicator for similar types of wait periods. Users recognize the distinctions we discussed in this pattern, and they will base expectations on the typically used graphic design patterns, so they may be misled by poorly selected indicators.

Do not allow indicators to stop animating. Users will perceive lack of movement as locking up or taking so long that they will not wait for it and will abandon the process.

Do not imply completion by the use of the wrong indicator graphic. Bars imply completion, and the common practice of repeatedly filling a bar (for each step of an install) is simply misleading to the customer. Rotating items must not be drawn so that they appear to be pie charts filling in, or the same can happen.

Avoid using a Wait Indicator as a mouse pointer, or otherwise mimicking modes from desktop computing platforms. Do not use the Wait Indicator as an error message; errors should be separate and distinct from fatal error messaging. As an example, the "Spinning Beach Ball of Death," which indicates all system resources are used up, is not a suitable way to communicate this situation, especially on a mobile device.

Reload, Synch, Stop

You must provide the user with controls for loading and synching operations with any remote devices or servers.

Reload, Synch, Stop features are simply represented as buttons, so at the interface level they can be easily built in any platform.

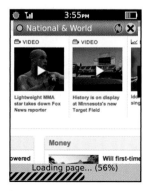

Figure 7-14. *Ghost progress indicators use the lack of completeness of the actual Refresh and Stop buttons that are key to all web browsers. Here, the two buttons are adjacent to each other. The Reload button is in use (and should be animated) but is grayed out, and cannot be selected again. The Stop function is available for selection. The status bar reinforces the behavior requested, as does the animation of the loading button, with additional details.*

Information from remote data sources provides the bulk of the functionality of interactive mobile devices. Due to specific user needs, accidental inputs, or simply system constraints, your users may sometimes have to manually start or stop data transfers.

Some examples:

- Loading web pages
- Synching with address books
- Synching with email servers
- Submitting transactions
- Transferring files to local devices
- Sending images over MMS

Although as much as possible should be done automatically, user override and user control should always be provided. These will be paired Reload and Stop buttons, either in the application directly or in settings.

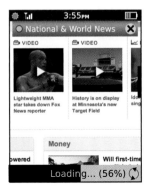

Figure 7-15. *Here, the superimposed style has only one button visible at a time. Since the page is loading, the Stop button is the available item. To ensure that this is clear enough, a different status row is used, with the icon seen in the Reload function animated here.*

Two very common methods are used to provide these functions. The selection of one over the other is largely up to design principles and space considerations; they are otherwise functionally equivalent.

Adjacent

Space is provided for both buttons to be visible at the same time. Generally, only one can be active, and the other is grayed out. Disabling inactive buttons and indicating this is usually preferable to removing the inactive button entirely so that users become familiar with the controls. Inactive buttons may also convert to status indicators; a Synch button can spin to indicate synching, while leaving the Cancel button location free to carry out its function. See Figure 7-14.

Superimposed

Only one space is provided for the two functions. Since they are mutually exclusive, only one function is needed. However, some users may become confused as to the location of the other function, and transitional states (e.g., stopping) require users to wait for their completion. Status may not be well communicated by the buttons, as the alternative action must take precedence. When synching, the button will convert to Cancel and status must be communicated elsewhere. See Figure 7-15.

Interaction details

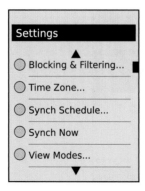 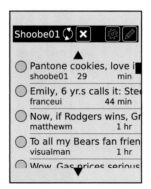

Figure 7-16. *Synch and Cancel buttons can appear conditionally, and in different screens. Here, for example, a Synch Now function within application settings returns to the main screen, with an animated status icon indicating that synch is taking place. Adjacent to it is a Cancel button that otherwise is not present.*

When the user selects any button, the action should begin immediately.

For either version, you should only allow access to buttons that can take effect in the current state. If a web page has completed loading, do not allow a Stop function to work.

Do not accept duplicate inputs. If a Stop command has been received, do not accept further Stop commands. If the system is poorly built, and might "forget" a previous command, fix the system instead of requiring the user to carry out a simple, single-point behavior multiple times.

For either case, it is best to assume that the request will take a brief time to take effect, and to account for accidental double-taps and other mistaken inputs. Especially if you choose the superimposed design, do not allow conflicting actions, such as Stop immediately followed by Reload. Make lockout periods brief so that a second selection within about half a second is disregarded.

Presentation details

For either version, only allow access to functions that are available in the current state. Suppress or gray out buttons that do not apply. See Figure 7-16 for more on conditional display.

For functions that have already been submitted, a subtler state change may take effect to imply that the button has converted to an indicator. For example, when a Synch button has been pressed, you should animate the icon to indicate that the synch operation is occurring. Remember that many remote operations, even just stopping a request, will take some coordination to take effect; the in-progress state will take a significant amount of time and so must be represented.

When a status area is present, such as those described in the Notifications pattern or the banner variant of the Tooltip, any current behavior should also be indicated there. This should be true for actions that will result shortly after the process becomes idle. A command to cancel loading should display "Stopping..." or something similar.

Labels, whether as a graphical Icon, a text label, or only as a Tooltip, must be clear and accurate. Do not use "Reload" for a synch operation, for example.

Antipatterns

Do not presume users will understand that selecting an action again will stop it. For example, an interface may have a Synch button. When selected, it animates. The user is then to understand that pressing it in the animated state is functionally "unsynch," but in practice this works poorly. Explicitly label all functions, both graphically and with text.

Never provide just one button of a pair. Do not provide a Synch button if you don't also provide a Stop Synching button.

Never display a status about load or synch behavior—especially not as a modal Pop Up—without a way for the user to cancel the operation. If canceling is technically impossible, clearly explain this, and indicate why.

Chapter 8

Information Controls

The Weilers, Version 1

Jack, Maggie, and their 5-year-old, Melissa, approach the entrance to the brand-new shopping mall that recently opened in their hometown. Melissa, thrilled with the opportunity to finally go to a Build-A-Bear Workshop, skips ahead in pure excitement.

Just inside the main walkway Jack sees the large vertical store directory and map. Rather than getting lost in such a large place through exploratory wandering, he decides to use the directory to figure out exactly where in the mall the Build-A-Bear store is.

The directory in front of him is typical. A floor plan is illustrated that labels each store using apparently arbitrary numbers. Those numbers are referenced and sorted by category to comprise the store directory.

Immediately, Jack's frustration begins to build. He struggles to determine what category Build-A-Bear falls into; he's looking under Gifts, then Baby, both without luck. Maggie chimes in and finally finds the store under Fun & Games with a label of "L34."

Once again, frustration builds when the family members can't find their current location, or determine where L34 is. They do not see a "You are here" indicator. Annoyed by this barrier, Jack and the family give up, and walk farther into the mall in hopes of eventually coming across the store.

The Weilers, Version 2

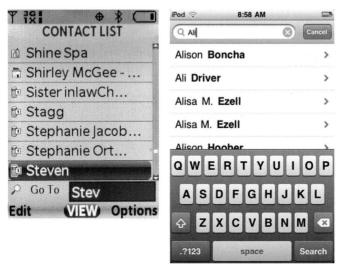

Figure 8-1. *Search within the address book is a modal behavior on some of the newest touch-centric OSes. The fact that it varies from all the other search capabilities on the device, and is more like the classic Search Within pattern, indicates how important and expected the function is.*

Jack, Maggie, and Melissa are excited about visiting the Build-A-Bear Workshop in the new shopping mall. As they enter the mall for the first time, they see a crowd of people gathering at the story directory kiosk.

As they walk toward the crowd, they are amused to see that the mall uses a multitouch interactive table to display its layout and directory. Jack places his fingers on a portion of the screen to begin. That portion of the interface lights up and generates a pop up with an option to locate a store or to begin a video chat with the mall's customer service department.

Jack is presented the option to filter his search by general category, with proxemics to his current location, or by alphabetized store name. Jack selects the alpha search, which reveals a vertical list of stores with location jump controls, as well as a text field with a touch keyboard.

Excited about this experience, Melissa engages with the table and uses the location jump control to find the stores that begin with the letter *B*. The "Build-A-Bear Workshop" label displays within the list. As she selects the name, an interactive floor plan of the mall immediately populates, illuminating the store's location.

The floor plan at first shows the entire mall's layout with callouts to the family's current location and the Build-A-Bear location. Then the display slowly zooms and reorients to the family's current position and animates an eye-level view of the walking route from their location to the Build-A-Bear store.

Having visually seen that the store is located on the second level next to the Food Court, one level above them and slightly to the right, the family heads in that direction, still excited by the engaging user experience.

The Difference

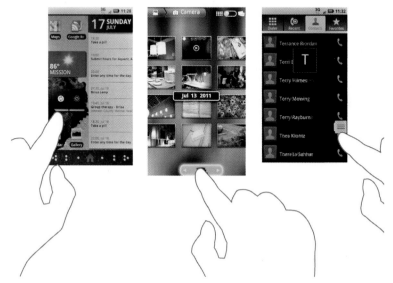

Figure 8-2. *Gestural interfaces, almost by their nature, have little or no affordance before use. Location Jump is useful to allow initial gestures to expose more functions, and allow the user to reach his goal faster.*

The two different scenarios provide polar examples of how a common task of locating information can lead to a frustrating user experience, or an enriching one. The solution was not just the power of the technology. It was also, and more importantly, how the content was organized, displayed, and made available to the user, in a manner useful to the user's immediate needs and current context.

In the first scenario, all the information was presented at one tier, without the user's ability to use controls to drill down, sort, and filter information for his current needs. This lack of control placed too much burden on the user, and resulted in a failed experience.

In the second scenario, the Weilers had access to a variety of information controls, such as location jump, search within, zoom and scale, and sort and filter, which made searching for relevant information much quicker. Each particular control provided a unique way of revealing different types of information.

Information Controls in the Mobile Space

In the mobile space, where limited display sizes constrain the amount of information presented at a given time, a user will require affordable functions that provide quick access to her intended goals.

The following are some guidelines for providing information controls in the mobile space:

Provide controls that afford their functionality by resembling their intended function

A control using a + or – button will be understood to zoom in and out, as opposed to using controls with arbitrary labels such as "A" and "B."

Make the controls visible

Users need to be immediately aware that controls exist to access and control the amount of information visible. Keep the control placement consistent across the OS or application. Because mobile screen real estate is valuable, consider using the Revealable Menu or Fixed Menu pattern.

Use appropriate metaphors to establish learnability through familiarity

As Dan Saffer explains, "Metaphor is not just about language; it's really about thought. We conceive of things in terms of other things" (Saffer 2005). When using metaphors, be aware that they may have various meanings across cultures, and match the metaphor to the content, not the content to the metaphor.

Consider how we understand the use of a magnifying glass to see detail. Some zoom and scale controls use this metaphor to communicate one's ability to see more or less information.

Provide immediate and appropriate feedback

A control must produce immediate feedback to communicate a change of state (contrast, color, shape, size, sound) that is measurably different from its initial state. Delaying feedback can result in the user performing other actions in the belief that the intended action failed.

Provide constraints

Provide only the control functionality that is needed to complete the user's task. Use a control for its purpose only. Do not assign it multiple, unrelated functionalities.

Follow wayfinding principles

Make sure your users know where they are in the control's current state, while providing information to communicate its range of control. For example, when using an alphabetical Location Jump widget, ensure that the selected letter is visible within the range of letters, while also communicating its relationship to other letters.

For more information on designing controls for people, refer back to Chapter 4. For a book on the subject, consider *The Design of Everyday Things* by Donald Norman (Basic Books).

Patterns for Information Control

Using the information control widgets allows users to quickly access the type and amount of information within the current state of the device. In this chapter, we will discuss the following patterns:

Zoom & Scale
> Provides the ability to adjust the level of detail of high-density information by changing the levels of zoom and scale

Location Jump
> Allows the user to quickly jump to a specific location within a list of information (see Figure 8-2)

Search Within
> Allows the user, via a search field, to quickly filter or jump to specific information the user knows exists within the page (see Figure 8-1)

Sort & Filter
> A method to aid exploratory search by progressively disclosing search options to narrow the relevant results

Zoom & Scale

Problem

You must provide a method for users to change the level of detail in dense information arrays, such as charts, graphs, and maps. Design a zooming function or metaphor to provide this control.

Certain platforms, especially gesture-driven OSes and those built for display of map data, will have built-in zoom controls which should be used when available. Custom zoom widgets are easy to build and interact with, but retrieval of information in a seamless manner, especially from remote sources, can be more challenging.

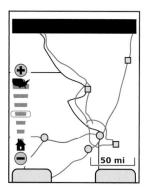

Figure 8-3. *The Interactive Scale combines the zoom control and zoom indicators. Often, they are accompanied by scale limit indicator icons as shown here in the country map and house (meaning neighborhood level). The scale can be used by clicking a zoom level to jump to it or dragging the control to a specific level. When included, zoom-in and zoom-out keys may be used to change one zoom level at a time.*

High-density information, such as that in charts, graphs, and maps (as in Figure 8-3), can be especially difficult to use on small-screen devices. Showing a sense of the whole space can preclude viewing sufficient details for analysis and other uses.

Zooming into (and out of) the information is the general solution. As this is similar to changing the depth of view, it is a different axis than scrolling, and so requires a unique control set.

In coordination with changing the zoom level, you must also communicate a sense of scale, whether relative or absolute.

Variations

The Zoom & Scale pattern has the following variations:

On-screen buttons

Buttons are very often provided to zoom in and out by single steps. There is no indication how large the step is, or how many are available. These may be placed at the corner of the view area, within a menu, or within an Annotation for a single point on the map or chart. See Figure 8-4.

Interactive scale

A bar may be provided for direct control that indicates the whole range of zoom. Generally, icons are used at the top and bottom of the control that indicate the range of the control, such as (for a map) the country for the maximum zoom out, and a house to indicate block level for zooming in. These controls may support direct selection of a zoom level or direct control of the zoom level by moving the slider dynamically. They are also often combined with on-screen buttons to offer all options.

Screen gestures

The most common of these is the two-finger "pinch to zoom" gesture. Other on-screen gestures are also used, including single-finger spinning actions (clockwise to zoom in, counterclockwise to zoom out). These may be useful for systems that cannot (for technical or user needs) support multitouch. See Figure 8-5.

Hardware buttons

Devices that do this a lot, such as GPSes, might have dedicated zoom buttons. In addition, you can repurpose unused hardware keys; if the five-way pad is used to scroll along the x- and y-axes (and select points), the volume rocker can be repurposed to control the z-axis and be a zoom control.

Additional methods may emerge in the future, such as the use of sensors to change detail level as the device is moved toward the viewer's eye.

You can combine some or all of these for interfaces that work on multiple devices without change, to appeal to different user types or to surmount lack of affordance in screen gestures.

For controls that do not integrate a scale, you should provide some sense of scale. This may be an explicit scale (labeling axes or a bar of distances), or it may be implicit, by showing items of a well-understood size.

Interaction details

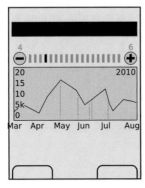 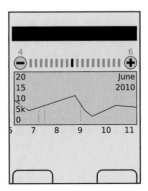

Figure 8-4. *On-screen buttons may also have a permanently visible indicator of zoom level. If it is not interactive, as this one is not, it does not become an interactive scale. Zoom may be for only one axis for some items, such as this one which zooms only the data. The scale indicators for graphs and charts will vary by axis. Make sure the scale is clear by providing all needed labels to understand the full context.*

Consider zoom to be a third axis of movement about the screen. Like the x- and y-axes are along the edges of the screen, the z-axis sticks straight out from the screen. You can think of the zoom control as moving to layers or levels along this axis.

The z-axis must be directly related to the request method, so the zoom does not break the user's expectations. There are two basic methods to establish this center point:

View center

> The zoom axis is about the center of the displayed graphic, usually the center of the viewport.

Activity centroid

> The zoom axis is centered on the indicator, pointer, or gesture used to indicate the zoom.

Zoom actions taken with no particular item in focus, without an on-screen cursor and without controls attached (visually or by use of gesture) to a specific point on the screen, use the view center method. All others use the activity centroid model, with the axis centered along the in-focus item or the gesture centroid. Note that these are not fixed conditions based on the device hardware or the overall interaction model; you can switch between them in a conditional manner. Zoom buttons (whether on-screen or hardware) will use the view center method when nothing is in focus, and switch to the activity centroid method when a virtual cursor is moved onto the screen or an item is in focus.

Any effect requested should show an effect immediately. Whenever possible, load a few additional zoom levels in advance, in a similar manner as described in the Infinite Area pattern.

For devices with keyboards and without dedicated zoom keys, you should provide access-key functions (and their corresponding labels) for at least the zoom-in and zoom-out controls (see the Accesskeys pattern).

Disregard input to zoom over the limits of the current information set, either too high or too low; for example, do not display errors for which insufficient information is available for the zoom level. The user will simply have to initiate another zoom control to back out until the information is again useful.

Presentation details

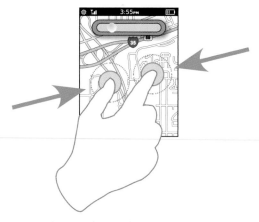

Figure 8-5. Even if the screen is otherwise devoid of zoom control elements, as with this touchscreen example, display the current zoom level during a zoom change. Place the indicator so that it is not occluded by the user's hand or fingers.

Regardless of the interaction variation, you must display a zoom indicator on the screen whenever the zoom level is changing. This indicator must immediately reflect the current zoom level setting; if there is a delay in loading the new data, display the requested setting, not the previous one. The indicator should remain visible for a short time after the zoom level has completed loading.

If there will be delays in loading, be sure to follow the principles in the Infinite Area pattern, and display a Wait Indicator, either for the entire application or for each frame of information being loaded.

Whenever practical, display a scale on top of the displayed information. For maps, this is a simple distance scale. The scale must automatically adjust to fit the available space, both in size and in units used. On maps, you may switch from hundreds of miles to hundreds of feet as the zoom rate increases.

When a zoom level is unavailable—such as when the end of the range has been reached—gray out the control to make it clear it is inaccessible. This is generally preferable to simply suppressing the control, as it may be confusing if it is usually present.

When there is a choice (such as on maps versus labeling individual axes of a graph), the scale bar should be horizontal. The scale bar should be large enough to be easily visible so that numeric labels are readable, and small enough that it takes up no more than about one-third of the width of the screen.

Label units of measure, and use standard abbreviations to label them. See the Ordered Data pattern for more discussion along these lines.

For gestural actions, such as pinch to zoom, use live transitions so that the user can see the effect of his gesture and adequately control the request.

For all zoom methods, if full-resolution imagery cannot be preloaded, you should use a variation of the ghost method described under the Wait Indicator pattern. You can scale the available imagery to match the requested panel size, giving a sense of the requested scale and information.

Antipatterns

Avoid delays in loading new zoom levels. Do not assume a Wait Indicator will solve the issue, but if one is needed, use it, as the Infinite Area pattern does when loading new data. Use the ghost method instead of explicit indicators such as short wait whenever delays cannot be avoided.

Don't use a toggle between two levels of zoom because your only convenient interaction does not support steps or gradations. Zooming via double-tap, for example, has value in some cases but is not a typical zoom and does not cover the majority of the needs for zooming into data.

Location Jump

Scrolling to items in a long vertical list is cumbersome. You must be sure to design an indexing system to assist in retrieval, and to design a method to allow easy access to key indexed portions of the list.

Location Jump controls are typically custom-built to integrate with their host interactive methods. Enough variations of this pattern exist to allow it to work on any device, but be sure to choose one that works best with your display and input features and standards of interaction.

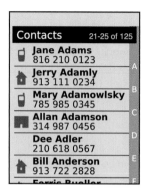 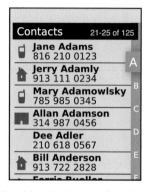

Figure 8-6. A very common use of this pattern is in the address book. It is often a long list, the data is easily divided into a simple alphabetical index, and contacts are referred to by name so that users can easily find them by using this indexing system. Note that the entire index often cannot fit on the screen. Some scrolling may be required to get to other parts of the index.

The problem of accessing long lists of ordered data is not unique to interactive systems, and the solutions are directly related to the methods derived in the machine era. The Rolodex (a portmanteau of "rolling index") is the prime example, a series of cards separated by tabbed label cards for letters of the alphabet.

This might seem to be akin to the Tab pattern, but the tabs here are different. Instead of providing access to different, parallel information sets, they are simply markers inline with the same, continuous list.

The variation on this, the flat Autodex with a single list of letters and a sliding control to select the section, is even more directly what is copied into digital Location Jump systems. An indicator of location, or indexing system, is visible on the screen, as in Figure 8-6, and this may be used to jump to the front of an indexed section.

These are used with the Vertical List pattern, but you could also use them in other types of displays as well. Specifically, the Location Within pattern used on Film Strip-like Home & Idle Screens can be interactive instead.

The Location Within pattern is very similar, but does not have interaction. Pagination is likewise similar, but page numbers are not an indexing method as they are unrelated to the content itself.

Variations

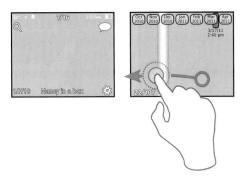

Figure 8-7. *The Location Jump widget can be used to reference or jump to any type of relevant index value. Here, a photo gallery is displayed in a single list, with index points by month and year. Without this pattern, a similar organization would have required a folder structure. The indicator shows index information with additional specificity.*

There are two key sets of variations. The first is in display:

Content indexing
> The strict Rolodex style, with letters, numbers, or whatever other key indexing item is in the list. Thumbnails generally do not work well.

General indicators
> Most often dots, sometimes in varying quantities or sizes, but also icons and numbers. These are only really suitable for smaller sets, such as Home & Idle Screens, but there is much room to expand the pattern here.

You can also allow interaction with each of these in one of two key ways:

Direct selection
> Click on the indicator label directly, or type it in. Press a P, and jump to the "P" section.

Drag selection
> Drag an indicator of the current position, much like a scroll bar, to the position you want. See Figure 8-7.

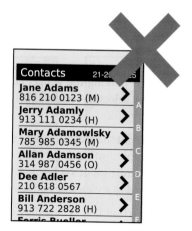

Figure 8-9. *Avoid placing functions in the list items next to the location bar, to avoid accidental input. Here, the indicator to see details on a contact will interfere with the Location Within widget. Allow for more space, or place the actions on the left side of the list.*

As with any scrolling method, do not jump to the new position. Always indicate that the new position in the list was scrolled to by actually showing the animated scrolling to the actual new position in the list.

Do not allow indicators to disappear during scrolling and other movement. They must also transition live whenever possible.

Use caution when placing items within a list. Do not allow critical labels to disappear under the position indicator. On touch devices, keep buttons and other interactive items away from the Location Jump controls side of the list. See Figure 8-9.

Search Within

Finding specific items within a long list or other large page or data array is cumbersome. You must provide a method for the user to find and display this information.

Though search forms are very commonly built-in functions, live or component updating may not work, or not work well, on all platforms. This pattern is especially difficult to implement well and consistently on the Web.

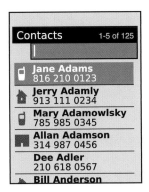 

Figure 8-10. *The live jumping Search Within style is almost ubiquitous on scroll-and-select address books such as this. Both the search field and the first line item (until the user scrolls) are in focus, one for typing and the other for selection. It is also often a good idea to indicate what about each result caused it to be displayed, here by making the search string in bold.*

Even the most common informational display on interactive systems can contain impressive amounts of information. Even if a user knows that a particular piece of information is present—from a web search or from memory—it may be very difficult to find by simply browsing or scrolling through lists.

It is also implausible to assume that all information finding may be solved by proper page design. The design of web pages, for just one example, is out of the hands of web browser designers. A method must then be provided to search for specific items within the displayed information set.

Search is, in fact, exactly what is used. You simply place a text search on the page, and the results of this search are jumped to or filtered within the space of the default results list. For mobile devices requiring this feature, the search must more often than not be visible by default, and very often have focus on entry to the page. When of secondary importance, it may be a feature that is explicitly summoned from a menu.

Variations of this pattern depend widely on the pattern used to display the information:

Live jumping

> The most common and the oldest style of the Search Within pattern is especially suited to long lists such as address books, and for scroll-and-select devices. When the page is loaded a search field is in focus. Typing will immediately result in a jump to the results. See Figure 8-10.

Explicit filtering

This is most often used as a replacement for live jumping when explicitly revealed by the user, and very often for touch or pen interfaces. Due to the perception of a conventional search and the familiarity of many users with web search, this behaves as a filter for the list data displayed, suppressing all other results. The search is typed in its entirety and then submitted by the user. This may also be more suitable for Infinite List pages, where live jumping may have odd-looking delays in the display of some results. See Figure 8-11.

Highlight results

This is a display method alone, though often explicitly operated as opposed to being opened by default. The search method should display as live results whenever practical but should use explicit submission when needed, as when the results set is not loaded. These are commonly encountered when searching within web pages, documents, or maps. The results only make sense when displayed in context, and so are left as such. See Figure 8-12.

A key and increasingly relevant variant of this is non-text-faceted search. Other, preset parameters are offered for selection and may be used alone or in combination with a text search. This is particularly suitable for maps; a preselected list may include restaurants, transit, and types of shopping.

Interaction details

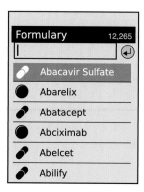
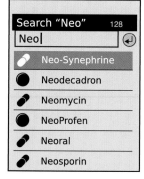

Figure 8-11. *Explicit filtering, especially in lists, may look almost exactly like live jumping. If both are used in adjacent lists, care must be taken to differentiate the two interaction methods. Users will usually have to deliberately select the search box to enter information, and then press the Search button (or OK/Enter key) to submit. This variation is particularly suitable for very large lists, especially if they are retrieved from a remote server and there is a time delay.*

The live-jumping style of list display is loaded with a Search Within entry box anchored to the top of the viewport. On scroll-and-select devices, this has an unusual, almost unique sort of in-focus relationship with the list. The top item in the list is generally in focus and is displayed as such. Pressing Enter (or sometimes using the Left and Right keys) will commit whichever action corresponds to the line item.

However, regardless of the list item in focus, typing from anywhere in the list, at any time, will enter characters into the Search Within field at the top of the viewport. Naturally, you have to disable Accesskeys for such interactions as the keys are dedicated to typing instead.

A variation is to make nothing appear in focus; the text field behaves in the same manner, but a downward scroll is required to bring the first item in the list into focus. When this is used, the top item in any result is still presented as though it is in focus. Yet a smaller subvariant of this—usually specific to an existing OS standard—is to have the same search either initiated by a dedicated search button or revealed when any text entry is made.

For the highlight-results variation, when more than one result is found, a simple Pagination widget will be provided, with the total number found, the current position in the results, and a method to move between individual results, such as Back and Next buttons.

Search methods—whose results are displayed from a particular entry—are left to the specific implementation, and must consider the way the data set is most naturally used.

Presentation details

Figure 8-12. Highlight results on web pages and within text documents simply highlight the text results and jump to the next relevant result when selected by the user. The current item is highlighted slightly differently, and may be in focus for other interface purposes. The map on the right is displaying in essentially the same manner, using one form of icon for all results, and the item in focus has a larger one. A button is provided for a list view, and Previous/Next buttons are provided.

Display of results must be immediate, and require as little additional user input as possible. Display the number of items found and the total number of items in the data set for all types of lists and results.

For the live-jumping variation, the list is unchanged and simply scrolls so that the first relevant result is at the top of the viewport. The remainder of the list is still there, and may be found by scrolling.

For the highlighted-results display variant, each individually found result will be indicated. Usually, as the namesake would indicate, they are highlighted with an underlying color that contrasts with the background and allows the text to be readable.

The current result, when multiple highlighted results are available, will be highlighted in a unique manner and must always be moved to appear within the viewport. Zooming and other techniques may be automatically employed to ensure that the result is readable in context if it is not of a readable size or position when the search is initiated.

When the information found is not explicitly visible or cannot be usefully displayed by simply zooming in, you can add a Tooltip or Annotation bubble to label the results. For multiple results, a small pinpoint is generally used; the item in focus is an expanded label.

Results are generally as close to instant as possible, as the information is already loaded. For certain uses, such as for Infinite List or Infinite Area displays, there may be a delay for some of the information. This may be confusing for live-jumping-type displays, so instead the explicit filter style should be used, and a suitable Wait Indicator used for any delays that cannot be avoided.

Antipatterns

Use the correct method for the information set available. Using an improper method can confuse the user or provide inappropriate or incomplete results.

Carefully consider what an "Enter" key action performs during a live-jumping search. If the user is not looking at the screen and does not notice the results are already displayed, he can assume an explicit search. If a result is in focus, pressing "Enter" may commit an action the user did not intend or which is destructive.

Sort & Filter

Problem

Large data sets often hold information with multiple, clearly defined attributes, shared among many items in the set. You have to provide a way for users to sort and filter by several of these common attributes at once, in order to discover the most relevant result.

Sort & Filter is an information processing and organizational pattern. Although some variant implementations may have special technical needs, the concept can be implemented on any platform, even on a website for the lowest-end, script-free browsers.

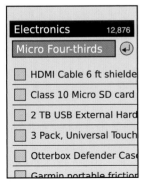
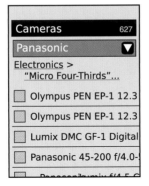

Figure 8-13. *In most cases, only one selection should be offered at a time, and the selection should be a simple single selection. This may be used without complex interactions, and works well with text entry or pull-down selectors. Note the title, revealing some of the categorization, and the counter of results, which assists the user in understanding if more filtering is required. Breadcrumb labels should be identical to the selectors, including category labels (such as "Brand" on the right), to preserve the user's mental model of the system architecture.*

Search can be considered as either information finding or information exploring. Simple tools such as Search Within are for finding specific information the user already knows to exist. This is clearly exemplified by that pattern's most common use, the handset address book. The user should know each of these individuals, has generally added each one by hand (though the rise of contact synching makes this less sure each year), and is looking for a specific one by name.

Exploratory search is used to find items the user suspects exist, or which the user cannot explicitly recall. Faceted search (searching on different attributes to approach the results from multiple angles, such as the cut facets or faces on a jewel) may be required to find such ill-defined items. Sorting is used to present the most relevant results first, from of a possibly large results list.

The keys to implementing Sort & Filter features on mobile devices are:

- Progressive disclosure
- Contextual relevance

Even when there is room, users may not understand that multiple selections are offered within a single selection area and may be confused or only make a single selection at a time.

Some options—and especially sorting options—may not have a clear value to the user before the results are displayed. Instead, offer a single key selection and then, when results are displayed, present additional facets for search, or the ability to sort the results.

To aid users in obtaining the best results, always make informed default choices for them. Sorting, for example, should be "best match" for most cases, but may need to be alphabetical for items such as people, due to the familiarity with names.

The exploration of information that Sort & Filter offers may be presented in one of two distinct manners:

Navigation
    The current filter set should appear as a similar navigational paradigm. The whole displayed set will be applied to the same level, and when the individual selection is made, others will be excluded and cannot be selected again. When no subsidiary facet is available (or when selection of another facet requires an explicit action), these may appear to be a structure like Tabs, and indicate which selection has been made. Otherwise, the selection is effectively a Button, and submits the change. This display method is not suitable for sorting, as it may be perceived as an exclusive set (a filter) and the results may confuse the user. See Figure 8-14.

Submission
    The current set of filters is offered as a series of links, or a set of Form Selections such as a pull-down list. When selected, these will be submitted immediately. Sorting should always use this pattern. When space allows, use a series of links as a breadcrumb (see the Link pattern) so that the entire set of options is visible to the user. See Figure 8-13.

As long as additional facets are available for selection, the next one in turn will be offered, generally at the top of the results page. Sorting is always offered, but in an area distinct from the filter, and with a clear label.

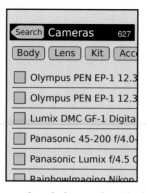

Figure 8-14. *Your site or application may benefit from making the filters appear as navigational elements, such as tabs or similar selectors. This exemplifies the pitfalls of any such system as text is inefficient in the horizontal plane. The "Search" label on the Back button indicates that the previous filter applied was a text search, and the user may step back to change the search parameters.*

As each facet is applied individually, whether of "navigation" or "submission" style, only the one set of items is offered, and the user must use a separate method to access and change other facets.

This will be presented as a history list, "breadcrumb," or "back" function. Allow the user to step back in the process, or jump to the relevant section. Additional, subsidiary choices should be retained and reapplied, or the previous settings should be the default values when subsequently offered within this session.

Though the initial selections should be offered individually, once they are selected, the user may have formed a mental model of the system, so the entire set may sometimes be presented as a series of selectors. The user may change as many as desired and submit these changes as a batch, or individual changes may immediately take effect.

User input should be preserved in general. Especially when some filters have been applied as text entry, some sort of retrieval, undo, or history function should be provided. Though it may not seem like user entry, the Cancel Protection and Exit Guard principles should be applied due to the possibility of the user spending some time creating a detailed, custom search.

Some filters should support multiple selections for a facet, or may already do so in the desktop iteration. An example is to view only certain brands; in this case, only being able to see one at a time may not help. If needed, this may require a large modal dialog or full-page selector and a Select List. A similar problem, also requiring a large dialog or full-page selector, may be encountered with range selectors.

Presentation details

Figure 8-15. *The bottom of the list should ideally have the same filter mechanism, or easy access to it (such as anchoring to the edge of the viewport). Try to make sure at least some results are still visible, so the user is aware he is still in the list view. As shown, this may preclude inclusion of some components; here, only the breadcrumb is shown. When space is short, sorting may appear only at the end of the list, and should always be explicitly different. This is centered, is labeled, and uses links, even though a pull down would work at least as well, to avoid confusion with the form elements used for the filter.*

Faceted search helps users who need or want to learn about the search space as they execute the search process. Facets educate users about different ways to characterize items in a collection. If users do not need or want this education, they may be frustrated by an interface that makes them do more work.

The search space is classified using accurate, understandable facets that relate to the users' information needs. As we've discussed before, data quality is often the bottleneck in designing search interfaces. Offering users facets that are either unreliable or unrelated to their needs is worse than providing no facets at all.

Display results in the most suitable method. Usually, you will find this to be a Vertical List. Sufficient space must be provided for the functionality described. When space provides, and when technically possible, it is desirable to have the Sort & Filter functions be docked to the top of the viewport. In this way, the user may decide additional or different selections are required at any time when browsing the results, without scrolling to the top. If this is not practical, place the selectors at both the top and bottom of the results.

Display methods, such as limited-size lists with Pagination or using an Infinite List, are up to the designer and the needs of the user. Consider the entire interaction, though, and ensure that the user can get to the Sort & Filter functions easily.

The currently applied filters should be presented in the page Title or just below it, as space allows, and repeated at the bottom of any scrolling page as in Figure 8-15.

Antipatterns

Many conventional sorting patterns, such as clicking on column headers in table displays, do not work well in small screens. Even when screen space is available, user expectations are not the same, and these generally will not work without explicit instructions.

Users can easily confuse the actions of sorting and filtering, so you should present these options in different areas or at different times. Use clearly differentiable labels such as "Show only" for filter and "Show first" for sorting. The terms *Sort* and *Filter* are largely meaningless to the majority of end users, and so should be avoided.

Do not simplify navigation or search systems for mobile due to the constrained interface. If the information is the same, the user's needs are the same. In many cases, the need for getting specific information quickly is greater due to the additional difficulty of browsing and reading content from many results.

Selected actions should always be performed on the entire results set. Do not, for example, just sort the items currently displayed on the screen, or that have been downloaded to the device cache.

Summary

Wrap-Up

As we've discussed, mobile widgets are highly reusable items and are used repeatedly, across the device's OS and applications. Widgets can be used to quickly access related levels of information, provide visual cues about the current state of the device, and control the amount of information and level of detail needed on a page.

Pattern Reference Charts

The pattern reference charts in the following subsections list all the patterns found within each chapter described in this part of the book. Each pattern has a general description of how it can apply to a design problem while offering a broad solution.

Cross-referencing patterns are common throughout this book. Design patterns often have variations in which other patterns can be used due to the common principles and guidelines they share. These cross-referenced patterns are listed in the following charts.

Chapter 5, "Lateral Access"

This chapter explained how lateral access widgets assist the user in quickly navigating to and accessing content in the same tier level. This is especially important on mobile devices because the potentially smaller screen sizes affect the amount and type of content presented, and the user's ability to successfully search, select, and read this information.

Pattern	Design problem	Solution	Other patterns to reference
Tabs	Access must be provided to a small number of items at the same level in the information architecture, while also clearly communicating this hierarchy of information.	Follow the file folder metaphor by creating distinct labels that separate three to eight sets of content.	Titles Pagination Location Within
Peel Away	Display a small amount of deeper information, or provide access to related controls or settings, in an organic manner.	For touch and pen displays, use a simulated 3D effect to represent a paper page, which can be peeled or flipped back to view a second page behind it.	Simulated 3D Effects Pagination Pop-Up Film Strip Fixed Menu On-screen Gestures

Pattern	Design problem	Solution	Other patterns to reference
Simulated 3D Effects	Display a small amount of directly related information, provide access to related controls or settings, or provide an alternative view of the same information, in an organic manner.	Use Simulated 3D Effects to pretend components on the page, or even the whole page, merely present one side of a physical object. Rotating the object, moving it aside, or looking past it can then be done without 3D technologies.	Stack of Items On-screen Gestures
Pagination	Location within a series of screens continuing display of a set of content should be clearly communicated and access provided to other pages in the stack.	Page numbers, and a sense of the relative position within the total, are displayed. Tied to this display is a method to move between pages easily and quickly. Methods to jump further than the previous and next pages are also usually offered.	Infinite List Search Within Location Within Peel Away Directional Entry Accesskeys On-screen Gestures
Location Within	Location within a series of screens with alternate views, or which continue the display of a set of content, should be clearly communicated.	When several screens of similar or continuous information are presented with an organic access method, an indicator is usually required so that the user understands his position within the system.	Slideshow Film Strip Pagination On-screen Gestures Tabs

Chapter 6, "Drilldown"

In this chapter you learned that drilldown widgets are used to support many functions that contain parent-child relationships. These widgets can be used in any context to provide access to related content, allow the user to submit information, create a change in the current state of the device, and provide related information in a glanceable manner. Drilldown widgets support the parent-child information architecture relationship.

Pattern	Design problem	Solution	Other patterns to reference
Link	A function must be provided to allow access to related content from arbitrary locations within a page.	Links are content-only items that provide access to additional information either by loading a new page, jumping to a piece of content, or loading a pop-up dialog.	Pop-Up Vertical List Indicator Icon Button
Button	Within any context, an action must be initiated, information submitted, or a state change carried out.	Use a Button to initiate an immediate action such as changing device modes and committing user selections.	Link Icon Indicator Mechanical Style Controls Pop-Up Exit Guard

Pattern	Design problem	Solution	Other patterns to reference
Indicator	Within any context, an action must be initiated, information submitted, a link followed to more information, or a state change carried out on the current page.	The Indicator pattern is a type of action initiator between a Link and a Button. Indicators are always used with text labels, and may perform any action: linking, state changes, and commit actions.	Link Button Icon Pop-Up Pagination
Icon	Provide access to disparate items or functions, in a glanceable manner.	Icon widgets provide immediate access to additional information such as target destinations and device status changes, and are easily understood by their graphical representation.	Link Button Indicator Home & Idle Screens Grid Carousel Avatar Thumbnail List
Stack of Items	A set of closely related items, which can be represented as icons or thumbnails, must be presented in a manner implying the hierarchy and providing easy display of the contents.	A set of stacked thumbnails are arranged with only the top one completely visible.	Hierarchical List Icon Grid Fixed Menu Revealable Menu Wait Indicator
Annotation	A data point in a dense array of information must be able to show additional details or options without leaving the original display context.	An iconic element points to the information selected, and presents (sometimes only after further selection, or in another area of the screen) a label and additional options.	Link Button Tooltip Fixed Menu Revealable Menu Simulated 3D Effects

Chapter 7, "Labels and Indicators"

In some situations, you may be required to use small labels, indicators, and other additional pieces of information to describe content. Mobile users each have unique goals. Some require instant additional information without clicking. Others may need additional visual cues to assist them while quickly locating information. In any case, the information labels must be presented appropriately while considering valuable screen real estate, cultural norms, and standards.

Pattern	Design problem	Solution	Other patterns to reference
Ordered Data	Present information, especially text and numerical data, in the most appropriate and recognizable format for the context and viewer.	Content types that are displayed frequently have become regularized in their display, and must be presented in specific formats to be easily recognizable to users.	None
Tooltip	A small label, descriptor, or additional piece of information is required to explain a piece of page content, a component, or a control.	A tooltip is a transient, contextual, informational assistance widget initiated by hovering over a target, or automatically presented when the system determines the user needs help.	Fixed Menu Revealable Menu
Avatar	A glanceable representation of a person should be provided, for use in various contact-listing contexts.	An avatar is an iconic image used to represent or support the label for an individual, such as a contact in the address book.	Icon Indicator Thumbnail List Link Tooltip
Wait Indicator	Processing, loading, remote network submission, and other delays must be clearly communicated to the user.	Use a Wait Indicator to inform users of delays which are imposed by technical constraints.	Interstitial Screen Infinite Area Reload, Synch, Stop
Reload, Synch, Stop	User control must be provided for loading and synching operations with remote devices or servers.	Due to specific user needs, accidental inputs, or system constraints, the user must sometimes manually start or stop data transfers.	Notifications Icon Tooltip

Chapter 8, "Information Controls"

This chapter explained how information control widgets can be used to quickly find specific items within a long list, information set, or other large pages or data arrays.

Pattern	Design problem	Solution	Other patterns to reference
Zoom & Scale	Data in dense information arrays, such as charts, graphs, and maps, must be able to change the level of detail presented by a zooming function or metaphor.	Use Zoom & Scale to zoom into and out of information while communicating a sense of scale, whether relative or absolute.	Annotation On-screen Gestures Ordered Data Wait Indicator Infinite Area
Location Jump	Scrolling to items in a long vertical list is cumbersome. The information must be indexed to assist in retrieval, and a method provided to allow easy access to key indexed portions of the list.	An indicator of location, or indexing system, is visible on the screen and may be used to jump to the front of an indexed section.	Tabs Vertical List Location Within Home & Idle Screens Accesskeys On-screen Gestures Search Within
Search Within	Finding specific items within a long list or other large page or data array is cumbersome. A method must be provided to find and display this information.	A text search may be placed on the page to search for information within the displayed information set.	Infinite List Accesskeys Pagination Annotation Wait Indicator Tooltip
Sort & Filter	Large data sets often hold information with multiple, clearly defined attributes, shared among many items in the set. Users must be able to sort and filter by several of these common attributes at once in order to discover the most relevant result.	Use Sort & Filter to present the most relevant results first, from of a possibly large results list.	Search Within Exit Guard Vertical List Infinite List Select List Pagination

Additional Reading Material

If you would like to further explore the topics discussed in this part of the book, check out the following appendixes:

The section "General Touch Interaction Guidelines" in Appendix D
> This appendix provides valuable information on appropriate sizes for visual targets and touch sizes for interactive displays.

The section "Fitts's Law" in Appendix D
> In this appendix, you will become familiar with Fitts's Law, which explains that the time it takes a user to either select an object on the screen or physically touch it is based on the target size and distance from the selector's starting point.

PART IV

Input and Output

The varying ways in which people prefer to interact with their devices highly depend upon their natural tendencies, their comfort levels, and the context of use. As designers and developers, we need to understand these influences and offer user interfaces that appeal to these needs.

User preferences may range from inputting data using physical keys, natural handwriting, or other gestural behaviors. Some users may prefer to receive information with an eyes-off-screen approach, and instead relying on haptics or audible notifications.

This part of the book will discuss in detail the different mobile methods and controls users can interact with to access and receive information.

The types of input and output we will discuss are subdivided into the following chapters:

- Chapter 9, "Text and Character Input"
- Chapter 10, "General Interactive Controls"
- Chapter 11, "Input and Selection"
- Chapter 12, "Audio and Vibration"
- Chapter 13, "Screens, Lights, and Sensors"

Types of Input and Output

Text and Character Input

Whether they are sending an email, sending an SMS message, searching, or filling out forms, users require ways to input both text and characters. Such methods may be through keyboards and keypads, as well as pen control. Regardless, these methods must work very efficiently in performance while limiting input errors.

General Interactive Controls

Functions on the device and in the interface are influenced by a series of controls. They may be keys arrayed around the periphery of the device, or they may be controlled by gestural behaviors. Users must be able to find, understand, and easily learn these control types.

Input and Selection

Users require methods to enter and remove text and other character-based information without restriction. Many times users are filling out forms or selecting information from lists. At any time, they may also need to make quick, easy changes to remove contents from these fields or from entire forms.

Audio and Vibration

Our mobile devices are not always in plain sight. They may be across the room, or placed deep in our pockets. When important notifications occur, users need to be alerted. Using audio and vibration as notifiers and forms of feedback can be very effective.

Screens, Lights, and Sensors

Mobile devices today are equipped with a range of technologies meant to improve our interactive experiences. These devices may be equipped with advanced display technology to improve viewability while offering better battery life, and incorporate location-based services integrated within other applications.

Getting Started

You now have a general sense of the types of input and output we will discuss in this part of the book. The following chapters will provide you with specific information on theory and tactics, and will illustrate examples of appropriate design patterns you can apply to specific situations in the mobile space.

Chapter 9

Text and Character Input

Slow Down, You're Too Fast!

Some say he was doing it to annoy the writers. He may argue that it was because the adjacent alphabetized keys kept jamming up due to interference when people were typing too fast. For whatever reason, in the early 1870s, Christopher Latham Sholes, a newspaper editor and printer in Milwaukee, continued to redesign his keyboard layout on his early writing machine into ways that seemed nonsensical.

After studying letter-frequency pairs, Sholes, along with the guidance and support of his backer, James Densmore, separated the most commonly paired letters in his latest layout. In 1873, with that layout, he sold the manufacturing rights for his Sholes & Glidden Type-Writer to the company E. Remington and Sons.

In order to impress customers, the workers at Remington made a slight change to the final key layout. They moved the letter R to the top row. This allowed their salesman to impress their customers by typing the brand name TYPE WRITER all from just one row. This became what is known today as the QWERTY layout.

When the QWERTY keyboard was introduced, writers struggled to learn its layout. The key-jamming problem was less problematic; however, typing speed, along with user satisfaction, was immediately reduced. Sales of the typewriter were poor. It wasn't until 1878, when the Remington No. 2 model was released, which incorporated both lower- and uppercase letters, that sales and performance increased.

An Improved Design?

For the next 60 years, the Remington typewriter maintained success with little competition. In the 1930s, August Dvorak and his colleagues at the University of Washington were determined to create a new and improved keyboard layout. Dvorak wanted to create a keyboard layout that improved typing efficiency and time.

In 1936, he patented his Dvorak Simplified Keyboard (DSK). His design targeted Sholes' problems, such as hand overload, unbalanced finger loads, excess finger movements, and awkward strokes (Parkinson 1972). Dvorak claimed through his testing that his design made typing faster, easier, and more accurate.

Failed Impact

Many critics have discounted Dvorak's findings that his keyboard improved performance. Some argued that experimental setups and statistical analysis done on the SDK were flawed. In what appeared to be a positive promotion for Dvorak, the Navy in 1944 published a document verifying an increase in typist productivity by an average of 74%. With these results, the US Navy Department had planned to order 2,000 SDK typewriters. But the request was turned down by the Procurement Division of the US Treasury Department, which felt there would be too much financial risk.

The Status Quo

Whether or not the Dvorak keyboard was more efficient in time and performance, it never gained the popularity the QWERTY layout achieved. People learned to use the QWERTY and dealt with its odd arrangement of letter placement. The QWERTY layout became the status quo.

Today, several cultural variations of the Latin-scripted QWERTY layout exist:

- QWERTZ, widely used in Central Europe and Germany

- AZERTY, used in France and Belgium

- QZERTY, used mainly in Italy on typewriters

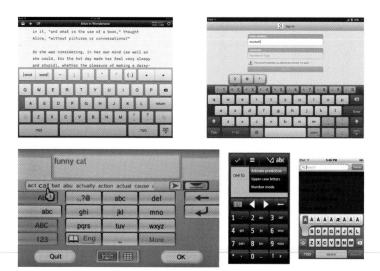

Figure 9-1. *A variety of keyboard layouts, including two tablet methods, a 10-foot UI using remote gestures and prediction, a virtual keypad with entry mode indicator, and a press-and-hold method to get to optional characters.*

Note that cultures that are not based on Latin script use keyboard layouts based on their own language alphabet. And even before the typewriter, specialized users had nonalphabetical layouts; typesetters pulled letters from drawers laid out by frequency and size (you need a greater quantity of the letter *e* due to its frequency of use, so the slot they fit in is bigger), making their layout apparently random.

Use What's Best for You

As we just discussed, even though more efficient ways to input text may exist, it's essential to understand that people will use what they are comfortable with. Some people are comfortable with handwriting, others with keyboard input (Figure 9-2). Some may prefer to use a pencil, pen, or stylus. Default to the most common method they can be expected to be familiar with, and provide options.

Text and Character Input on Mobile Devices

When designing for mobile devices, understand that your users have different skills, preferences, and expectations when it comes to entering text on a device. If possible, provide multiple input options that they can choose from:

- Use input controls and layouts already familiar to users. See Figure 9-1. Don't introduce a new layout that requires them to learn an entirely new process.

- Consider the context. Will your users be using the device outdoors, while wearing gloves? If so, a capacitive touchscreen becomes useless. Consider providing pen input or a hardware keypad as well, or instead.

- Consider your target audience and their habits. According to the Neilson Company, teens in the United States in 2010 sent and received an average of 3,339 text messages per month. A study done by Harris Interactive found that 47% of teens can text with their eyes closed (mostly using 10-key devices and triple-tap entry). Another study, done by PEW, found that 72% of adults text, but they average only 10 texts per day, compared to 50 per day by teens.

- Use functions that promote efficient and quicker input. Use assistive technology such as auto-complete and prediction during text entry. On mobile devices, limit the amount of unnecessary "pogo sticking" when using key controls. For example, when you begin entering a URL, consider the benefit of having a key or button labeled as ".com" predicatively appear.

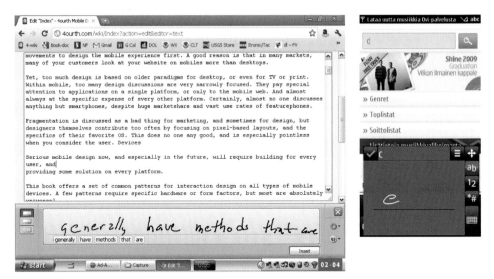

Figure 9-2. *Writing methods can be useful for inconvenient environments, users not accustomed to typing, annotations or nonlinear editing, or ergonomic reasons. Writing with the finger is available on some devices, and may become a pattern eventually.*

Patterns for Text and Character Input Controls

Using text and character input functions appropriately will increase efficiency, performance, and user satisfaction. In this chapter, we will discuss the following patterns:

Keyboards & Keypads
> Provides guidelines for text and numeric entry on mobile devices that use physical and virtual keyboards and keypads

Pen Input
> Provides an alternative input method for users who are less familiar with keyboards or are more comfortable writing in gestures

Mode Switches
> Expands the capabilities of a keyboard or keypad without using up additional screen or hardware space

Input Method Indicator
> Uses explicit indicators to communicate the current mode of input the user has selected

Autocomplete & Prediction
> Provides assistive technology to reduce effort, errors, and time during text entry

Keyboards & Keypads

You must provide a method of text and numeric entry that is simple, is easy, and should be so predictable in behavior that it may be performed by any likely user with little or no instruction.

Keyboards and keypads are built into the device hardware or are provided by the OS. Many platforms, such as the Web, cannot influence input methods, but all must support the available input devices correctly. For some devices, add-on virtual keyboards can be built as well. Some applications may demand a custom keyboard; a typical one is the Dialer, covered separately.

The typewriter or computer keyboard, and the telephone keypad, have become so ubiquitous that they are, to most populations, the expected method of entering information. This extends even to devices which are, on their surface, unsuitable for such entry, such as membrane panels and on-screen displays.

Figure 9-3. A variety of slide or fold-out keyboards found on mobile phones. These include conventional individual keys with backlight, transparent keys with ePaper labels, and membrane panels. The arrangements of each are different, in a search for the best possible way to use the space, but all remain approximately reflective of their regional desktop computer keyboard.

Though several other input methods exist, none are remotely as common, and this is expected to remain true for the foreseeable future. Therefore, keyboards and keypads will continue to be key components of all interactive devices.

The details of any particular implementation must meet user expectations as closely as possible, within the technical limitations imposed by the device and its interaction methods. See some hardware examples in Figure 9-3. A touchscreen virtual keyboard on a small screen cannot faithfully reproduce a full-size physical keyboard, but adhering to the basic principles will generally impose sufficient order that users will be comfortable and able to enter information without any cognitive burden.

Be aware of regional and cultural norms. Avoid imposing unexpected keyboard or keypad layouts on your users. If you have a global product, make provisions for as many input types as possible. Recognize that some users will have wildly different layouts and character sets, and this can influence your interface in many ways. Accesskeys, for example, if not modified for each language, may work unexpectedly or not at all.

Instead of variations, you can consider Keyboards & Keypads to have three axes of variation. Any one implementation combines the attributes from each axis to define the category. Most devices will use multiple modes, which are switched between as the state of the device changes. For more discussion on mode switching, see the Input Method Indicator pattern. For additional details on keypad entry for telephony, see the Dialer pattern.

Hardware/virtual

> The keyboard or keypad may exist as a series of physical buttons, or be represented on the screen of the device. Devices with hardware keys that can be relabeled digitally fall in the middle of this axis and will use attributes of both.

Keyboard/keypad

> Keyboards are those with direct mapping to characters in the native alphabet. Keypads, as shown in Figure 9-4, are for numeric entry and often dialing when on mobile devices. Chording keyboards and some specialized key entry panels fall in the middle of this axis and may use attributes of both types.

Direct/multitap

> Keyboards allow typing words and phrases via (mostly) direct entry; an *a* is typed by pressing a key marked "a" (or "A"). Multitap is an indirect entry method; an *a* is typed by pressing the "2" key, which is also labeled "abc"—sequentially until that letter appears. Certain keyboards combine attributes of both of these, but none are currently commercially important.

Figure 9-4. *A typical mobile device numeric keypad is coordinated with direction keys and other functions such as Talk and End, soft keys, and so on. This keypad is for use in North America, as the letter labels comply with the NANP standards. Other regions will demand other labels or layouts.*

Although these descriptions and the remainder of this pattern focus on the standard implementation, each axis is actually more of a continuum. A single implementation may fall in the middle, such as some nonstandard keyboard/keypad combination layouts. This simply means the same principles apply to even those implementations.

Predictive systems, although associated with 10-key entry, are actually just methods that can be applied to any input method. These are addressed in detail in the Autocomplete & Prediction pattern.

You might think virtual keyboards and keypads are purely for touch interfaces, but they can also be used on scroll-and-select devices. You may find this an especially useful solution if occasional text or character entry is required on devices with no dedicated keyboard or keypad, and no touch or pen input.

When a field which allows free input is selected, display the virtual keyboard largely as described in this pattern. One key or button is always in focus while the keyboard is visible, and scrolling with the five-way Directional Entry pad will move in the corresponding direction. Keys should be in a grid, not offset, so that directional movement is more predictable. Selection will "press" the key and enter it in the field provided. The keyboard or keypad should be "circular" on both axes so that scrolling to the edge will move the selection to the opposite edge, to reduce the already large number of clicks to get to a character. A method must be provided (see Figure 9-5 and Figure 9-6 with the "Done" key) to complete entry and close the virtual keyboard.

Interaction details

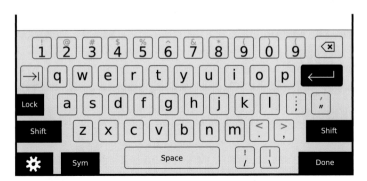

Figure 9-5. *Numerous keyboard layouts can be devised to fit the space available, and still provide a meaningful experience for the user. This virtual keyboard is fairly full-size, with number keys, and symbols assigned to each of those. A few have been removed to save space, and a few have been moved slightly. The Mode Shift keys and the Submit key are along the edge, to allow space around them while increasing the effective contact space. Note that the keycaps are in lowercase, since the Caps Lock key is not on.*

Interactivity of a key press is generally very straightforward. For hardware, the user simply presses the button. For pen and touch devices, each key is functionally a Button and should have hover and activated states when available. Some input methods may be more difficult. When Directional Entry or Remote Gestures is used, pointing and other focus or effect indicator standards from those patterns must be applied.

Hardware keys should comply with human factors standards for pressure, size, and spacing. Virtual entry systems should have targets no smaller than 10 mm. Certain keys, especially the "Enter" key, space bar, and modifiers, may be placed along the edges of the viewport on flat bezel devices, to increase their effective size. High-use keys, such as the space bar, should be larger to make them easier to find and activate. Keys that can have consequences past a single entry, such as the "Enter" key and modifiers, should have spacing around them to prevent accidental activation. Figure 9-5 shows several of these principles.

The layout and key arrangement should, whenever possible, use common and expected formats. For keypads, use regional standards promulgated by the local telecommunications authorities. In North America, for example, this does not just mean the layout is 1 at the top left, but that the 2 key has the letters "ABC" assigned to it.

Figure 9-6. *For more compressed spaces, keyboards often lose the dedicated number row. You can make up for this by switching modes to a number entry or function key mode. This is a virtual keyboard, so alternative labels are not needed, as long as the mode switch key is clear. The primary symbols on the keyboard are still present, and in mostly the same place; the less-used slash and question mark are moved slightly for space purposes. Each of these has the shift mode labels above, so the user knows she does not need to enter the symbol mode and hunt, but simply presses Shift to get these characters.*

You may have to compress your keyboard to fit into the available space, and combine the number row with the top row of letters, as shown in Figure 9-6. Make sure the numbers align to their normal location, so Q will also be *1* when a function modifier is used. Likewise, place all conventional symbols as close as possible to their typical location. The !, @, #, and $ symbols should be alternatives of the 1, 2, 3, and 4 keys—or the closest equivalent when compressed—for US English keyboards.

Although most keyboards are displayed or built as a single block, a sometimes-seen variant is to provide a split in the middle. Sometimes this is just a structural and visual break, which otherwise does little to the interface, whereas other times it may be so wide that the screen sits between the two halves. Virtual keyboards may also benefit from this, both for space considerations and by considering the ergonomics of the device. It may be better for users to grasp the device firmly; if your device is very large, like many tablets, they will not be able to reach the center of the screen, so a split layout may have advantages.

Multitap keypads are capable of very high-speed text entry for populations familiar with their use. They must be designed to follow standards and work without delay. Pressing a key will cause the first valid character to appear (see the Input Method Indicator pattern for more details). Pressing the same key within a short time frame will display the next valid character in turn, as shown in Figure 9-7. This display is a circular list, and will eventually return to the first character. After a brief pause, when another key is pressed, or when a "next" key is used, the character displayed is committed and further key presses will display in the same manner, for the next space in the field.

Figure 9-7. *An example of how the multitap (or "triple-tap") text entry method works.*

Only enable key repeat on fairly high-fidelity devices. For smaller devices, and those expected to be used in difficult environments, such as most mobile devices, do not allow a key press to enter multiple instances of the character. This will help to prevent accidental multiple inputs due to the user being in a vibrating or bouncing environment, such as typing while in a vehicle.

The behavior of mode-switching keys is covered in the Mode Switches pattern. Despite being on the keyboard, they behave very differently and so must be considered separately. Also be sure to see the Press-and-Hold pattern for details on another way to gain access to alternative entry for a particular key.

Directional Entry keys should almost always be included with any hardware keypad. A five-way pad is naturally included in the device design for a typical keypad-based device. For slide-out keyboards and directional keys, an enter function should be included. This will allow users to scroll and submit entries without leaving the keypad. The enter function is often useful on even virtual keyboards, to prevent accidental inputs from the user shifting his grip or reorienting to change the entry mode.

Be sure to accept input from hardware direction keys and enter keys, even for devices that are otherwise touch- or pen-driven. Don't be that one application that works differently, and confuses users.

Presentation details

Key labels must be clear, large enough for users to read, and easy to read in all lighting conditions. Alternative labels (Shift or "Fn" characters) should be in a different (less prominent) color, or in a tint of the primary label color. Alternative labels should be above the primary label. There is no need to display uppercase letters as alternative labels to lowercase letters.

On virtual keyboards, use the label for the current state, to ensure that the state is clear. Display lowercase labels unless the Shift or Shift Lock key is activated.

Keys must perform their labeled behavior. Note, for example, that backspace and delete are different functions. Users accustomed to Windows will think "Del" means "forward delete" and not backspace as it is labeled on the Macintosh and some other devices.

Especially for hardware keyboards, the use of symbols may be helpful for special keys, such as "Tab" and "Enter". However, these icons are by no means universally recognized. Use caution when selecting and designing labels, and use simple text descriptions of the function—or text along with an icon—whenever possible.

Antipatterns

Figure 9-8. *This keyboard has a "Back" button immediately adjacent to other keys, where it can be accidentally hit. Without an Exit Guard this can cause catastrophic failures, discarding large amounts of user-entered data.*

Avoid accidental input:

- Lock keypads when sensors determine they are in pockets, against the user's face (on a call), and so on.
- In any context, if multiple adjacent keys are pressed within a few milliseconds, this can be assumed to be incorrect, and only the most likely key—by physical mapping or Autocomplete & Prediction—should be accepted, or none at all.
- Staggered keyboard layouts can also help to prevent accidental input or increase the accuracy of filtering.
- Be careful when placing functional keys on the keyboard, and provide sufficient space to avoid accidental input. If required, use an Exit Guard to reduce the risk of losing user-entered data, unlike the keyboard shown in Figure 9-8.

On the other hand, be careful going further to avoid accidental input, such as instituting lockout periods. It can be hard to differentiate fast typing from accidental input.

Use caution with backlight in marginal lighting. It is easy to have the backlight illumination match the key material reflectivity, resulting in the label being entirely invisible. Besides careful coordination of the backlight algorithm, using a color that is different from the key material for the label (e.g., yellow instead of white light on black keys) can alleviate this issue.

Do not use conventional insertion point indicators (such as pointers or vertical lines) during multitap, especially when a character is not committed, as it may not make it sufficiently clear where key presses will take effect. Highlight, underline, or otherwise indicate the actual character space being currently affected.

Be careful when implementing layouts, labels, or input methods that are not absolutely standard. Be sure to test this robustly with users. Inertia (by users and operators) will still impede adoption, even if it is a superior idea.

For physical keyboards, a mode in which a 10-key pad is imposed over some of the keys is rarely well understood by the user. When the keyboard is staggered it is also difficult to use. Usually, the best solution when no dedicated number pad is provided is to simply use the number keys along the top row, and lock them on for the user during numeric entry, such as when dialing a number. Be sure to inform the user with an appropriate Input Method Indicator.

Pen Input

Problem

You must provide a text and graphical input method that is simple, is natural, requires little training, and can be used in any environment.

Pen Input requires hardware to be dedicated to the input method. When designing for pen devices, you must consider what input methods will need to be adjusted to work best with them. Many pens have hover states, so this would need to be added to existing touch designs.

Solution

Writing is a function inherently performed by the use of a tool. Until the development of typing, this was exclusively done with the stylus, then the pen and pencil.

Pen interfaces can be an excellent alternative to keypads or touchscreens when touch is unavailable due to environmental conditions (gloves, rain), when the device is used while on the move, or just to provide additional precision and obscure less of the screen. Pen input used to be the norm for touch-sensitive screens, when relatively low-resolution resistive input was the norm; capacitive touchscreens have moved to direct touch and multitouch inputs, but pens as an add-on to touch are making a slow resurgence.

Text input is particularly suited to pen devices, especially for populations (or cultures) with little or no experience with keyboard entry, when there is no standard keyboard for the data type, or for the aforementioned environmental conditions. In situations where only one hand may be used to enter the text, character entry by pen can often be much faster than the use of a virtual keyboard.

Other systems, including newly developed capacitive screens, can use specialized pens to isolate touch input, allowing the user to more naturally interact with the device by resting his hand on the screen. Mode switching is also available, either automatically or manually, to allow either input method.

Pen input devices may be used for multiple types of interfaces, and will often be the primary or only method for the device. This pattern covers only character and handwriting input and correction. Although any pen input panel should also support a virtual keyboard mode, this mode is covered under the Keyboards & Keypads pattern.

Variations

Figure 9-9. Word entry with pens will appear in an input panel, occupying part of the screen. Options such as the mode switches and special characters appear as buttons within this panel. Translation candidates appear beneath the written word, and may be explicitly accepted or automatically loaded after a pause.

Pen input for text entry may fall into one of two modes:

Word entry

Provide an area where users may input gestural writing, as they would with a pen on paper. Printed characters are the most common, but script styles should also be supported. The device will use the relationships between characters to translate the writing to complete words, based on rules associated with the language, and will automatically insert spaces between words. The input language, and sometimes a special technical dictionary, must be loaded. See Figure 9-9.

Character entry

The input panel is broken up into discrete spaces in which individual characters may be entered. This may be accomplished by simply dividing the word entry area with ticks or lines, or presenting a series of individual input panels. The device will translate individual characters, and will generally disregard the relationships between adjacent characters. The user is responsible for spelling words correctly and adding space between words. The language must be set correctly for the fastest recognition, but alternatives within the language family will also be available (e.g., all accented characters for any Roman language). See Figure 9-10.

A subsidiary version of the character entry mode uses a shorthand, gestural character set instead. This is not required for most languages with current input and recognition technologies, but it may have advantages for certain specialized applications or for the input of special characters or functions.

All available input modes should be made available to the user at all times. The user will generally have to select which mode to use, so you will have to provide a method of switching. This does not preclude development of a hybrid approach, or even simply appropriately selecting which mode is most suitable to the current situation. If a gestural shorthand is available, it should be offered as a part of the natural character input method; both will be accepted, and the user may change the input method per character.

In certain cases, such as where the note taking must be as fast as possible or the user is combining images, formulas, and other data with words, you may provide a mode (or a whole application) that will store the gestures without live translation. Later, the user may select all or part of the gestures to translate. This behavior should follow the pattern covered here when similarities arise, in order to preserve a consistent interface. Portions will, of course, be different.

The finger may also, on some devices, be used to input gestural writing without a pen. Many of the principles outlined in this pattern will apply, but the size of the finger may obscure the screen and make this suboptimal.

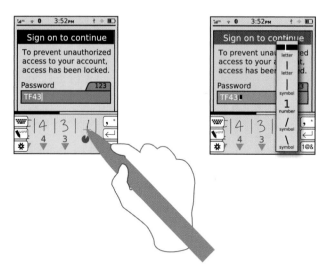

Figure 9-10. *Letter input divides the input panel into discrete entry areas. Each character displays a translated candidate. One way to handle correction is for optional translations to be offered in a list. If the best guess is incorrect, the user may open the list and select another option.*

When a field that can accept character entry becomes in focus, the input panel should open to the last-used mode. This mode may be a virtual keyboard, but as the user may switch between the modes, the principle applies.

Whenever the user makes a gesture in the input panel, the path taken is traced as a solid line in real time, simulating pen on paper. As soon as a word (or, for character entry, a character) is recognized, it should be offered as a candidate. The translated characters and words are the same as the candidate words, as discussed in Autocomplete & Prediction; see that pattern for additional details on presentation, the interaction to select alternatives, and user dictionaries.

To provide sufficient space for the user to write, you must allow scrolling within the input panel. When the panel is filled, the system may automatically scroll, or the user may be allowed to scroll manually instead, to enable review and correction. Avoid dynamically adding additional space to the input panel, as this will just obscure more of the screen.

At least two methods are available for actions when input is completed:

Manual

> The user completes his entry in the input panel, and presses a button to load the content into the field. The next field may be entered, or the form submitted.

Automatic

> The system loads characters or words into the input field as they are completed, or after a brief pause, to allow the user to make immediate corrections. This may be used to eliminate the need for scrolling the input panel, as it only needs to support one long word, at most.

Mode switches may either commit all completed characters and words to the input field, or switch them to the new entry mode. As discussed in the Cancel Protection pattern, never discard user-entered information.

You must make some provision for functions which are not visible characters, such as line feeds, the "OK/Enter" key, and the "Backspace" and "Tab" keys. These keys should be visible as part of the input panel. See the Keyboards & Keypads pattern for additional information on key placement; the handwriting area of the input panel may be considered equivalent to the character keys, in this sense. Place these other keys in the proper relationship to one another so that the user may find them easily.

On-screen Gestures may also be supported for the user to more quickly input key functions. However, these will require learning, and so should be used as a secondary method in most cases; continue to provide keys for these functions as well.

Some pens will have buttons that the user can activate to bring up contextual menus or which may be programmed to perform other functions. These are not usually noticed and so should be considered like gestures, and are only shortcuts to functions which can be found otherwise.

A robust method for correction must be designed into the system. Do not require the user to manually delete and rewrite, but build in approval steps, or gestures to correct existing phrases, even when loaded to the input field. A good path is to allow easy conversion from word to character input, to allow correction of characters that could not be recognized correctly. Avoid training, but do make the system learn the user's input method and accept new words in the dictionary so that accuracy improves over time.

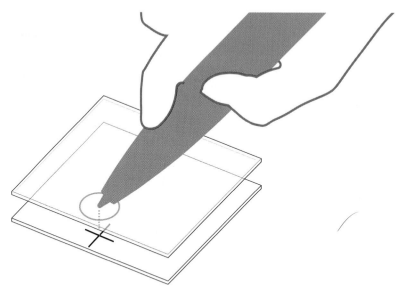

Figure 9-11. *Parallax for input on screens manifests as a perceived offset between the surface where contact is made and the actual display area. As shown here, this is because protective overlays, the sensing panel, and other materials cause a gap between the two. When viewed straight on, they line up and there is no problem. However, users will not always view directly perpendicular to the screen, and will notice the difference with pen inputs. This is especially troublesome on larger devices, such as tablets and kiosks, as the user cannot view the entire screen straight on.*

Writing is entered into an input panel that appears as a part of the screen. For small screens, this should always be a panel docked to the bottom of the viewport and should automatically open when an input field is placed into focus. For larger screens, the input panel may be a floating area displayed contextually, adjacent to the current field requiring input. This panel must always be below the input field, to prevent obscuring the information.

For very small screens, the input panel may take up essentially the entire screen. In this case, you must take special consideration to display the text that has already been entered in order to provide context to the user.

When input is completed the input panel should disappear to allow more of the page to appear.

Any delays in translation or conversion, especially for Autocomplete & Prediction methods, should be indicated with a Wait Indicator so that the user is aware the system is still working. Besides the usual risk of the user thinking the system has hung, the user may believe the phrase cannot be interpreted, and will waste time trying to correct it.

If the pen can be detected as a unique item, you may place the input panel anywhere on the screen. Otherwise, you must always place the panel to avoid accidental input, usually along the bottom of the viewport. Resistive and most capacitive screens will perceive multiple inputs and may not be able to tell the difference between the pen and the user's hand.

Be sure to continue providing a cursor within any input areas on the screen, regardless of the method used. Within the handwriting input panel, you should use a small dot or crosshair to indicate the current pen position. This is crucial due to miscalibration and parallax, as explained in Figure 9-11. Both are always present in screen-input devices. For touch devices, the finger is so large that the errors are generally not obvious, but pens are very small and so reveal these errors and other phenomena.

When the pen leaves the input area, this cursor will revert to the normal cursor. When the pen leaves the screen and cannot be detected, the cursor should disappear. Do not use the desktop metaphor of "cursor always present," as that is based on the pointing device used. Mice are *relative* devices, and always provide input from the last location, so the cursor is critical. Pens (and fingers) are *absolute* devices, and point wherever they are placed on the screen.

If a hardware Directional Entry key is provided which can move the cursor, this is again a relative pointer, and the cursor may reappear from its last location when this input is detected.

Antipatterns

Be careful when using a pen for large stretches of input on kiosks or where the user cannot rest his hand or arm. The "gorilla arm" effect can impose fatigue or even repetitive stress injury. Allowing the user to rest his hand or arm on the input device, as well as more casual use or regular changes in the type of entry, can avoid this.

The space taken by the input panel must be considered by the page display. Do not allow important information to be obscured by the panel. For example, do not make it so that the panel must be closed to scroll to the bottom of a form so that the Submit button may be activated. Consider the input panel to be a separate area instead of an overlay so that items may be scrolled into view.

Handwriting recognition can be very fast and may even be approximately as error-free as typing on a keyboard. However, correction methods and other parts of the interface can serve to make the overall experience much slower. Use the methods described here to reduce clicks and provide automatic, predictive and otherwise easy access to mode changes.

Respect user-entered data, and do whatever is possible to preserve it. Use the principles outlined in the Cancel Protection pattern whenever possible. Especially when user input must be committed before being populated, do not allow gestures to be discarded by modal dialogs or other transient conditions.

Separate entry areas are generally no longer required technically and so are not suggested unless there is some specific technical or interface reason that the screen cannot or should not be used. If only one screen or portion of the screen is to be used for pen input, ensure that it is as optimized as possible for this task. Generally, this area may be dedicated to input, and may take on all the behaviors of the input panel as described earlier, when in pen input mode.

Mode Switches

Problem

You must provide access to additional and alternative controls, without taking up more hardware or screen space, through the use of mode switching.

Mode switches are simple buttons or similar menu items that are easy to implement on any platform that can support the primary input method, such as a virtual keyboard.

Solution

Mode switching is a method almost universally employed to expand the capabilities of a keyboard or keypad. A default condition exists and is very usable, but to include variations of characters, to access the full range of symbols, and to repurpose entry (such as numeric keypads to text), mode switches provide almost unlimited extra capabilities.

Mode switches are often poorly implemented, with arbitrary and conflicting controls. These can impede text entry enough that text-related functions are underutilized or alternative entry methods are used, and they can be so obscure that users cannot access certain modes at all.

Be sure to implement the same methods across the entire interface, and whenever possible inherit existing standards in the OS. For example, if press-and-hold is used as a shift lock, do not use a double tap to lock the symbol key instead. Be especially careful to avoid drastic differences between virtual and hardware keyboards, when both are available on a single device.

You can also apply the same principles to switching modes in any interface. However, Tabs are the most common implementation outside of character entry. The types of modifiers discussed here only truly become a pattern when used with Keyboards & Keypads, Pen Input, and related functions such as the Dialer.

Figure 9-12. *The neutral condition for all virtual keyboards should display lowercase letter keys, to indicate this is the character that will display when a key is pressed. Mode Switch keys should be clearly different from character entry keys, but not so different that they may be confused as being in a selected mode.*

Variations can be discussed best by categorizing them using machine-era analogies. Consider the Mode Switch to be a replacement for a hard-wired electrical switch.

Single-throw

> An individual button activates an alternative mode. When inactive, the input panel returns to a neutral state. This concept is important because multiple single-throw switches may be added to the panel, such as "Shift", "Caps Lock", and "Symbol" keys for a keyboard. This type of entry is subdivided further; the subdivisions are discussed in the "Interaction details" section.

Multithrow

> An individual button switches sequentially between multiple modes. The number is preset and will not vary arbitrarily, but certain modes may be skipped when locked out. Switching is *circular*—once the last item has been reached, the next switch activation will return to the neutral state. This switch type is very commonly used on keypads with text entry. One key will be a mode switch, moving between multitap, predictive text, and numeric-only entry, for example. In this manner, the key works much like a "triple-tap" entry key, but as it does not enter data itself, it is a Mode Switch instead.

Both of these variations perform equally well on both virtual and hardware keyboards and keypads, and can be used as the mode selector for Pen Input methods as well.

Figure 9-13. *When a mode is entered in which some keys have no function, these keys should retain their neutral position label, but be grayed out to indicate that they are unavailable. This labeling will keep the user oriented through the change. All other keys should remain unchanged. In most states, this will include punctuation, and in all states, the "Return" or "Enter" key.*

Single-throw Mode Switches operate in two modes. "Shift" keys (as shown in Figure 9-14) and the like are activated with a single press and are *sticky* in the accessibility sense. When the mode switch key is pressed, it remains active for a single character entry keystroke, after which it deactivates and the entire entry panel returns to the default state.

The other mode is for locked modes, such as Caps Lock. When activated, these keys will remain active until the panel is dismissed or entry is no longer occurring. In some cases, these will be dedicated keys and will activate as soon as they are selected or pressed. The other option is to make the lock mode an alternative function of the key. To avoid a mode switch for the key, a secondary activation method using only the mode switch key itself must be used. The most typical are:

Press-and-hold

The mode will be entered immediately upon being selected or clicked. If the key is pressed for a short time, the mode will switch from sticky single activation to locked.

Press-twice

The mode will be entered immediately upon being selected or clicked. If the key is pressed again in quick succession, the mode will switch from sticky single activation to locked.

You should determine which method to use by following which practice is most common (if any) in the enclosing OS. You should determine the speed of double-clicks or time for press-and-hold by best practice of the OS.

When the user simply selects the same mode switch (or any another mode switch), it will disable the current mode lock.

Figure 9-14. *When Shift (or "caps") is selected, display the capital letters on each key. Though this is the primary method of communicating the mode switch, the switch button itself should also indicate it is selected. If a full-size keyboard is emulated and two Shift keys are present, be sure to indicate that both are active.*

The last mode switch activated takes priority, and disables any other switches that may have been activated before. When deactivated, return the input panel to its neutral position, not to the previous activated alternative mode.

When reactivated, input panels governed by single-throw switches will always open to their neutral mode and disregard the last condition.

Multithrow switches always behave as though they are locks. The input mode will remain in the selected mode until manually changed, or the mode becomes unavailable.

For either switch type, mode switching will be disabled for any required entry modes. For example, if a full keyboard is provided but only numeric input can be accepted, it should be locked to numeric input. A virtual keyboard under the same conditions should display a numeric keypad instead.

Multithrow switches should not allow inappropriate entry types. For example, predictive text should not be allowed on password fields (they are unlikely to auto-complete a secure password, it might encourage easily guessable words, or it may even store a unique password in the user dictionary). The unavailable mode should be skipped; if the normal progression is Abc→ABC→T9→123, for a password field, it would simply be Abc ABC 123.

Presentation details

Figure 9-15. *Just some of the many ways to communicate a shift function. Note that single-character capitalization is not precisely the same as Shift. Be sure to use the expected behavior, and make sure your label matches your behavior.*

You must always clearly indicate the current mode. If it is not displayed as part of the OS or integral to a virtual keyboard, you must display it as a part of the field. The lock key should change display in some manner. For hardware keyboards, this is generally only displayed as an Input Method Indicator. Ideally, the key itself would illuminate or otherwise change, but the extra cost in hardware is so often deemed prohibitive that it is not particularly common.

For virtual keyboards and keypads, the mode switch key should always change. This may follow the general OS guidelines for a currently pressed key. The locked mode should be different and may change the icon of the key, or change the state of the "pressed" indication.

In addition, virtual keyboards and keypads should change the individual key labels. When in neutral mode, letters should appear in lowercase (a) as in Figure 9-12. When the "shift" mode is active, they should change to uppercase (A). Lock modes have no effect on these. When a key becomes inactive due to a mode change, the label should gray out or otherwise indicate that it is unavailable, as shown in Figure 9-13.

Labels must match in both *reflective* (ambient light) and *transmissive* (illuminated backlight) modes. If the "Fn" key is entirely yellow, to match the yellow function key mode labels on the rest of the keyboard, when illuminated this key must be entirely yellow and the labels on the individual keys must also be yellow. See the improper way to do this in Figure 9-16.

Labels must also be clear and unambiguous. Use caution with iconic labels, and test when possible; some user populations will not understand the symbols, and may only determine the intent by position and familiarity with keyboards. See Figure 9-15.

Mode switches for alternative character entry on hardware keyboards, such as numbers or symbols, should use a color to differentiate them. For example, if letter characters are in white, symbols may be in yellow and numbers in blue. The symbol mode switch key should be labeled in yellow and the number mode switch in blue.

Differentiate the mode keys from character input keys by position, shape, and label. A common method to differentiate secondary modes such as the symbol key is to reverse the colors. In the preceding example, the label text would be in black on a yellow background.

Figure 9-16. *This keyboard uses yellow to identify the function modifiers. The reflective labels, for daylight mode, work very well. The function key is all yellow, and the modifier labels are yellow symbols. But under backlight, while the modifier labels remain yellow, the function key itself is black, with the "Fn" label in white. This disconnect is bad, but the change between view modes is even worse.*

Do not implement a "Shift Lock." Older typewriters had no concept of a Caps Lock, but allowed the Shift key to be locked. This also would type the alternative special characters for symbol and number keys. It is not the expected behavior today on any device, and should never be implemented.

Avoid using multikey combinations for routine controls. Despite their prevalence on desktop computing, they are poorly understood and the smaller keypads of mobile devices (even with a sticky key paradigm) do not support their use well. This does not preclude their use for obscure functions. In fact, the lack of any reason for users to routinely use multikey combinations means they are a good way to support highly specialized or technical functions, such as for system resets.

Use caution with backlight in marginal lighting. It is easy to have the backlight illumination match the key material reflectivity, and the label cannot be seen at all. Besides careful coordination of the backlight algorithm, using a color different from the key material for the label (e.g., yellow instead of white light on black keys) can alleviate this issue.

Input Method Indicator

Problem

You must make the user aware of the current mode of the selected input method, and of any limits on selecting modes for a particular entry field.

The Input Method Indicator may be integral with the virtual keyboard, or provided as part of the OS, or may need to be implemented within your application as an individual component, within the page or for each field. When it is not available at the OS level, you are encouraged to build a custom version when input types are critical or misunderstanding would cause errors.

Solution

When not otherwise immediately apparent, an indicator should be placed on the screen to explicitly indicate the current input mode or method.

On virtual keyboards and keypads, the Input Method Indicator is generally not required as the keyboard itself changes modes and serves as an implicit indicator. No "symbol mode" indicator is needed when the keyboard is filled with symbols. There may be value in attaching field-level indicators.

Explicit indicators are valuable with most hardware keyboards and are most common with hardware keypads, which provide numerous entry modes and methods. Only explicit indicators are discussed here. See additional information about virtual keyboard labels under Keyboards & Keypads and Mode Switches.

When additional types of input-field control are offered, such as selection (for cut/copy/paste) or highlighting, you can use the same indicator field to indicate this condition. Since character input is not valid during these operations, this is a good use of the space, without having to remove the indicator area entirely.

Figure 9-17. *Input Method Indicators may be anchored to the individual fields, as on the left, or attached to the viewport frame, usually in the annunciator row or notification strip, as on the right. Both methods have advantages.*

The key variation for explicit indicators is of position. Two key options are available, both of which are compared in Figure 9-17:

Field

The indicator is adjacent to, and usually attached to, the current input field. This associates input with the area on the screen where input will take effect. Each field may have an indicator, so the user may plan ahead, and have a better understanding of what is a field and expose any variations in advance.

Viewport

The indicator is locked to the edge of the viewport. This associates the input mode with the device, as only one mode may be used at a time. This may seem to indicate that the edge closest to the keyboard or keypad should hold the indicator, but this is not always true. The area in which alerts and other status messages already appear is instead usually the best choice, and the user will learn to glance here for all status information.

Pen Input panels will have a different set of available modes than are discussed here, but they follow the same principle and always indicate through implicit interaction and layout or explicit indicator which mode is currently selected. Their mode indicator may be best attached to the pen input panel. More on this is discussed in that pattern.

Input Method Indicators are exactly as the name indicates, and are only indicators of changes carried out elsewhere, such as with Mode Switches. No direct interaction with the indicator is possible, or it becomes another type of mode switch.

Indicators are dual-purpose; they always indicate the current input mode and sometimes are a flag to indicate when a particular mode must be used. When a field with a restricted entry (such as numeric only) is in focus, the indicator will switch to the available mode as well as indicate this.

If a field input restriction exists but more than one mode can meet the needs, the automatic selection can become more complex to design. If the keypad is already in allowed mode, no change will take place. If not, the first or most similar available mode will be used. For example, if symbols are disallowed, the Abc mode should be the first selection, but the ABC, Word, and 123 modes can also be used and switched between.

ABC	(ALL CAPS)	ABC
Abc	(Initial cap)	Abc
abc	(lower case)	abc
123	(numeric)	123
!@# or Sym	(symbol)	!@#
T9 word	predictive lower case	abc
T9 Word	Predictive initial cap	Abc

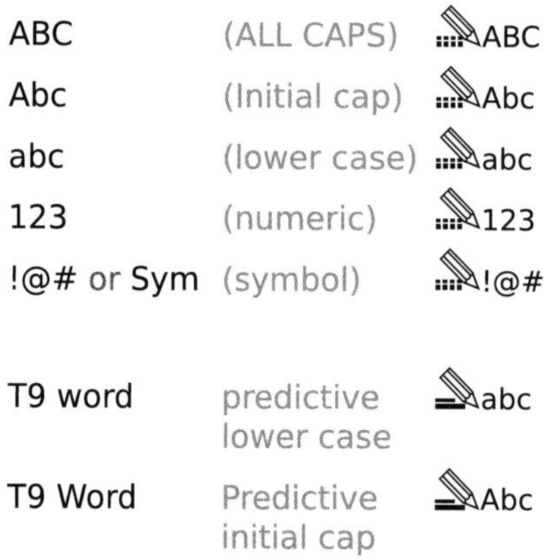

Figure 9-18. *A series of indicators, all characters on the left and with a graphic on the right. Note that predictive text is a subset of other methods, and can be input with its own capitalization. Each text label is internally descriptive. The graphics use the pencil as an anchor, and an icon for individual on continuous entry to indicate character or predictive text. Numerous other schemes are available or can be devised.*

Make sure these labels are clearly readable and are obviously associated with the input method. Field-level labels should be attached to the edge of the field in some manner, so they are clearly associated. Tabs hanging off the field are most typical.

You can apply this principle within any interactive design to indicate fields with exceptional entry (such as all-numeric) even if the OS has no universal indicator itself.

Viewport-level label location and design is much more associated with the overall OS design, and especially with the design of the Annunciator Row and any Notifications strip.

Labels must not only clearly explain the current mode, but also should be designed to associate with one another so that changes do not confuse the user. Employ the same symbology and typography, at the same size, color, and overall contrast; if outlined, do not have one mode be a solid. For graphical labels, consider an anchoring icon from which all the indicators are built. Figure 9-18 shows two related languages of icons.

All labels must change immediately upon a switch of the input mode.

Mode labels may disappear when irrelevant, such as when not in focus for field-level indicators, or when no input is available on the page for viewport-level indicators. This may be advantageous if space is at a premium and additional information may be presented. However, there can be advantages in offering the information, so carefully consider the downsides to decluttering techniques such as this before implementing them.

The phone Dialer keypad and some other specialized keypads, such as calculators, generally will not need such indicators due to the lack of a mode change. Dialers, for example, support Accesskeys and sometimes use other shortcuts to return results considering both text and numeric input, but this is not a mode switch. When text entry is allowed for a Search Within pattern on the address book, or for recent calls, these are usually switched to at a higher level within the application, and are not just an entry mode switch.

On the other hand, within the Dialer, do go ahead and use an Input Method Indicator if any confusion may arise from restricting multipurpose input methods. For example, if a hardware keyboard is used but only numeric entry is allowed, the lock to numeric input should be indicated on the screen.

You can extend the indicator methodology to related selection mechanisms if relevant to the interaction model, as shown in Figure 9-19.

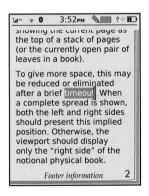

Figure 9-19. *Other symbols that replace input methods in function, such as select and highlight, shown here, may also replace the Input Method Indicator when active.*

Never display a mode label that is unavailable, or prevent input by allowing the user to select an improper mode and then blocking input or displaying errors. Always switch to the best available mode instead.

Do not allow the indictor to scroll off the screen, or be obscured by input panels or modal dialogs. If the indicator cannot be locked to the viewport and made to be always on top, use a field-level indicator instead.

Do not use both field- and screen-level indicators. If field-level exceptions are needed, they should be displayed as field "hints" or as contextual content, not as an interactive control or indicator.

Avoid jargon or brand names for labels unless they are very well known. T9 is an almost-generic shorthand for predictive word entry. As long as your device uses Nuance/Tegic T9, you may use that label. If you use a homegrown solution (and so are legally barred from using the label "T9"), do not use your abbreviation, and use a generic word or symbol instead.

Autocomplete & Prediction

Whenever possible, you should build in assistive technology to reduce the user's text entry effort, and to reduce errors.

Even when provided by the OS, you may wish to build your own Autocomplete & Prediction methods specific to the operation of your application, or due to the use of information outside of dictionaries, such as URLs in browser history.

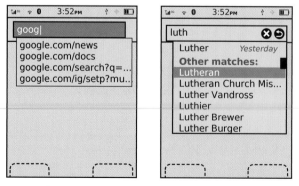

Figure 9-20. *Two versions of the suggestion variation. On the left is a simple list. On the right, the most likely match is at the top of the list, and others are in a scrolling area below.*

Auto-completion, predictive entry, and related technologies such as spelling correction have become well-established, highly expected assistive technologies for day-to-day computing. For mobile devices, they have proven to be especially valuable due to their relative difficulty and reduced speed of text entry and especially for complex, technical character entry such as for URLs.

Though many current high-end devices either disregard these features or do not consistently implement them, the same pressures apply to any mobile device. At least one method of assistive entry should be employed universally, across the entire device. If it is not available on the targeted OS, consider if adding this feature is possible and if it would assist or distract the user due to the different interface, and determine the best method for the type of entry taking place.

The actual functionality of any of these features, such as the algorithms used to perform the prediction, is outside the scope of this pattern, but is also crucial to its good operation.

Variations

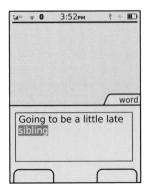

Figure 9-21. *Completion enters the word directly in place, as with this predictive text example.*

A range of related functions, associated with all types of text entry, can be considered under the Autocomplete & Prediction pattern:

Completion
> Entry is automatically completed or modified with the goal to create valid entry with as few keystrokes as possible. This is common in search or URL history, when a list of recent or common results is presented during typing. See Figure 9-21.

Suggestion
> One or several suggestions are displayed for an entry. An action is available to change to the suggestion, often by picking from a list. This is encountered in triple-tap text entry (a Next key is provided to scroll through options). See Figure 9-20.

Replacement

A word or phrase which is recognized as likely to be a misspelling of an entry in a dictionary will be replaced with the correct version shortly after user entry has been completed. No options are offered.

Error notation

A word or phrase which is not recognized as an item in the dictionary or does not meet the input criteria (e.g., a URL without a TLD) is indicated as such inline. The most common is underlining misspelled words in free-form text entry. Alternatives may be offered after manual selection (at which point it switches to being function-ally a "suggestion" variation), but no automatic action is taken by the system. See Figure 9-22.

In some cases, multiple variations may be encountered at once. This is especially common when an entry is completed in the field with the best match, but suggestions are offered adjacent to the field.

Coupled with this variety of functions are a number of interactive methods. This pattern outlines some of the more fixed points and provides general guidelines, but it cannot be prescriptive about all aspects due to the degree of variation.

Interaction details

Figure 9-22. *Error notation, such as this miskeying of a triple-tap entry, simply indicates suspect words without changing them.*

The suggestion variation requires explicit acceptance by the user. A selection must be made from the list of candidates for any to be applied. No selection will simply use the characters as typed.

Multiple candidate matches should be offered whenever possible. Allow the user to scroll through them, usually by presenting them as a drop-down-style list, and by using On-screen Gestures or with the up and down Directional Entry keys. In some cases, and most notably with multitap text entry methods, a key is dedicated (while in the mode) to displaying the next available candidate.

Lists of candidates should often be circular, so scrolling past the end will load the first item again. This is especially critical when a single key is used to display the next option. For certain results sets, such as predictive search or URL completion, the results may be too large for this, or a circular list may imply too few results found. For these, use some principles of the Infinite List to present the arbitrarily large, remotely fetched data. Remember, it's still a Vertical List, even if a very small one.

For lists, a method to select the candidate in focus must be made available. Avoid using methods that have other meanings when a selection is not being made. For example, if OK/Enter is used to submit the form, it may be confusing or lead to errors if it is used for selection of suggested words also.

"Completion" is an implicit method, and any exit of the word, phrase, or field will commit the last candidate displayed. This includes simply pressing the space bar, period key, or comma key at the end of a word, or submitting the entry to a field. On the other hand, if the user keeps typing, his character entry will be used instead, and the completion disregarded or switched to the next best match.

Many of these methods may be used in concert. Figure 9-23 shows one example.

Presentation details

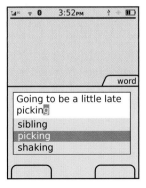

Figure 9-23. *The variations may also be combined. In this example, words are completed inline with the best match, and options are shown in a list below much like the suggestion variation. Scrolling is used to select options instead of the "next" key of a conventional word input completion variation.*

All automatic behaviors, even inline items such as replacement, must take a brief time to take effect, and have a visual cue to attract the user to the behavior. Many of these behaviors, although helpful, are also destructive and may change the intent of the user entry. You must be careful to inform the user of the change.

Selectable completion text may appear in one of two basic places:

In place

Text is automatically entered and modified in the actual field where entry occurs. This is almost always used with multitap keypad text entry, but is also the way that auto-correction works.

Adjacent

The suggestion appears adjacent to the word, either right below the entered text or within the input panel itself. Suggestions for search, and matching candidates for Pen Input systems, are typical examples.

A rare variation is to complete "in place," but in a field, or location in the current field outside the current focus or cursor location. This is best exemplified by code completion, but can be encountered in certain types of form filling. For example, an appointment creator with start, end, and length-of-event fields will find the three closely coupled; completing any two will automatically fill the third. This may seem to be a simple, one-off interaction trick, but it must follow the principles outlined in this pattern to ensure that the user observes the behavior and gets the benefit from the auto-entry.

In place display should be highlighted while still a candidate, such as for multitap text entry. The entire word is still being changed by the user's character entry, so this highlighting is a sort of extension of the text insertion point.

Candidate options displayed adjacent are typically offered as multiple selections. One item is always in focus, but others may be made available. These may be presented as a list, which can be scrolled through, or only one may be visible as a "bubble" (much like a Tooltip, or Annotation if directly selectable) before selection is made.

Error notifications, such as spellchecks, should appear as a dashed or dotted underline in a very contrasting color. Red has become the typical color, but you should avoid this if the background, text, or overall theme uses red routinely, and especially if it is used as a link color.

Antipatterns

Avoid automatic replacement without a method for the user to opt out or disable the feature, either session- or device-wide.

Do not violate the principles of Cancel Protection and allow destructive automatic behaviors without a method to revert or correct the change. For example, whenever automatic replacement is used, be sure to allow a method to revert to the spelling as typed. This may be as simple as automatically disabling auto-correction for any second entry of the same word in the same field (or location in the field), so the user may simply rekey the phrase as intended. This is suboptimal as it requires additional typing, but it is better than rejecting the second entry as well, and frustrating the user further.

Chapter 10

General Interactive Controls

Darkness

It's pitch-black outside. The air is cold and wet, yet it carries a lingering sweet smell. Sporadic beams of light dance in the night, casting an eerie glow on the landscape. Giggles, whispers, and even the occasional scream carry through the streets, a reminder that others are out and about.

Through the eyes of one of these figures, a house is seen. The figure changes course and heads in the direction of the house. The house is unlit and looks unoccupied. In one hand the figure holds a large sack; the other yields a blunt sword.

As the figure makes his way up the porch and to the door, the hand that is holding the sword points forward. The hand is not a human's hand. It's about twice as big as a man's hand. Coarse, dark fur covers its skin, while jagged claws extend from the aged fingers.

The creature now stands directly in front of the door, its purpose clear. It only wants one thing and that thing remains inside the house. With a blast of energy the hand with the sword raises, lunges, and slams into the house.

A chime echoes. The front door opens. The man who opens the door smiles happily while looking down, hardly frightened by the four-foot tall, hairy monster screaming "Trick or treat!"

That Sounds Like a Great Idea

Take a moment to catch your breath and slow your heartbeat. Despite my enjoyment of Halloween, it's not my focus for the remainder of this chapter. However, the doorbell that so frequently sounds during that annual holiday is.

The doorbell is an outstanding example of an effective interactive control. If a 10-year-old dressed as a monster with oversize latex hands can use it effortlessly in the dark, and anyone listening in the house (even the family dog) instantly understands what it means, it must work well!

Let's examine why the doorbell is an effective control, using Donald Norman's Interaction Model.

Make It Visible

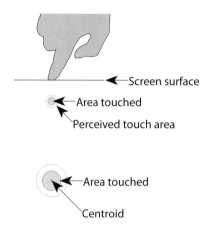

Figure 10-1. *Touch operates over an area, but defines the contact as a point. Understand how touch works to use it correctly.*

Any control needs to be visible when an action or state change requires its presence. The doorbell is an example of an "always present" control. There aren't that many components on the door anyway (doorknob, lock, windows and peepholes, maybe a knocker), so the doorbell is easy to locate. In many cases the doorbell is illuminated, making it visible and enticing when lighting conditions are poor.

Cultural norms and prior experiences have developed a mental model in which we expect the doorbell to be placed in a specific location—within sight, within easy reach, and to the side of the door. So our scary monster was quickly able to detect the location of the doorbell from even his relatively more limited prior knowledge, as well as see it in the dark.

Having an object visible doesn't have to mean it can be seen. It can also mean the object is detected. Consider someone who is visually impaired. She still has the prior knowledge that the doorbell is located on the side of the door, at eye level and within reach. Its shape is uniquely tactile, making it easy to detect when fingers or hands make contact with it. Most doorbells are in notably raised housing, and although it's not clear if this is an artifact of engineering, it serves to make the doorbell easy to find and identify by touch as well.

In mobile devices this is a very important principle. Many times we don't have the opportunity to look at the display for a button on the screen, but we can feel the different hardware keys. Consider people playing video games. Their attention is on the TV, not the device controller, yet they are easily able to push the correct buttons during game play.

Mapping

The term *mapping* within interactive contexts describes the relationship established between two objects, and implies how well users understand their connection. This relates to the mental model a user builds of the control and its expected outcome. When we see the doorbell, we have learned that when we push it, it will sound a chime that can be heard from within the house or building to notify the person inside that someone is waiting outside by the door.

Our 10-year-old monster mapped that pushing and sounding of the doorbell will both notify the people inside that a trick-or-treater is there and that he will receive a handful of free candy.

So mapping relates heavily on context. If the boy pushes the doorbell on a night other than Halloween, the outcome will differ. The sounding chime will remain, which indicates a person is waiting outside, but the likelihood of someone being there to answer and, what's more, to hand out candy, is much less great.

On a mobile device, controls that resemble our cultural standards are going to be well understood. For example, let's relate volume with a control. In the context of a phone call, pushing the volume control is expected to either increase or decrease the volume level. However, if the context changes, for example, on the Idle screen, that button may provide additional functionality, bringing up a modal pop up to control screen brightness and volume levels. Just like call volume, pushing up will still perform an increase and pushing down will perform a decrease in those levels.

But you must adhere to common mapping principles related to your user's understanding of control display compatibility. On the iPhone, in order to take a screenshot, you must hold the power button at the same time as the home button. This type of interaction is very confusing, is impossible to discover unless you read the manual (or otherwise look it up, or are told), and is hard to remember as the controls have no relation to the functional result.

On-screen and kinesthetic gestures can be problematic too if the action isn't related to the type of reaction expected. Use natural body movements that mimic the way the device should act. Do not use arbitrary or uncommon gestures.

Affordances

Affordances describe that an object's function can be understood based on its properties. The doorbell extends outward, can be round or rectangular, and has a target touch size large enough for a finger to push. Its characteristics afford contact and pushing.

On mobile devices, physical keys that extend outward or are recessed inward afford pushing, rotating, or sliding. Keys that are grouped in proximity afford a common relationship, many times a polar functionality. For example, four-way keys afford directional scrolling while the center key affords selection.

Yes, affordances are just like other understandings of interactive controls and are learned. Very few of these are truly *intuitive* (meaning innately understandable) despite the common use of the term, and may vary by the user's background, including regional or cultural differences.

Provide Constraints

Figure 10-2. *Pen tablets are reasonably common in industry, for sales and checkout (as shown here), and for various types of data gathering or interchange. The interaction methods allow the user to remain standing, or to use the device in situations where a laptop would take too much effort or attention. Touch is already making some headway into these markets, and gesture interfaces may be even more suitable for some fields that are dirty and/or dangerous.*

Restrictions on behavior can be both natural and cultural. They can be both positive and negative, and they can prevent undesired results such as loss of data, or unnecessary state changes.

Our doorbell could only be pushed and not pulled (in fact, a style of older doorbell did require pulling a significant distance). The distance which the button could be offset is restricted by the mechanics of the device; it cannot be moved in any direction except along the z-axis.

Despite the small surface and touch size, the button can still be pressed down with a finger, an entire hand, or any object, even one much larger than the button's surface. In this context, that lack of constraint was beneficial. A user doesn't have to be entirely accurate using the tip of the finger to access the device (see Figure 10-1). Our trick-or-treater, with huge latex hands holding a sword, was still able to push the button down, allowing for a quick interaction. More commonly, the doorbell can be pushed by someone wearing mittens or by an elbow if the person is carrying groceries.

On mobile devices, however, it's often necessary to define constraints on general interactive controls. Some constraints that should be considered involve:

Contact pressure
> This means having a threshold of contact pressure that initiates that the control should be used to prevent accidental input. Touchscreens and similar controls without perceptible movement will have a lower threshold, as opposed to hardware keys which move to make contact.

Time of contact
> Similar to contact pressure, requiring that keys remain pressed for longer periods may provide a useful constraint to accessing buttons that can lose user-generated data. Pushbutton power keys, for example, do not immediately turn devices on or off but must be held for a few seconds.

Axis of control
> The amount of directional movement can be restricted to a single axis or a particular angle or movement. Gestural interactive controls must use appropriate constraints so that they are not activated out of context, even though the user's finger is not constrained itself. Unlocking an Idle screen may require a one-axis drag to access additional features on the device. This is not different for hardware keys; the user may "nudge" controls by using scroll controls at an angle. Only the input hardware or sensing may be constrained.

Size of control
> The ability to interact with a button can heavily depend on its physical size and its proximity to other buttons or hardware elements. With on-screen targets, you must carefully consider this when designing UIs that have these interactive controls. Small touch targets must have larger touch areas, and must not be too close to a raised bezel. Consider as well the use of other sensors, and how Kinesthetic Gestures can change the effective control size, such as for the card reader in Figure 10-2. Refer to the section "General Touch Interaction Guidelines" in Appendix D.

Use Feedback

Feedback describes the immediate perceived result of an interaction. It confirms that action took place and presents us with more information. Without feedback, the user may believe the action never took place, leading to frustration and repetitive input attempts. The pushed doorbell provides immediate audio feedback that can be heard from outside as well as inside the house. Illuminated doorbells also darken while being pressed, providing additional feedback to the user as well as confirmation in loud environments, for more soundproof houses, or for users with auditory deficits.

Use feedback properly. Unlike the doorbell's immediate indicator of an action, an elevator call or floor button (both work the same, as they should) remains illuminated as a request-for-service, until it has been fulfilled. As an antipattern, crosswalk request buttons generally provide no feedback, or beep and illuminate as pressed. There is no assurance that the request was received, or is properly queued.

On mobile devices, when we click, select an object, or move the device we expect an immediate response. With general interactive controls, feedback is experienced in multiple ways. A single object or entire image may change shape, size, orientation, color, or position. Devices that use accelerometers provide immediate feedback showing page flips, rotations, expansion, and sliding.

Gestural Interactive Controls

A growing number of devices today are using gestural interactive controls as the primary input method. We can expect smartphones, tablets, and game systems to have some level of these types of controls.

Gestural interfaces have a unique set of guidelines that other interactive controls need not follow:

- One of the most important rules of gestural controls is that they need to resemble natural human behavior. This means the device behavior must match the type of gestural behavior the human is carrying out.

- Simple tasks should have simple gestures. A great colleague and friend of mine who is visually impaired was discussing her opinions on gestural devices. She provided me an example of how gestures actually make her interactions quite useful. If she types an entire page, and finds later that she is unhappy with what she wrote, she can shake her device to erase the entire page. A simple shake won't activate this feature (constraint), so she is confident she won't lose her data accidentally. But rather than her having to delete the text line by line, this simple gesture quickly allows her to perform an easy task.

Dan Saffer (Saffer 2009) points out five reasons to use interactive gestures:

More natural interactions
   People naturally interact with and manipulate physical objects.

Less cumbersome or visible hardware
   Mobile devices are everywhere: in our pockets and hands, on tables and storefront walls, and in kiosks (see Figure 10-3). Gestural controls don't rely on physical components such as keyboards and mice to manipulate the device.

More flexibility
   Using sensors that can detect our body movements removes our hand-eye dependence and coordination normally required on small mobile screens.

More nuance

> A lot of human gestures are related to subtle emotional forms of communication, such as winking, smiling, and rolling our eyes. The nuance gestures have yet to be fully explored in today's devices, leaving an area of opportunity in user experience.

More fun

> Today's gesture-based games encourage full-body movement. Not only is this fun, but it also provokes a fully engaging social context.

Figure 10-3. *Working or playing in groups encourages other types of interactions, such as the use of remotes and gestural interfaces.*

Patterns for General Interactive Controls

The patterns in this chapter describe how you can use general interactive controls to initiate various forms of interaction on mobile devices:

Directional Entry

> With controls used to select and otherwise interact with items on the screen, a regular, predictable method of input must be made available. All mobile interactive devices use lists and other paradigms that require indicating position within the viewport.

Press-and-Hold

> This mode switch selection function can be used to initiate an alternative interaction.

Focus & Cursors

The position of input behaviors must be clearly communicated to the user. Within the screen, inputs may often occur at any number of locations, and especially for text entry the current insertion point must be clearly communicated at all times.

Other Hardware Keys

Functions on the device and in the interface are controlled by a series of keys arrayed around the periphery of the device. Users must be able to understand, learn, and control their behavior.

Accesskeys

These provide one-click access to functions and features of the handset, application, or site for any device with a hardware keyboard or keypad.

Dialer

Numeric entry for the dialer application or mode to access the voice network varies from other entry methods, and has developed common methods of operation that users are accustomed to.

On-screen Gestures

Instead of physical buttons and other input devices mapped to interactions, these allow the user to directly interact with on-screen objects and controls.

Kinesthetic Gestures

Instead of physical buttons and other input devices mapped to interactions, these allow the user to directly interact with on-screen objects and controls using body movement.

Remote Gestures

A handheld remote device, or the user alone, is the best, only, or most immediate method to communicate with another, nearby device with a display.

Directional Entry

Problem

To select and otherwise interact with items on the screen, you must provide a regular, predictable method of input.

The hardware for Directional Entry is inherent to the device, but it must be well understood to ensure that you are building interfaces and interactions that take best advantage of the type of control and pointing available. Discrete clicks are very different from continuous scrolling functions.

Figure 10-4. *One-axis buttons may be paired or appear as a single rocker, though the function is the same. A now-uncommon variation is the roller or scroll wheel, this one only shown here in profile for clarity. These often used their larger range of motion to accept multiple speed inputs or a range of speeds.*

All mobile interactive devices use lists and other paradigms that require indicating position within the viewport. Two separate solutions work hand in hand to meet these needs: the hardware input devices themselves and the behaviors that govern this Directional Entry.

Although the two are related and one cannot work without the other, there is not an absolute relationship between their capabilities. For among the most extreme and most common examples, touchscreens can be restricted to a single axis of movement on a particular device, application, screen, or mode.

Hardware for input can be broadly categorized as:

Paired buttons
    Up/down or left/right pairs, often repurposed volume controls which may also be used for scrolling in certain modes.

Four buttons
    The very common four-way or five-way rocker control. These are four discrete inputs: up, down, left, and right. Even if shown as a contiguous circle, there are four micro-switches, and they cannot be combined to yield diagonal movement or change direction.

Arbitrary input controls
    Usually trackpads or trackballs (both of which may be much smaller than a user's finger), but also input pads that look like five-way panels. These all accept input on any angle, and should allow the user to change the angle and speed during the movement.

Direct entry controls
    Touch and pen entry methods directly on the screen. Just like arbitrary inputs, these accept input at any angle, can change speed and angle during movement, and may initiate movement from any position on the screen. The only difference is that they

have a "direct" 1:1 correlation with the screen, as they are overlaid. Note that this covers only direct and directional entry by touchscreens. On-screen Gestures are discussed separately.

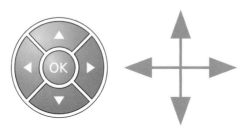

Figure 10-5. *A typical two-axis hardware controller is often built as a five-way pad with four directional keys and the OK/Enter key in the middle. Only discrete directional movements are possible, not diagonals.*

You can consider Directional Entry to be your best option if it is itself broken down into at least three types of interaction:

One-axis
> The interface (or the portion in focus) only accepts movement in one axis, usually up/down or left/right. See Figure 10-4.

Two-axis
> The interface only accepts movement along the up/down and left/right axes. Most scroll-and-select devices are like this, though some can accept diagonal inputs as well. See Figure 10-5.

Planar
> Input of any angle within the viewport may be accepted. See Figure 10-6.

These approximately correspond to the hardware types, if arbitrary input and direct entry are grouped together. Each of these can be applied to essentially any input hardware used. Although a touchscreen is clearly "planar" at a device level, specific screens or modes may limit you to one-axis movement.

No universally applicable three-dimensional hardware or interactive control exists. Certain gestures, discussed later, do account for movement in three dimensions, but they do not involve direct user control of a cursor in three dimensions.

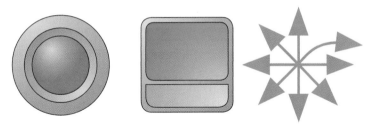

Figure 10-6. *Aside from direct input mechanisms such as touch and pen, planar input methods use hardware such as trackballs and touchpads. These may be limited to two-axis input, but they should not be. Any direction may be input, and the input direction may change during the course of a single movement. An OK/Enter function is often built into the trackball (press to enter) and may be built into the trackpad, or must be immediately adjacent to it.*

Interaction for directional control is integral to the bulk of the patterns in this book, and of course to almost every electronic device you use on a daily basis or might design for. See especially the Focus & Cursors and Scroll patterns.

When a single axis of entry is allowed, such as scrolling a vertical list, consolidate the inputs from the other axis into the active axis. Up and right are generally paired, as are down and left. For arbitrary input and direct input methods, all other angles are consolidated to the nearest active axis. A lockout range may be implemented, so inputs close to the opposite axis will be disregarded. Without this, dragging directly sideways on a vertical list could cause the scroll to jump between up and down as the angle changes slightly.

If an item is locked to a single axis of movement, make sure input movement starts with the item selected. However, once movement has begun in the correct axis, off-axis input is irrelevant. For example, if the user is vertically scrolling an item with a touchscreen, only the vertical component of input is accepted once the drag has begun; when the horizontal component inevitably drifts to the side, this has no effect on the drag action.

Many input variations are functionally restricted from use by certain types of hardware. However, there are workarounds. A key one is to extend off-axis consolidation to allow scrolling perpendicular to the two-way keys. Vertically oriented keys will scroll sideways when that is the only available scroll action.

Though cumbersome, two-axis scrolling may be implemented with only a one-axis control by use of a modifier key. This is not a mode switch and must be held down to create a key combination. Directional pairing is the same as for off-axis consolidation.

Speed and acceleration are important to build into your Directional Entry design; they are discussed in the Scroll and Focus & Cursors patterns.

In general, this pattern is about the input method of Directional Entry. Presentation details for directional entry on-screen are best discussed within the context where they are used, such as the various display elements on the page. Refer especially to the Focus & Cursors and Scroll patterns.

Some trackpads and all touch and pen devices also vary from all other methods in that they are *absolute* pointing devices, with no inherent sense of a cursor location when the user is not implementing the device. Wherever the user provides input is the starting point. For all others, there is a distinct sense of focus, and movement always begins from that point.

Relative input usually demands that a cursor be visible at all times. Absolute pointing almost requires that no cursor be visible, though there are exceptions for when the pointing device is in use, as described in Pen Input. See Focus & Cursors for a general discussion of communicating cursor locations.

Antipatterns

Make sure that reaction to scroll is always immediate. If the content cannot be loaded as quickly as the scrolling occurs, you may use a Wait Indicator for the various components on the page, but never delay the scrolling behavior itself.

Avoid needless extra axes of movement. Lists, for example, should function in a single axis only, regardless of the capabilities of the input device.

Press-and-Hold

Problem

In certain contexts, you must provide an alternative function centered on the cursor or focus area. A simple, universal method of applying this mode switch should be provided.

Press-and-Hold is built into many OSes as a standard method for some interactions. It can generally be detected, even by web pages, and so can be built into almost any interaction if desired.

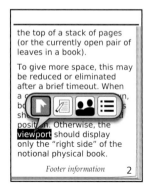

the top of a stack of pages
(or the currently open pair of
leaves in a book).

To give more space, this may
be reduced or eliminated
after a brief timeout. When
a ... n,
b... s
sh... ...
posit... . Otherwise, the
viewport should display
only the "right side" of the
notional physical book.

Footer information   2

Figure 10-7. *This contextual menu appears based on selection of a piece of text, then the user pressing and holding the OK/Enter key. A small menu appears, and the first item is in focus. Selecting an item will commit that action on the selected text.*

No matter how cool, gestural, and interactive your new mobile application is, most of the interactions are still relatively simple, and are about pointing and selection. In some situations you will find an alternative selection method to be useful. In many cases a mode switch on the keypad or in an on-screen menu is unsuitable due to space or in order to preserve the contextual function of the pointing and selection behavior.

In the desktop computing world you might be familiar with the *right-click* on the mouse, by the use of buttons on the pen, or by using keyboard Mode Switches to change the pointer effects.

In mobile, none of these are really very suitable. There is insufficient space on the input area for alternative inputs, and touch devices cannot provide a secondary button on a user's fingers. Meta keys work poorly in all platforms due to the need to educate users, but they are particularly unsuitable on mobiles for ergonomic reasons; it's easy to drop the phone while messing with keys at the same time as tapping the screen.

If conventional click actions take effect on *mouse up* (on release of the button or removal of the finger from the screen), you can use Press-and-Hold of the selection mechanism to initiate an alternative interaction. Within this pattern, we will use the web conventions of mouse up and mouse down to refer to these conditions regardless of the pointing device.

By far the most common use of Press-and-Hold behaviors is to present a contextual menu of options, or other contextually relevant optional selections. For example, pressing and holding the keys for virtual Keyboards & Keypads offers a list of special characters such as accented versions that cannot be easily accessed via Mode Switches.

The same behavior may be used for other functions such as a temporary zoom to aid in selection of a text insertion point, placing a pinpoint on a map, or options on selected text, as shown in Figure 10-7. These behaviors are generally either immediate actions (although a confirmation or undo may be provided) or function only while the mouse is down. These behaviors are not entirely settled, so the remainder of this pattern will discuss only the functionality and appearance of the contextual selection variation.

Interaction details

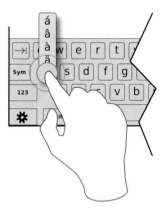

Figure 10-8. *Pressing and holding a character on this keyboard presents a list of optional accented characters. Dragging over the list and then releasing will select and type the character on the screen. The "Sym" modifier allows entry of common symbols; this is a good way to offer even more characters without adding modes or entering the settings and switching languages.*

Press-and-Hold works equally well with any selection hardware. Buttons, pens, and fingers all may press and hold equally well.

The length of time needed to hold before the action occurs should be about one-half of a second. You can vary this greatly by testing how users interact with the interface. The menu or other function must appear immediately after this timing is reached, and with absolute consistency in all locations.

For direct entry methods especially, you have to allow for a certain amount of jitter (or inadvertent movement by the user), instead of detecting the mouse-down time as all in-motion. This should be around 10% to 20% of the minimum target size for touch devices.

Your menu may work in one of two modes:

- The mouse is kept down while scrolling to the item to be selected. Selection is made by mouse up when the desired selection is in focus, at which point the action occurs and the menu is dismissed. Releasing the mouse without moving after the menu appears, or via mouse up outside the menu area entirely, will dismiss the menu without selection. This is typical of virtual keyboards, as in Figure 10-8.

- Once the menu appears, the mouse is released and the menu remains in place as a semimodal dialog. The user performs conventional scroll or point and selection activities. Once a selection is made, it is committed and the dialog is dismissed. If no selection is to be made, the user may select any area outside the menu; this first selection will not activate any items, and just dismisses the dialog. The use of a close function is acceptable, but is not suggested due to space concerns and the additional risk of accidental activation on most devices.

For semimodal dialogs, actions outside the menu will dismiss the dialog but will not perform any other action, either from within the menu or on the parent screen.

The decision as to which to use should be based on the manner in which the remainder of the OS functions generally. Avoid mixing the two methods within a single interface.

Menus should follow the patterns already established, and will usually appear similar to any other second-tier menu, as described in Fixed Menu and Revealable Menu. They may follow the principles of the Annotation instead, and offer selectable buttons or icons, instead of a Vertical List.

Presentation details

Before the hold time has been reached and the alternative action occurs, show that Press-and-Hold is about to take place. The timing when this appears is variable, but about halfway through the time is a reasonable starting point. The indication must be visible past any on-screen indicators. If you are designing for a touchscreen, it must be larger than the finger, or presented adjacent to it.

This indicator should indicate progress to count down to the activation time. A common symbol is to animate the drawing of a circle around the current in-focus item or the centroid of the contact area. This is timed so that the context menu appears as the circle is completed. Even though this countdown timer appears halfway through the true time, consider the moment it appears as the start of the clock.

Make the timing indicator disappear as soon as the context menu appears.

Menus resulting from Press-and-Hold behaviors must appear adjacent to the in-focus item or the contact point. Ideally, the menu will emerge directly from the option selected, such as an Annotation. If there is no distinct item on selection, you should usually also create an object from which it may emerge.

The menu must visually be clearly separate from the rest of the interface so that it cannot be confused with any objects in the parent, no matter how complex or interactive they appear. Use border and shadow effects to indicate that the menu is associated with the pointing device, and not just the screen. Typically, these will appear to be floating above the interface.

Figure 10-9. *This dialog is too complex for a Press-and-Hold menu. If data must be entered, launch it from a location that is more robust and integral to the application.*

Avoid overly complex dialogs offered via Press-and-Hold behaviors, as shown in Figure 10-9. Stick with individual selectable items, and avoid complex selectors, text entry, and submit buttons. If such dialogs are needed, present them via a Revealable Menu, Fixed Menu, or link within an Annotation.

Do not absolutely rely on use of a Press-and-Hold behavior. As the function has no affordance, users may not discover or remember it. Attempt to instruct users in some manner as to this functionality, or provide other paths to presenting this same information. Those methods may go ahead and use the same display method (of a layered menu) to try to instruct users that there is contextually relevant functionality they are missing.

In addition, ensure that this function works in all included input methods. If a touch device also has hardware Directional Entry keys, make sure the function works as well for both input methods.

Use these behaviors consistently, and as pervasively through the device UI as possible. Once learned, if it fails irregularly, this will violate the user's mental model and may lead to general disuse of the feature.

Focus & Cursors

You must always clearly communicate the position of current and future input behaviors to the user. Within the screen, inputs may often occur at any number of locations, and especially for text entry the current insertion point must be clearly communicated at all times.

Focus & Cursors are built into every OS, but there are many opportunities to use them in-appropriately. Be sure to use the correct methods, and to actually design easily perceived indicator states in your application or website.

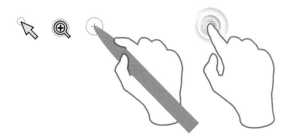

Figure 10-10. *Some examples of the variety of position cursors. In general, the simple pointer is most useful with the indicated space at the tip of the arrow. Very few state changes are helpful, but magnifiers are one that can work well. Pens should have high-precision pointers. Certain other interactions can use subtle touch indications, such as the splash shown here, to indicate that contact has been made.*

Generally, you must allow the user to only immediately interact with one element of the screen at a time. Areas and items for which immediate interaction is possible are consid-ered *in focus*. When an exact point is in use within this area, this is indicated by a cursor.

Expressions of Focus & Cursors serve to indicate the area and exact position so that the user is aware that his Directional Entry took effect correctly, and so he may take appropri-ate actions based on the item to be affected.

You can use a similar indicator to indicate certain types of connections. This can be seen with charge indicator LEDs that are adjacent to the port, and should be more common with near-field communications such as mobile payment schemes. A light, icon change, or other effect either on-screen or not occurs at or adjacent to the point of communica-tion to indicate successful linking. This overlaps heavily with the LED pattern; more clar-ity should emerge as these are implemented more often.

The most basic type of focus indicator or cursor simply indicates position. The specificity with which this is communicated can vary widely. Focus is general, and area- or item-level, while a cursor indicates down to the pixel level.

Cursors always indicate position, but they may also indicate the current state, as described in the Input Method Indicator pattern.

Though these two indicators are similar, users perceive them as being different, so they may be inconsistently employed, with position-only focus indication and state-indicating cursors.

Pointing cursors indicate position in two dimensions. See Figure 10-10. Text entry cursors are almost a mix of focus and pointing; precision on the vertical axis is only to the line level, whereas in the horizontal it is to the pixel.

Although scrolling input devices of all sorts are *relative*, and all input starts from the co-ordinates of the last input, most of the purely touch and pen devices are different. They are instead *absolute* pointing devices. When the finger or pen is not in contact, there is no focus point or area. Any point may be selected and then immediately becomes the cursor point, and the area it is in is within focus. There are certain exceptions, such as modal dialogs, which may restrict this "arbitrary" selection to specific areas.

Figure 10-11. *Text cursors should be made of the thinnest visible lines, with caps—making it an "I-beam" style—to make it more visible. Entry mode indicators may be either vague, such as the diamond in the center, or specific, such as the lowercase "a" on the right.*

Focus indicators and cursors are purely display elements. They react to changes made as a result of general interaction, but are not themselves interactive elements.

Devices that have both touchscreens and other Directional Entry methods such as a five-way pad will need to support focus and cursor pointing at all times. Even they are if not displayed, the system must have an internally declared location for the focus and cursor at all times.

One exception to the relative pointing discussed here is that you will usually impose a cursor position on the interface when a page change is implemented. This is unlike a desktop system, where the mouse pointer generally does not move about the page without user input.

The focus model, on the other hand, is very similar to the desktop application or website. New applications or pages automatically seize focus, and you may control this to the point of placing focus within a specific scrolling area or text field, allowing entry without any use of a Directional Entry device.

For a related discussion of state changes based on position cursors, and interactivity with these changes, see the Press-and-Hold pattern.

Focus indicators and cursors must always be contextual, and not just reside within the interface but work with the existing design.

You should almost always represent focus by changing either the background or the border of the element. For the background, use a color or saturation change, with the item in focus getting the highest contrast state. Sometimes additional effects are useful, and only the item in focus has gradients to make it appear dimensional or shinier. Borders may change color, saturation, or weight. Both borders and backgrounds may be changed together, of course.

You can use additional effects, such as shadows, but be careful not to make the focus element float off the page, or it may look like a modal dialog instead. Any number of other variations may also be used, but they should always be in conjunction with borders or backgrounds.

Don't forget to take advantage of other elements or widgets. State indicators may also be used to emphasize that the element is in focus. If the Input Method Indicator is attached only to the side of the field in focus (and no other fields), this can help clarify the focus, and attract attention to the field or area. See Figure 10-11. Understanding how the page will actually be presented may save you from having to work too hard to express focus.

State indicators, aside from indicating focus, are often attached to focus areas or cursors to provide additional, highly contextual information about the current state.

Cursors must be able to indicate to pixel-level precision. The position pointed at is usually the pixel immediately under or adjacent to the tip of the pointer; avoid obscuring user inputs with the cursor. Text entry cursors are derived directly from typewriters (and Linotype machines, etc.); the next character entered will begin to be entered at the current cursor location, and extend to the right. The cursor will then move to the right side of the character just entered.

Make sure cursors are visible, regardless of the background. Use shadows and/or contrasting borders to ensure this. Cursors are typically white, with a black hairline border. Some text cursors for controlled areas such as form fields can be solid hairlines, as long as you make sure the field background is a solid, contrasting color. Use the I-beam shape to make the cursor easier to pick out from the surroundings, with the crossbars above any ascenders and below any descenders. Text cursors must never fall outside the bounds of the text entry fields.

Scroll-and-select devices, with only a five-way pad and no other pointing device, often have no cursor pointer. These rely on line-by-line or other item-level focus, as shown in Figure 10-12. Naturally, the cursor exists during text entry as a subset of the field in focus.

Figure 10-12. *Focus denotes areas of the page visible within the viewport that are currently able to accept input. These are designated with borders, as in the example on the left, or with highlights, as in the button and line item in the examples in the middle and on the right. Tiers of focus may also be presented, as in the left two examples where the modal dialog and individual fields are both in focus. On the right is a special case where two areas are in simultaneous focus, one for text entry and the other for selection.*

Text cursors are at the insertion point for character entry, and they highlight the affected area much like a focus area, when in predictive text modes. State changes should be depicted to indicate when a mode switch such as caps lock has been activated. These indicators may be vague and require learning such as adding to or modifying the shape of the cursor, or may be specific and indicate the mode switch symbolically.

Arbitrary modifications to the cursor may be sufficient to indicate a state change when an Input Mode Indicator is displayed elsewhere, but otherwise you should try to make it understandable by itself or at least have a related visual language. This minor change, especially if some user-deciphering is required, may seem to not be worth the effort, especially when another indicator may already exist. However, users are often very focused on the cursor and areas immediately around it, and so are very likely to see changes in this which they may not otherwise become aware of immediately.

Position cursors should always be used for pen input devices. Due to the precision of the pointer, small parallax changes may induce errors, and presenting a cursor will help to avoid this.

Position cursors are rarely used persistently for touch, but they may sometimes appear to indicate that single contacts have occurred. This can be especially helpful to ensure that the device has accepted the input, such as in sleep states, or where the device may otherwise not respond immediately but a Wait Indicator would be unsuitable or intrusive.

Devices that have both touchscreens and other Directional Entry methods such as a five-way pad may choose to not show the cursor when in touch or pen mode. However, as soon as a directional key is used, the focus area and cursor (if applicable) must appear. This should generally not require a key press to "activate" the mode; the first key press will move the focus or cursor in the direction selected.

Avoid expressing focus only by pointing to the area or field, such as with an arrow in the left column. Although such indicators may help to guide step-by-step processes, they are insufficiently contextual to be the only method of indicating focus.

Avoid under-indicating. Focus and cursors work hand in hand. Make the focus area clear, to draw the user's eye to the cursor or text insertion point, even for touch or pen devices which may seem to not need traditional focus indication.

Do not follow desktop cursor examples strictly. For example, avoid the use of the "hour-glass" pointer mode, and use the more mobile-appropriate Wait Indicator instead.

Other Hardware Keys

Problem

Functions on the device and in the interface are controlled by a series of keys arrayed around the periphery of the device. Users must be able to understand, learn, and control their behavior.

Hardware is part of the device, but the control can often be overridden, modified, or simply ignored by individual applications or even web pages.

Solution

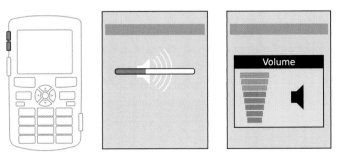

Figure 10-13. *On-screen actions, such as these common volume overlay widgets, should be loaded as soon as a hardware action is performed, unless something else that is obvious and immediate occurs, such as launching an application.*

Setting aside Keyboards & Keypads and Directional Entry controls, practically all mobile devices have numerous additional keys. Even the increasingly button-free devices generally have volume rockers, lock keys, power buttons, rotation control buttons, and so forth.

Regardless of the number of keys or their individual functions, all of these must comply with some basic behavioral standards in order to be useful, usable, and valuable to the user.

There are two basic types of hardware keys, based on their effect:

Hardware
> The effect is to the device hardware functions. Any on-screen interactive display is incidental to the function. These are functions such as volume and radio toggles.

Interaction
> The primary effect is only to interactive, software-driven elements, generally resulting in a principle, unmistakable interaction. Back, search, and camera keys are representative of this type.

In either case, the attributes and behaviors discussed in this pattern are applied in the same manner.

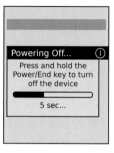

Figure 10-14. *Optional functions activated by Press-and-Hold require either an explicit Exit Guard or presentation of an options menu. Note that both of these use the symbol for the power key to denote how the menu was activated. The righthand one is also pointing at the location of the key on the device.*

You must make sure the use of any hardware key has an immediate, noticeable response, regardless of the type of function being performed. This applies even if the action takes some time to take effect, such as time to load the camera application.

There should always be a visible effect. Generally, this should be an on-screen display of some sort. If the primary effect either is not an interactive element or cannot be launched within about one-tenth of a second, a secondary display must be used. Though many options are available, some frequently used variations are as follows:

- Use a Pop-Up dialog, as in Figure 10-13. Sometimes a variation is for display only, has no opaque bounding box or background field, and cannot be interacted with.

- Use an Annunciator Row icon change, or other Notifications indication. An icon can be added, or it can change in an obvious manner.

- Prelaunch the application. If camera hardware takes a few seconds to load, the camera application can still begin launching immediately, and open in a state that indicates this delay.

Pop ups will generally disappear after a few seconds. Annunciator or notification icons may need to remain in place, to indicate the change of state permanently, such as enabling Bluetooth features, which displays the Bluetooth icon in the Annunciator Row.

You can also use direct responses when they are technically feasible. If individual-key backlight is available, the backlight should dim momentarily during the key press. Rarely, this may be used successfully as an indicator with all key backlighting changing based on certain key presses.

Haptic Output or Tones are also valuable as a way for the user to confirm the key has been activated, or for multistep keys (such as volume) to give a better sense of the number of steps entered. Subtle clicks, whether through vibration or the speaker, can serve this need.

Don't forget that every key can perform at least two functions. Enable Press-and-Hold actions, or double-click behaviors as detailed in the Mode Switches pattern (see Figure 10-14). This may be difficult to explain or learn, so it should be used sparingly and only when relevant to the original key label. Holding down the Home key, for example, may load a list of currently running applications.

However, most of these are not common. The only relatively regular uses involve power.

- Devices with "Talk" and "End" keys tend to place the power toggle on the "End" key. A press and hold of the "End" key will activate the power-off function.

- Devices without an "End" key tend to have a dedicated "Power" key, which acts as a lock function when pressed. To power off the device, a long press is required.

Actions such as power should be protected from accidental activation by an Exit Guard or similar function, such as revealing a menu of options, only one of which is the power off function.

Presentation details

The key must be unambiguously labeled with the function. Volume keys may get a pass as they are expected, so any single key pair may be understood to be a volume control. However, the difference between the camera and power buttons, for example, should be made clear by labels.

Labels should be visible in all conditions. See the Mode Switches pattern for additional details on this, and especially for risks of backlight under certain lighting conditions.

Labels and descriptions for on-screen effects such as notification icons should comply with the hardware label to draw a clear parallel between the two.

Consider key position carefully. Keys should meet established best practices whenever possible. Camera buttons are very often arranged like a shutter, and so are at the right side of the top edge when in conventional (landscape) orientation.

On-screen behaviors and their keys should be adjacent to each other whenever possible. Soft keys, for example, must be near the bottom edge of the screen, and aligned near the left and right edges. If the position of more obscure keys is known to the software, it may be pointed to as an aide to the user, either routinely or during first-run tutorials or help topics.

Secondary functions, such as power toggle on the "End" key, should also be labeled on or adjacent to the key. Typically, these are smaller, and above the primary label as for hardware Keyboards & Keypads.

Antipatterns

Figure 10-15. *This handset does at least two things poorly. First, the keys are only labeled with impressed symbols and words, which are largely unreadable; the illumination is only there for some keys, and only for a brief time. Second, the soft keys are far removed from the screen, and they have no obvious relationship to soft keys by shape, position, or symbol. Note that the topmost key in the circle is not even a key, and is there purely for decoration.*

Do not block informational dialogs about a key press for any reason. Even in video playback, where other status messages or Notifications may be suppressed, hardware key presses should almost always display the results on-screen.

Do not cast labels into keys without highlighting them with color and/or providing a backlight. Impressed key labels are unreadable under all but the best lighting conditions, and the user will not understand them.

Do not make up new symbols to denote key functions. Use common, universally recognized icons or other labels.

Preserve relative key functions when switching modes. For example, when in portrait mode, the top key of the volume rocker increases volume. When switching to landscape, make sure the same key performs the same function. When a secondary set of keys is provided for the new mode (such as another set of soft keys to align with the screen location), both should function at the same time, to support habituation or eyes-off use.

Do not attempt to use lines or other guides to make up for soft keys that are far removed from the screen. If this cannot be surmounted, users will eventually learn it, but guides do not tend to help. The handset in Figure 10-15 let the pretty circle get in the way of clear soft-key placement.

Avoid using multikey combinations for routine controls. Despite their prevalence on desktop computing, they are poorly understood and difficult to label. This does not preclude their use for obscure functions. In fact, the lack of any reason for users to routinely use multikey combinations means they are a good way to support highly specialized or technical functions, such as for system resets.

Accesskeys

Problem

You must provide single-click access to an arbitrary set of key functions, which may change for each page or view.

Accesskeys are almost universally supported by any platform that has a keyboard or keypad. Even web browsers usually can handle Accesskeys, allowing shortcuts to be added to any web page as well.

Solution

Figure 10-16. *Typical use of Accesskeys to provide quick access to list items. Also note that the title, serving as a breadcrumb, has a key to quickly go back up a level. The search button cannot have one, as the only time it would be helpful is while the text field is in focus, when typing the character would just enter into the field. The user may press "OK/Enter" to get the same result.*

Accesskeys provide one-click access to functions and features of the handset, application, or site for any device with a hardware keyboard or keypad. A number or letter is assigned to a particular link or other function, and when selected it activates this function as a shortcut.

The primary value is in the arbitrary assignment, and larger number of keys available than you can get with soft keys, or other built-in "quick access" keys. Not only each application or website, but each page, view, or state can have its own set of shortcuts. The other advantage, if you build it right, is that the labels are immediately adjacent to the function. The user can glance at the page and then select a function with a single key, without the need to scroll to the function. See Figure 10-16.

A secondary value is in learnability. Users may become accustomed to certain features of the device or a specific application always having a particular accesskey. For example, if you always assign 0 (zero) to the "home" function for a site, this may be learned and it will be used even if the link to home is in the footer, and not visible at all times. This adds speed and efficiency to the expert and other highly accustomed or trained users.

Variations

By far the most common type of accesskey uses the number (and the symbol characters # and *) from the telephone number pad. This is applied to the basic 10-key device, which also very often has a simple scroll-and-select interface and can use the additional speed provided by the shortcut to avoid excessive scrolling.

Full keypads may use letters as Accesskeys, though this is rarer, and therefore is not as well understood. They may also use numeric characters, allowing a single set of keys to be assigned for all devices, though the mode switch to access the numbers may reduce the value of the Accesskeys for these devices.

This pattern cannot really be used on devices without a hardware keyboard or keypad. It is even often impossible for a user to launch the virtual keyboard or keypad without being in a text entry area, so Accesskeys cannot be used at all. On-screen Gestures used as shortcuts fulfill a different need and are not implemented in a manner that is consistent enough to qualify as a distinct pattern just yet.

Figure 10-17. *Accesskeys can be used to perform one-click functions on scroll-and select devices, anytime they might be needed. Here, pressing "OK/Enter" on the in-focus item will load a details page. The # accesskey provides a shortcut to see the same item on the map immediately. This sort of labeling can also be used to make users become accustomed to Accesskeys, and can encourage their use in the rest of the site or application.*

When the user presses a key mapped to a function in the current state, you should make sure the system will immediately perform the function.

Text or numeric entry will always take over the keyboard or keypad, and Accesskeys will not function while a form or other text field is in focus. The key labels cannot usually change their state to indicate that they are disabled, so use caution when providing Accesskeys on pages where text entry is the primary function.

Accesskeys should be assigned in a logical method. For keyboards, shortcuts are the best guidance, and following principles of desktop shortcuts works well. Abide by desktop heuristics whenever possible (e.g., Cut, Copy, Paste as X, C, V).

You can assign directional keys that are subsidiary to the built-in Directional Entry method. For example, if you want to be able to move a map within a page, but not move the page itself, assign the up/down/left/right keys as though the keypad is a virtual direction pad, with 5 at the center. Use the same logic even if only one axis of movement is required. This sort of behavior is useful for interfaces such as web pages, where you have no control over the overall page size and scroll behavior, and subsidiary scrolling might otherwise be troublesome.

Use mnemonics if available, such as * for favorite. The asterisk is commonly called and perceived as a star, an already common icon for favorite marking.

When possible, exercise bilateral symmetry. Zoom should be 1 and 3, never 1 and 2. This may be more difficult on keyboards, due to more variable layouts, staggered keys, and so on.

The following is a set of the most common assignments for numeric keypads:

Key	Suggested functions
1	Zoom out
2	Scroll up
3	Zoom in
4	Scroll left
5	Home, recenter or switch view
6	Scroll right
7	
8	Scroll down
9	
*	Favorites, bookmarks, etc.
0	Help
#	

Unassigned items have no discernible pattern to their use, and so are open to your needs. Of course, if your application has no need for zooming, those keys are free for the taking.

If you do not have complex interaction and are simply displaying options in a Vertical List, a typical method is to simply assign the Accesskeys to the functions in order, from the first available function down.

Key assignment consistency is key for learnability. Try to follow OS standards when available and possible. Design accesskey assignments at the same time as the overall application or site design. When determining what components are visible in the header and footer of each page, for example, assign keys to those so that they are fixed and repeatable across the service.

Try not to reuse keys that are common controls. If you have a weather map, and in many states the zoom function is important, try not to use 1 and 3 for some other purpose when on a page without the map.

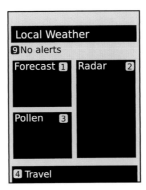

Figure 10-18. *Accesskeys can be a way to provide easier navigation for multicolumn and other complex layouts. Scroll-and-select users who are accustomed to scrolling only vertically or whose devices make column switching cumbersome can still access all parts of the page instantly. The downside is that not all users understand Accesskeys.*

You must be sure to label each accesskey-enabled function. Don't just provide these instructions in a help document, as they will be forgotten. See Figure 10-17 for more on labeling and describing Accesskeys.

Luckily, it's easy to place these labels. The general method is to display the labels as a graphical representation of the physical button. This emphasizes the intent of the label and reduces confusion with other items in the page. You can, if needed, use any label style that is different from other items on the page, such as gray and italic.

Labels should usually be much less prominent than other items so that they do not interfere with those who are unfamiliar with the concept of the accesskey. Low prominence may be attained via simply being very small. Accesskey labels can often carry text much smaller than the normal minimum suggested text size.

For lists, the labels should usually use the left margin, where no other elements are located. If numbered labels for line items are required, whether adjacent to the accesskey label or not, they appear in the normal place, slightly to the right of this, in the primary content area.

When items to be labeled are not all in a single list, place the labels in the same relative position, either to the left or to the right of the title, as in Figure 10-18, for the object or function. Place them within containing objects, such as title bars. Avoid switching label locations, even if symmetry or other needs seem to make design sense. If you are providing an accesskey for a button, place it in the same relative position (left or right) but inside the button itself.

Do not attempt to overuse the accesskey label, by also making it serve as a list identifier. Even if it functions in one instance, this may imply additional importance to the labels in other instances, or may be interpreted as only position labels and not be understood to be accesskey labels.

Do not rely on accesskey use. Even when 10-key devices were ubiquitous, it was not universally understood. Only a subset of the population used it regularly. Do include it for those users, but consider it as a secondary or shortcut method unless you know you have trained or expert users.

Avoid using Accesskeys for subtle interactions, such as granting focus to a search field. The user may not note the change and will become confused, or may even make entry errors due to misunderstanding a state change.

Dialer

Problem

Most mobile devices are still centered on voice networks connected to the PSTN. You may have to provide access to a phone dialer.

The dialer is provided with devices, but you may have to recognize and enable phone numbers to launch the dialer. Voice communications outside the voice network, using various VoIP implementations, also open the possibility of building part or all of a phone dialer into other products.

Solution

Figure 10-19. *The dial pad, whether virtual or (here) physical, must be laid out in a conventional-enough format that it can be easily used. Letters and symbols must be assigned per local standards to ensure that numbers can be dialed correctly. This is a keypad for the NANP, used in the United States, Canada, and the Caribbean.*

Numeric entry for this dialer application or mode varies from other entry methods, and has developed common methods of operation that users are accustomed to. You have to follow these standards to provide easy access to the voice network. There are also notable technical requirements, and failing to address those—even at the interface design level—will make it difficult or impossible to accurately connect.

This pattern is concerned with only the numeric entry and certain aspects of the overall interaction of the Dialer application. Numerous additional features and details of the operation are not discussed here, largely for space concerns. Follow the guidelines from the operator, especially as they regard specifics of the network, local regulatory requirements, and interoperability needs.

Variations

For most of the history of mobile devices, telephony was the key feature and was integral to the device at all levels. This is still the method employed for almost all devices with a dedicated numeric keypad, as shown in Figure 10-19. From the Home & Idle Screens, keypad entry will result in switching to the Dialer mode, and the entered character being typed onto the screen.

In these cases, the Dialer application does not need to be deliberately activated as the handset is considered a phone when it is not otherwise engaged. Naturally, when in other applications the keypad is used for text or numeric entry within that context, or to activate Accesskeys and does not immediately launch the Dialer.

A variation employed first by PDA phones and now by most smartphones is to make the Dialer just another application on the device, with few or no special UI conditions. In these cases, dialing cannot be initiated directly from the Idle screen. The Dialer application must instead be deliberately launched before any characters can be entered.

The Dialer application is very often closely coupled with—or a part of—address books, call history, and other features. However, only the Dialer itself will be discussed here.

Figure 10-20. *On mobile handsets with dedicated number pads—today, mostly feature phones—pressing a key from the Idle screen will launch the dialer and enter that number. Each number entered narrows the matching results from the internal address book. If no match is found, the best region guess from the area code entered is made. When sufficient characters are entered, the number is formatted for the user, with appropriate characters added.*

When users enter characters, they will appear in the phone number field, which is generally near the top of the viewport. Character entry is much like any other text entry, with a cursor and the ability to delete characters using the "back" or "delete" function. If the handset does not include a hardware back or backspace button, you will need to provide it on the screen, usually next to the input field itself.

Display any matching results below the numeric entry field. *Matching* is the practice of providing extended information about the number entered so far. A live search of the address book will be performed as characters are typed. Users may select from the list displayed, or simply confirm that the dialed entry is correct.

For example, if the characters "210" are entered, the following items (already in the address book) may be listed:

- Jane Adams 816 210 0123
- Dee Adler 210 618 0567

All typed characters are always entered into the phone number field, and the matching results are in a separate space, as shown in Figure 10-20. This is the *suggestion* variant of the Autocomplete & Prediction pattern. This generally results in a "split focus" where vertical scrolling changes the item in focus in the match list, and horizontal scrolling changes position in the entry panel. Pressing the "Talk" button will dial the last selected number; if the last entry is scrolling it will be an item in the match list, and if the last entry is typing (or scrolling up and out of the list) it will be the entered value.

If no address book match is found, you should display the region (such as the US state) that corresponds to the number format. For example, as soon as "816" is typed, with no matches, "Missouri" may be displayed (as this area code is entirely within that US state). This will increasingly be troublesome, as mobile phones allow the handset to physically be anywhere, and number portability allows any number to be associated with any region.

Errors and null-value information will not be displayed when address book or regional matching is not available or not found. Instead, you simply display no matches at all.

Pressing the "Talk" button is generally required to dial a call or select a match from the list.

There is no significant variation for the use of a virtual, on-screen keypad. When visible, the same behaviors take place. Virtual keypads are usually presented as only one of several options, and so may not be available by default when the Dialer application is open or when a call is initiated or received. Easy and unambiguous access to the keypad must be provided at all times.

If the user presses and holds a numeric key, this will initiate a shortcut and dial the corresponding number. If no shortcut is assigned, no special action should be taken and no errors displayed. The key will simply be entered as though a short press was performed.

The shortcut character should appear on the screen, to confirm it has been entered. These may display inside the phone number field (usually with a special character to indicate the difference, such as a preceding pound sign) or may be displayed in a different manner, such as a momentary overlay, as in Figure 10-21.

The 1 key is most often reserved for voicemail, though this may vary based on operator standards.

Whenever a call is active—whether sent or received—a version of the Dialer is generally made available in order to allow selection of extensions or entering information in inter-active voice response (IVR) systems. The same principles will be followed as in the Dialer, with keypad entry displayed immediately. The tones will also be sent over the network immediately. The display should show all numbers entered since the original connected number. Do not display these additional values appended to the dialed number, as it will become a long, nonsensical string that is hard to decipher.

Whenever possible, increase the time that the screen backlight is entirely removed while in calls, in order to allow the user to see the current state of an ongoing call without handling the device. Use sensors, as described in the Kinesthetic Gestures pattern, to au-tomatically lock the screen or keypad, and turn on and off the backlight to reduce battery use or prevent unintended input.

Presentation details

Figure 10-21. *Dialing shortcuts are activated by pressing and holding a number on the keypad. The contact to be dialed should be populated as if dialed and matched, but the shortcut should be indicated in some manner as well. Here, a large numeral pops up and fades rapidly away.*

The Dialer screen will be fixed, with elements remaining in the same location and always visible. Selectable items which must scroll (such as name matches) will do so only within a small area and will not scroll the entire page. The number pad itself must never scroll off the page.

The number entered must display in a common, readily recognized format for the locale. In the United States this would mean the characters 8162100455 display as 816-210-0455, for example. Avoid wrapping the primary number to a second line; if needed due to space, break at the end of the dialed number so that extensions and IVR entries are on the second (and third...) lines. Format characters are only for appearance, and cannot be edited directly by the user; they will disappear before they can be scrolled to.

Special functions for use while the call is active, such as mute and speaker, may be presented as on-screen buttons for touch or pen devices. Whenever possible, display these alongside virtual keypads instead of requiring switching from keypad to control mode.

Additional functions, such as three-way calling, should be placed under options menus, such as soft keys. The exact implementation will vary depending on the OS standards of the device. Avoid hiding important and frequently used in-call features, such as mute, speakerphone, and headset controls under options menus, even for scroll-and-select devices. Avoid placing these functions only on hardware keys, especially if the labels are small, if the labels are not illuminated, or if the keys are on the side of the device. Place these functions on-screen whenever possible. Feel free to provide informational graphics to indicate where a hardware key can be found that provides the same function.

Display the current state of the device when on a call. The time on call, connected status, and audio features (mute, speaker, headset) should be clearly visible on the screen, without scrolling or searching inside options menus.

Keys must be clearly labeled, and must comply with the layout and labeling of the local region. Although innovative layouts or graphic design may be used to differentiate the product, the user must be able to dial the handset without undue difficulty. Secondary key labels—discussed in the Keyboards & Keypads pattern for their use in multitap entry methods—were originally and are still used as mnemonic shortcuts for dialing, and must also comply with the locale in which the device is to be used.

When the Dialer is displayed for address book entry, formatting hints or optional entry methods should be provided. Hard pause, soft pause, and international dialing codes are not directly supported by the numbers on the keypad, so they will require special methods (such as menus or custom buttons) or informing the user of keypad shortcuts.

Antipatterns

Characters entered must be typed immediately and each character entered as typed. Do not require a "wake-up" character to open the Dialer but then be ignored, even if all subsequent characters are entered correctly.

Do not include unnecessary functions. For example, pause characters are not needed when live dialing, just when saving entered information to the address book.

Avoid hiding virtual keypads behind ambiguous labels. Whenever practical, place a Keypad button, with a suitable iconic label, on the screen or as a primary soft key.

The "OK/Enter" button should not perform unexpected or out-of-domain functions. For example, when selecting a match from the list, it should either dial or display additional options. However, a default condition (or pressing the "Talk" key from here) should dial the call. Loading the address book would be confusing and make it difficult for the user to dial the call quickly and without error.

Do not require the user to dial using PSTN standards when they can be deciphered and solved. In the United States, you must dial 1 before the area code for calls outside the local area, but may not dial the 1+ area code for local calls (from and to numbers in the same area code). Even if the mobile network is not intelligent enough, make the handset software interpret any dialed number so that it does not display these simple errors, and sends the appropriate codes. Never allow the user to hear a voice saying "Please dial a one or zero to complete this call…"

On-Screen Gestures

Instead of relying on physical buttons and other input widgets being mapped obtusely to interactions, you should allow the user to directly interact with on-screen objects and controls.

If any device you are designing for uses a touch or pen interface, you should support the existing On-screen Gestures and can consider building your own. On many platforms, enough touch information is available that you can build custom gesture controls even for individual web pages.

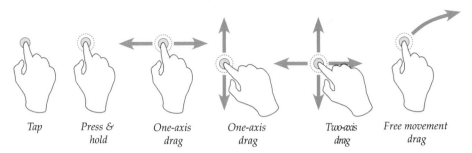

| Tap | Press & hold | One-axis drag | One-axis drag | Two-axis drag | Free movement drag |

Figure 10-22. *A selection of single-point gestures and actions.*

Gestures, as a whole, attempt to circumvent the previous state of the art where your controls are somewhat arbitrarily "mapped" to actions. Instead, this is replaced with a direct interface concept. Items in the screen are assumed to be physical objects that can be "directly" manipulated in realistic (if sometimes practically impossible) ways.

Gesture-based interfaces should always keep this in mind, and seek to use natural or discoverable behaviors closely coupled to the gestures. Avoid use of arbitrary gestures, or those with little relationship to the interface-as-physical-object paradigm.

Technology may limit the availability of some gestures. Two key pitfalls are poor or no drag response, and lack of multitouch support. Many resistive or beam screens do not support drag actions well, so they must be designed to use only click actions, such as buttons instead of draggable scroll bars. Capacitive screens are considered the absolute best, but optical, IR, and acoustical position sensing can detect just as well under some conditions, so there is no telling what will be used in the future.

See the Directional Entry pattern for additional details about the use of gestural scrolling, especially in constrained inputs such as single-axis scrolling.

Variations

Gesture, within this pattern, is used to refer to any direct interaction with the screen, including simple momentary tapping. A variety of gestures are available, but not all complex actions are universally mapped to a behavior. Instead, gestures are listed by class of input alone:

Tap

> The contact point (finger, stylus, etc.) is used to activate a button, link, or other item with no movement, just momentary selection.

Press-and-hold

> The contact point is held down for a brief time, to activate a secondary action. This may be an alternative to a primary (tap) action, or to activate an optional behavior for an area that is not normally selectable. See the Press-and-Hold pattern for additional details.

One-point drag

> This refers to the entire series of actions where a single finger (or stylus) is down, and moves items or scrolls the screen. See Directional Entry for details on interaction, such as behavior for restricted axes of entry (Figure 10-22).

> *One-axis*

>> Scrolling or dragging of items (repositioning) may be performed in only a single axis, either vertical or horizontal. The other axis is locked out.

> *Two-axis*

>> Scrolling or dragging of items may be performed in only the vertical and horizontal axes, but usually only one at a time. A typical use would be a series of vertical lists which are on adjacent screens, and must be scrolled between in a Film Strip manner.

> *Free movement*

>> Objects or the entire page can be moved or scrolled to any position, at any angle. For movement of areas, see the Infinite Area pattern.

Two-point actions

Generally, using two fingers, these use movement to perform gestures on larger areas, where enough room is provided for both points to be on the area to be affected. See Figure 10-23.

Both move

Two-point gestures are similar to Press-and-Hold, in that they are usually used to perform a secondary action, different from a one-point drag. These are often used to deconflict from single-point gestures, as shown in Figure 10-24. For example, a single-finger gesture may select text, and a two-finger gesture scrolls the page.

Axis and move

One point fixes the item to be rotated, providing an axis, while the other moves about this axis. To be perceived as this action, the axis finger must move as little as a typical tap action.

Multipoint input

This may use three or four (rarely more) fingers to perform simple gestures that generally impact the entire device or the entire currently running application. This is often used to deconflict single-point drag actions. For example, a single-finger drag may scroll the screen, but a four-finger drag will open a running applications list.

Interaction details

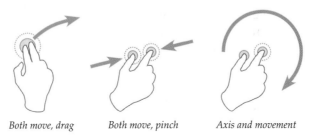

Both move, drag   *Both move, pinch*   *Axis and movement*

Figure 10-23. *A selection of two-point gesture types.*

Whenever possible, you must attempt to use gestures to simulate physical changes in the device. Pretend the screen is a physical object such as a piece of paper, with access only through the viewport. If it helps—and we think it does—you can even very roughly prototype the interactions by moving paper on a desk. Drag to move the paper, rotate by fixing it with one finger and moving the other, and so on.

Some actions require a bit more imagination. The "pinch" gesture assumes the paper is elastic and contains an arbitrary amount of information; spreading stretches the paper, showing more detail, and pinching squishes the paper, showing less.

You can also use inertia when it would help the interaction, such as scrolling of long lists. This must also follow an arbitrary set of reasonable physical rules, with the initial speed being that of the drag gesture and deceleration to simulate friction in mechanical systems. If you try simulating this by scooting paper across your desk you will find that the convention for interactive spaces presumes the screen is greased well, or is an air hockey table.

Whenever possible, follow these principles, and do not use arbitrary gestures to perform tasks. This may be reflected in the response. For example, if a running-apps list is opened with a four-finger swipe, do not have a Pop-Up dialog simply appear. Instead, have it drag in from the direction (and at the speed of) the gesture.

All on-screen gesture interactions follow a basic protocol:

1. A contact point touches the screen.

2. Any other input, such as dragging, is performed.

3. The screen reflects any live changes, such as displaying a button-down or drag action.

4. Input is stopped, and the contact point leaves the screen.

5. The action is committed.

Although many actions will be reflected while the contact point (e.g., finger) is down, they are generally not committed until the point leaves the screen. This can be used to determine when input is unintentional, or gives the user a method to abandon certain actions before committing them. Areas that only accept tap actions, such as a button, should disregard drag actions larger than a few millimeters as spurious input.

For multipoint gestures, the device generally senses so quickly that it can tell which finger makes contact first. When the second finger makes contact, any actions performed to date (including things such as drags) will be disregarded and returned to the start position, in preparation for the rest of the multifinger gesture to come.

For multipoint gestures, all contacts must be removed within a very short time frame, or the gesture may convert to a single-point gesture with the remaining contact point.

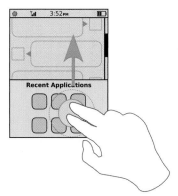

Figure 10-24. *Multifinger gestures can be used to access features which would otherwise be in conflict with primary interactions. Here, a single finger drag scrolls the discussion thread; a two-finger scroll gesture loads other features such as the running-applications list.*

Since gestures are not committed until all contact points are removed, a sort of "hover" or "mouse-over" state must be included for all links and buttons. While the point is in contact, this should behave exactly as a web mouse pointer, and highlight items under the centroid. An additional "Submit" state should also be provided to confirm the action has been committed.

Position cursors may be used for simple tap behaviors in order to provide additional insights and confirmation of action to the user for more complex actions. See the Focus & Cursors pattern for some details.

Instead of a conventional cursor, you can show a gesture indicator adjacent to the contact area, such as a rotate arrow when rotating.

When on-screen controls allow indirect input (via Form Selections) of the same action as the gesture, these should be changed to reflect the current state. Examples are scroll bars moving as a list is dragged, or an angle change form displaying the current angle in degrees as a rotate action is performed. All status updating must be "live," or so close to real time the user cannot see the difference.

Be careful using gestural interfaces for large stretches of input on kiosks or in other environments where the user cannot rest his hand or arm. The "gorilla arm" effect, well
understood with pen interfaces, can impose fatigue or even repetitive stress injury when
actions such as this are used. Allowing the user to rest his hand or arm on the input device
(when technically possible) as well as more casual use or regular changes in the type of
entry can avoid this.

Use caution with design, and test the interface with real users in real situations to avoid
overly precise input parameters. If tap actions are only accepted within an overly small
area, the user will have trouble providing input.

Do not commit actions while the contact point is still on the screen. At best, if additional
actions are blocked to prevent accidental second input, users may be confused as to how
to continue. At worst, no actions are blocked, and whatever is under the contact point will
be selected or acted upon without deliberate user input.

Devices without multitouch support will either accept the first input, or (depending on
the technology) determine a centroid based on all contact areas. Since touch devices do
not generally display a cursor, this makes the device unusable. Instead, you can just block
all inputs whose total contact area is much larger than a typical fingertip (larger than
about 25 mm). This avoids not just confusion over users who think it may be multitouch,
but also palm activation, pocket activation, or even babies grabbing the phone from doing
as much damage.

Kinesthetic Gestures

You should always make mobile devices react to user behaviors, movements, and the
relationship of the device to the user in a natural and understandable manner.

Sensors of all sorts are provided with most mobile devices, and can be used by applications and the device OS. Devices without suitable sensors, those where the sensor output is limited, may not be able to take advantage of these. Despite several attempts, web
browsers so far generally will not pass sensor data to individual pages, though some can
be made available with operator-level agreements.

Roll handset
(face up to
face down)

Figure 10-25. *Orientation changes when directly accounted for by switching screen orientation are covered under their own pattern. However, other types of device orientation can be used to signal that the device should perform an action, such as the relatively common facedown to lock. Orientation is also useful when combined with other sensors to perform other activities; raising a phone toward the user is very different when faceup versus facedown.*

Kinesthetics is the ability to detect movement of the body; although usually applied in a sense of self-awareness of your body, here it refers to the mobile device using sensors to detect and react to proximity, action, and orientation.

The most common sensor is the accelerometer, which as the name says measures acceleration along a single axis. As utilized in modern mobile devices, accelerometers are very compact, integrated micro electro-mechanical systems, usually with detectors for all three axes mounted to a single frame.

In certain circles the term *accelerometer* is beginning to achieve the same sense of conflating hardware with behavior as the GPS and Location. Other sensors can and should be used for detection of Kinesthetic Gestures including cameras, proximity sensors, magnetometers (compasses), audio devices, and close-range radios such as NFC, RFID, and Bluetooth. All these sensors should be used in coordination with one another to cancel errors and detect their own piece of the world.

Kinesthetic gesturing is largely a matter of detecting incidental or natural movements and reacting in appropriate or expected manners. Unlike on-screen gestures—which form a language and can become abstracted very easily—kinesthetic gesture is exclusively about the environment and how the device lives within that context. Is the device moving? In what manner? In relation to what? How close to the user, or other devices?

When designing mobile devices, or applications that can access sensor data, you should consider what various types of movement, proximity, and orientation mean, and behave appropriately.

Specific subsets of device movement such as Orientation have specialized behaviors and are covered separately. Location is also covered in its own pattern, but both of these are only broken out as they are currently used in specific ways. Subsidiary senses of position such as those used in augmented reality and at close range (e.g., within-building location) may have some overlaps, but are not yet established enough for patterns to emerge.

The use of a secondary device for communicating gesture, such as a game controller, is covered under the Remote Gestures pattern.

Variations

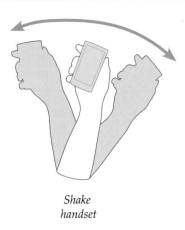

Shake
handset

Figure 10-26. *Device gestures are generally true gestures or shorthand for another action instead of attempts at direct control by pointing or scrolling. Actions best suited to these are those that are hard to perform, or for which there is no well-known on-screen gesture. Shaking to randomize or clear is a common, easy-to-use, and generally fun one.*

Only some of the variations of this pattern are explicitly kinesthetic, and sense user movement. Others are primarily for detecting position, such as the proximity of another device.

Device orientation

Relative orientation to the user or absolute orientation relative to the ground; either incidental or natural gestures. For example, when rotated facedown and no other movement is detected, the device may be set to "meeting mode" and all ringers silenced; as the screen is the "face" of the device, facedown means it is hidden or silenced. See Figure 10-25.

Device gesturing

> Deliberate gestures, moving the device through space in a specific manner; usually related to natural gestures, but may require some arbitrary mapping to behaviors. This may require learning. An example is the relatively common behavior of shaking a device in order to clear, reset, or delete items. See Figure 10-26.

User movement

> Incidental movement of the user detected and processed to divine the user's current behavioral context. For example, gait can be detected when the device is in the user's hands or in a pocket; the difference between walking, running, and sometimes even restless and working (e.g., at a desk at work) can also be detected. Vibration generally cannot be detected by accelerometers, so riding in vehicles must be detected by location sensors. Other user movements include swinging the arm, such as moving a handset toward the user's head.

User proximity

> Proximity detectors, whether cameras, acoustical, or infrared, for detecting how close the device is to a passive object, such as the user's head. A common example is for touch-centric devices to lock the screen when approaching the user's ear, so talking on the handset does not perform accidental inputs. Moving away from the user is also detected, so the handset is unlocked without direct user input. See Figure 10-27.

Other device proximity

> Nonpassive devices, such as other mobiles, or NFC card readers that can be detected by the use of close-range radio transmitters. As the device approaches, it can open the appropriate application—sharing or banking—without user interaction. The user will generally have to complete the action, to confirm that this was not accidental, spurious, or malicious. See Figure 10-28.

Proximity to signals or other environmental conditions farther than a few inches away is not widely used as yet. When radio, audio, or visual cues become commonly available they are likely to be used in the same way as these, and simply added on as another type of signal to be processed and reacted to.

These methods of gesturing can initiate actions in three categories:

State change

> The entire device reacts to a conditional change with a state change, such as the handset screen and keys locking when against the user's ear while the user is engaged in a phone call.

Process initiation

> This is similar to state change, in that the gesture or presence of another device causes a new state to be entered. However, this is manifested as an application launch or similar action which is not a device-wide state change that must be reversed to continue; the application can be set aside temporarily at any moment.

Control

Akin to the use of On-screen Gestures a kinesthetic gesture is associated with a specific action within the interface, such as shake-to-clear. Pointing and scrolling is also available, but is rarely used outside of game environments.

Interaction details

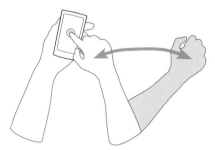

Gesture toward handset
(non-handset hand moves)

Figure 10-27. *Various types of user proximity can be detected so that the device can behave appropriately. The most common is to lock against accidental activation (when in the pocket, or against the user's ear when on a call) and to unlock when possible. The variation shown here presumes a more stringent lock condition, and uses sensors to determine when the user is about to perform on-screen input.*

The core of designing interactions of Kinesthetic Gestures is in determining what gestures and sensors are to be used—not just a single gesture, but which of them, in combination, can most accurately and reliably detect the expected condition. For example, it is a good idea to lock the keypad from input when it is placed to the ear during a call. If you tried to use any one sensor alone, that would be unreliable or unusable. If you combine accelerometers (moving toward the head) with proximity sensors or cameras (at the appropriate time of movement, an object of head size approaches), this can be made extremely reliable.

Kinesthetic Gestures reinforce other contextual uses of devices. Do not forget to integrate them with other known information from external sources (such as server data), location services, and other observable behaviors. The device can surmise a meeting from traveling to a meeting location at the time scheduled, and then being placed on a table. The device could then behave appropriately, and switch to a quieter mode even if it is not placed facedown. You should also consider patterns of use, and adjust the reactions according to what the user actually did after a particular type of data was received.

Context must be used to load correct conditions and as much relevant information as possible. When proximity to another device opens a Bluetooth exchange application, it must also automatically request to connect to the other device (not simply load a list of cryptic names from which the user must choose), and open the service type being requested.

All state-changing gestures should be reversible, usually with a directly opposite gesture. Removing the phone from the ear must also be sensed so that it reliably and quickly unlocks and does not cause a delay in its use.

Some indication of the gesture must be presented visually. Very often, this is not explicit (such as an icon or overlay) but switches modes at a higher level. When a device is locked due to being placed facedown or near the ear, the screen is also blanked. Aside from saving power when there is no way to read the screen anyway, this serves to signal that input is impossible in the event that it does not unlock immediately.

Nongestural methods must be available to remove or reverse state-changing gestures. For example, if sensors do not unlock the device in the earlier examples, the conventional unlock method (e.g., "Power/Lock" button) must be enabled. Generally, this means that Kinesthetic Gestures should simply be shortcuts to existing features.

For process initiation, usually used with transactional features such as NFC payment, the gesture toward the reader should be considered to be the same as the user selecting an on-screen function to open a shortcut; this is simply the first step of the process. Additional explicit actions must be taken to confirm the remainder of the transaction.

Kinesthetic Gestures are not as well known as On-screen Gestures, so they may be unexpected to some users. Settings should be provided to disable (or enable) certain behaviors.

Presentation details

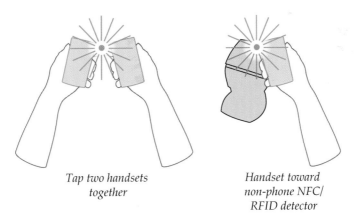

Tap two handsets together

Handset toward non-phone NFC/ RFID detector

Figure 10-28. *Two types of proximity to other devices exist. In the first, the devices are broadly equal and activity may be one-way (sending a file) or two-way (exchanging contact information). In the second, the user's mobile device is one side of a transaction, and the other is a terminal such as a payment or identity system.*

When reacting to another device, such as an NFC reader, make sure the positional relationship between the devices is implied (or explicitly communicated) by the position of elements on the screen.

When gesture initiates a process, use Tones, Voice Notifications, and Haptic Output, along with on-screen displays, to notify the user that the application has opened. Especially for functions with financial liability, this will reassure the user that the device cannot take action without the user initiating the action.

When a sensor is obvious, an LED may be used on or adjacent to the apparent location of the sensor. For the NFC reader example, an icon will often accompany the sensor, and can be additionally illuminated when it is active. This is used in much the same manner as a light on a camera, so surreptitious use of the sensor is avoided. This is already common on the PIN Pad readers for NFC cards, and when handsets carry NFC, they should use similar methods.

When using Kinesthetic Gestures for control, such as scrolling a page, you should include an on-screen display indicating that scrolling is deliberately active. You can use this to reinforce the reason the device is scrolling, and for the user to interact with it. Especially if the indicator shows the degree of the gesture, the user can balance the output (scroll speed) as well as directly have a way to control the gesture.

Kinesthetic Gestures are not expected in mobile devices, so users will have to be educated as to their purpose, function, and interaction. This can be done with:

- Advertising.
- Printed information in or on the box (when available).
- First-run demonstrations.
- A first-run (or early-life) Tooltip hint.
- Describing the function well in settings panels.
- Leaving it on by default, and explaining the function via a Tooltip. An Annotation may be preferred for some cases, so the user can immediately change or disable the function.

Antipatterns

Whole-device Kinesthetic Gestures can also be used as a pointer, or method of scrolling. These are not yet consistently applied, so they require instruction, icons, and overlays to explain and indicate when the effect is being used, and controls to disable them when not desired. Their use in games (think of the rolling-a-ball-into-a-pocket type of game) is due to the difficulty of control; this should indicate that using gestural pointing as a matter of course can be troublesome.

Avoid using graphics that indicate position when they cannot be made to reflect the actual relationship. For example, do not use a graphic that shows data transferring between devices that are next to each other, if the remote device can only connect through a radio clearly located on the bottom of the device.

Use caution when designing these gestures. Actions and their gestures must be carefully mapped to be sensible, and not conflict with other gestures. Shake to send, for example, makes no sense. But a "flick away" to send data would be OK; like casting a fishing line, it implies something departs the user's device and is sent into the distance. Since information is often visualized as an object, extending the object metaphor tends to work.

Remote Gestures

Problem

You must build interactivity so that a handheld remote device—or the user alone—is the best, only, or most immediate method to communicate with another, nearby device with a display.

Remote Gestures rely on hardware solutions that are not very common. However, the increasing use of sensors and of game systems specifically designed to capture this input may make them very common in the mid-term.

Solution

Remote control has been used for centuries as a convenience or safety method, to activate machinery from a distance. Electronic or electro-mechanical remote control has generally simply mapped local functions such as buttons onto a control head removed from the device being controlled.

Ready availability of accelerometers, machine vision cameras, and other sensors now allows the use of gesture to control or provide ambient input.

Combining the concepts and technology of Kinesthetic Gestures with fixed hardware allows remote control that begins to approach a "natural UI" and can eliminate the need for users to learn a superimposed control paradigm or set of commands. Although these have so far mostly been used to communicate with fixed devices such as TV-display game systems, they may be able to be used as a method of interacting in contextually difficult scenarios such as driving, walking, and presenting, or in dirty or dangerous work environments.

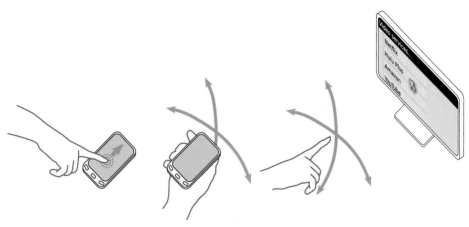

Figure 10-29. *Touch and pen devices, pointable devices, and hand gestures all display their position on the screen as a cursor. All conventional interactions, such as acceleration, are supported.*

You will find Remote Gestures may be used for:

Pointing
> Direct control is provided over a cursor or other position-indicating item on the remote screen.

Gesturing
> Direct but limited control is provided over portions of the interface, via gestures that map to obvious physical actions such as scrolling.

Sensing
> The user, or the device as a proxy for the user, is sensed for ambient response purposes, much as the user movement and user proximity methods are used in Kinesthetic Gestures.

Figure 10-29 summarizes all of these. Additional methods, such as initiating processes, may emerge, but as implemented now they are covered under Kinesthetic Gestures.

Methods are very simply divided between device pointing and nondevice gestures:

Remote device
> A handheld or hand-mounted device is used to accept the commands and send them to another device some distance away. The handheld device may be a dedicated piece of hardware or a mobile phone or similar general-purpose computer operating in a specific mode. This is most clearly exemplified by the Wii, though many other devices use some portion of this control methodology, including some industrial devices, and 3D mice.

User

> User gestures, including hand, head, and body position, speed, attitude, orientation, and configuration, are directly detected by the target device. This is most exemplified by the Xbox Kinect, but it is also used transparently to provide Simulated 3D Effects on some mobile and handheld devices.

This pattern is mostly concerned with gestures themselves, and does not explicitly describe the use of buttons, Directional Entry pads, or other interactivity on a remote control. These interactions, in general, will be similar to those on self-contained hardware, but an additional interaction will occur on the remote device, as described shortly. For example, if On-screen Gestures are used as the pointing control, the handheld device will accept the input and display output as appropriate on the device in hand, and a cursor and output will appear on the remote display as well.

For certain devices, or certain portions of the control set, gestures will simply not be suitable or will not provide a sufficient number of functions. You might need to use buttons, Directional Entry pads, Keyboards & Keypads, or other devices. Use the most appropriate control method for the particular required input.

Interaction details

Pointing may be used to control on-screen buttons, access virtual keyboards, and perform other on-screen actions as discussed in the remainder of this book. When touch or pen controls are referred to here or elsewhere, a remote pointing gesture control can generally perform the same input.

Remote pointing varies from touch or most pen input, in that the pointer is not generally active but simply points. A mouse button or OK/Enter key—or another gesture—must be used to initiate commands, much as a mouse in desktop computing.

Make control gestures interact directly with the screen. When the hand is used (without a remote device), a finger may be sensed as a pointer, but an open hand as a gesture. In this case, if the user moves up and down with an extended finger, you can make the cursor move. However, moving up and down with the hand open will be perceived as a gesture and will scroll the page. Two methods are provided without a need for traditional Mode Switches or addressing specific regions of the screen.

Decisions on the mapping of gesture to control are critical. You must make them natural, and related to the physical condition being simulated. Scroll, for example, simulates the screen being an actual object, with only a portion visible through the viewport. Interactions that have less-obvious relationships should simply resort to pointing, and use buttons or other control methods.

Some simple controls may still need to be mapped to buttons due to cultural familiarity. A switch from "day" to "night" display mode may be perceived as similar to turning on the lights. Even if required often, a gesture will not map easily, as the expected condition is one of "flipping a light switch."

Sensing is very specific to the types of input being sensed, so it cannot be easily discussed in a brief outline such as this.

Presentation details

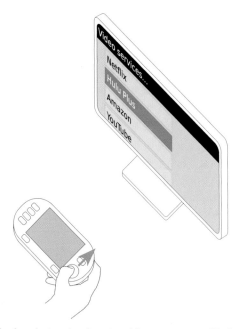

Figure 10-30. *Scroll-and-select devices (or directional keys on a pointable device) while being used with conventional interactive elements such as icon grids and lists use the focus paradigm for the remote screen and generally should not display a cursor. These same keys can be used to control a cursor or avatar when in other interfaces, such as when maneuvering a character in a game.*

For navigating or pointing, always provide a Cursor. Focus, as shown in Figure 10-30, may also be indicated as required, for entry fields or other regions. Just use the scroll-and-select paradigms. The nature of the pointing device requires a precise cursor to be always used, even if a fairly small focus region is also highlighted. See the Focus & Cursors pattern for details on the differences, and implementation.

When a Directional Entry control is also available (such as a five-way pad on top of the remote), you may need to switch between modes. When the buttons are used, the cursor disappears. Preserving the focus paradigm in all modes will help to avoid confusion and refreshing of the interface without deliberate user input.

For direct control, such as of a game avatar, the element under control immediately reflects the user's actions, and so acts as the cursor. You are likely to find that a superimposed arrow cursor is not important. If the avatar departs the screen for any reason, a cursor must appear again, even if only to indicate where the avatar is relative to the viewport, and that control input is still being received.

When sensing is being used, an input indicator should be shown, so the user is aware that sensing is occurring. Ideally, input constraints should be displayed graphically—for example, as a bar of total senseable range, with a clearly defined "acceptable input" area in the middle.

When a gestural input is not as obvious as a cursor, you have to communicate the method of control. The best methods of this are in games, where first use presents "practice levels" where only a single type of action is required (or sometimes even allowed) at a time, accompanied by instructional text. The user is generally given a warning before entering the new control mode, and until that time either has conventional pointing control or can press an obvious button (messaged on-screen) to exit or perform other actions. This single action is often something like learning to control a single axis of movement, before any other actions are introduced.

When game-like instruction of this sort is not suitable, another type of on-screen instruction should be given, such as a Tooltip or Annotation, often with an overlay graphically describing the gesture. These should no longer be offered once the user has successfully performed the action, and you must always allow the user to dismiss them. The same content must either be able to be turned back on, or be made available in a help menu.

Antipatterns

Be careful when relying on Remote Gestures for the primary interface, and for large stretches of input. Despite looking cool in movies, user movement as the primary input method is inherently tiring. If it is required for workplaces, repetitive stress injuries are likely to emerge.

Do not attempt to emulate touchscreen behaviors by allowing functions such as gesture scroll when the mouse button is down. Instead, emphasize the innate behaviors of the system; to scroll, for example, use overscroll detection to move the scrollable area when the pointer approaches the edge.

Be aware of the problem of pilot-induced oscillation (PIO) and take steps to detect and alleviate it. PIO arises from a match between the frequency of the feedback loop in the interactive system and the frequency of the response of users. Although similar behaviors can arise in other input/feedback environments, they lead to single-point errors. In aircraft, the movement of the vehicle (especially in the vertical plane) can result in the oscillation building, possibly to the point of loss of control. The key facet of many remote gesturing systems—being maneuvered in three dimensions, in free space—can lead to the same issues. These cannot generally be detected during design, but you must test for them. Alleviate this via reducing the control lag, or reducing the gain in the input system. The unsuitability of a response varies by context; games will tolerate much more "twitchy" controls than pointing for entry on a virtual keyboard.

Mode changes on the screen, such as between different types of pointing devices, should never refresh the display or reset the location in focus.

Avoid requiring the use of scroll controls, and other buttons or on-screen controls where a gestural input would work better. Such controls may be provided as backups, as indicators with secondary control only, or for the use of other types of pointing devices.

Chapter 11

Input and Selection

The Wheels on the Bus Go Round and Round

I remember as a kid singing the song "The Wheels on the Bus Go Round and Round" on those long bus rides up to summer camp. It was the adults' secret weapon to pass the time and keep the kids out of trouble, I presume. It went something like this:

The wheels on the bus go round and round,

round and round,

round and round.

The wheels on the bus go round and round,

all through the town.

The baby on the bus says, "Wah, wah, wah;

Wah, wah, wah;

Wah, wah, wah."

The baby on the bus says, "Wah, wah, wah,"

all through the town….

Well, the rest of the song is outside the scope of this book. But imagine if the song was describing today's city bus commute instead. It might go something like this:

The wheels on the bus go round and round,

round and round,

round and round.

The wheels on the bus go round and round,

all through the town.

The teen texters on the bus tap "LOL, LOL, LOL;

LOL, LOL, LOL;

LOL, LOL, LOL."

The teen texters on the bus tap "LOL, LOL LOL,"

all through the town.

The businessmen's emails go "Clicky, click, click;

Clicky, click, click;

Clicky, click, click."

The businessmen's emails go "Clicky, click, click,"

all through the town.

The traders on the bus type "Buy, buy, buy;

Buy, buy, buy;

Buy, buy, buy."

The traders on the bus type "Buy, buy, buy,"

all through the town.

Mobile Trends Today

Figure 11-1. *Forms can reduce the number of steps in a process by adding multiple selections, but they can also end up being dense and unusable. Many select lists, especially, confuse the paradigm and pull the selection mechanism out of the context as well as change the expected behavior.*

The landscape of mobile use is defined by user-generated input. In his blog post titled "Data Monday: Input Matters on Mobile," Luke Wroblewski points out the following:

> Web forms make or break the most crucial online interactions: checkout (commerce), communication & registration (social), data input (productivity), and any task requiring information entry. These activities are taking off in a big way on mobile. So getting input on mobile devices matters more each day (Wroblewski 2010).

As Wroblewski notes, the process of filling out fields and forms makes up a large part of our mobile experience, which is why it's essential for UI designers to make interfaces capable of handling the way users input and submit information.

Slow Down, Teen Texters!

Take a moment and reflect on some of your most aggravating moments filling out fields or forms on a mobile device. Here is a common scenario that frustrates me:

First, I'm not a savvy texter. I'm amazed by the accuracy and rate at which some people, mainly teens I've seen, can engage in a text conversation at a supernatural rate. I'm embarrassed to text whenever I'm next to them. Though when I am forced to input text and characters in a field, such as an SMS, it's very likely I'm going to make errors.

I constantly enter the wrong characters because my fingers extend past the target size. Positioning the cursor to edit my mistakes is even more maddening. I end up clearing the wrong part of the word or deleting everything I didn't want to delete. The only reason I continue with these tasks, despite my buildup of dissatisfaction, is because many times a better UI is not available.

If these basic functions of inputting data fail due to an unusable interface, the user will likely not bother with the site or service. That's a huge risk to take, considering the emerging trends users are engaged in with their mobile devices.

Input and Selection in the Mobile Space

Figure 11-2. Gestural interfaces can benefit from simulating physical objects, even for form selections. Make switches actually slide, and use thumbwheel-like spinners to replace short select lists.

For many of us, having to always type characters on a small mobile keyboard is a challenging task. The process is quite error-prone and the keyboard takes a lot longer to use. Luckily, there are solutions we can implement in our designs that will offer a more valuable user experience with less frustration and loss of data.

Here are some recommended tactics to promote quicker, more efficient and less error-prone input:

- Consider using assistive technology such as auto-complete and prediction during text entry. Requiring your user to only type the initial part of the word before the device recognizes the rest can allow for quicker text entry and fewer errors.

- When possible, and only if appropriate, consider using a drop-down list rather than a text field to complete an input. For example, selecting a country or state from a list can be more efficient than having to type it and risk misspelling. Remember, though, to never take your user out of the current context when using a drop-down list.

- Limit the amount of unnecessary "pogo sticking" and selection of key controls. For example, when you begin entering an email, consider the benefit of having a key labeled as ".com" predicatively appear to save the user from typing those additional four characters. Such functions can be offered as on-screen options even when hardware keyboards are used for entry.

- Always keep in mind the principles of Fitts's Law. The more clicks it takes to make a selection, the longer the process takes to complete. Use appropriate controls to limit the number of steps a user has to take to achieve a desired result. In this chapter you'll become familiar with spinners and tapes, and when to use them based on the number of incremental choices available.

- Take advantage of any sensors or stored data to assist the application in prepopulating information that can be detected. If the location can be determined, do not ask for it. And remember that absolute precision is not always required; weather reports are regional, so they can use vague senses of location and do not require GPS access.

Patterns for Input and Selection

Using input and selection functions appropriately provides users with methods to enter text and make selections within a list or field. In this chapter, we will discuss the following patterns:

Input Areas

Provides a method for users to enter text and other character-based information without restriction.

Form Selections

Provides a method for users to easily make single or multiple selections from preloaded lists of options. See Figure 11-1.

Mechanical Style Controls
> Provides a simple, space-efficient method for users to easily make changes to a setting level or value. See Figure 11-2.

Clear Entry
> Provides a method for users to remove contents from fields or entire forms without undue effort and with a low risk of accidental activation.

Input Areas

Problem

You must provide regions and methods for your users to easily enter text and other character-based information.

Forms and other types of input areas are common features of every platform. Many use web paradigms, so they are simple to understand whether in application or web mode.

Solution

 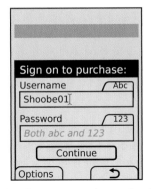 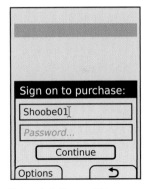

Figure 11-3. *These three examples show how space can be saved with efficient labeling and hint techniques. However, each compressed item also risks reduced readability and confusion. Field labels inside the form are suitable for single-use forms such as this, but should be avoided when prepopulated information or revisiting of the form is likely. The user may not know what each field is for once the information is entered.*

Text and textarea input fields are long-established principles and are heavily used to accept user-generated text input in all types of computing. Mobile is no exception, and employs these elements in several methods, with variations to meet the needs of mobile devices.

The Input Areas pattern is concerned largely with differences between the typical implementations of these web or desktop computing form fields and their most common practices within mobile OSes and applications.

For guidelines on using forms for the mobile web, see existing standards from organizations such as the W3C for the presentation, interaction, and design of these form fields. Of course, the principles outlined here still apply.

Variations

There are three variations for Input Areas. The differences are based on the amount of required text input by the user as well as limitations of available display space on the mobile device.

Text

Text fields are single-line entry methods. For mobile, these should usually be restricted to only accepting as much input as can be seen in the form, though exceptions may be made, such as for URL entry.

Textarea

Textarea fields are multiline entry methods, with fixed display heights. They are often configured to accept an arbitrary (though not infinite) amount of text, and are provided with a vertical scrolling mechanism to display text which does not fit in the field.

Convertible

In certain cases, such as the numeric entry field for the Dialer, mobile devices may display an entry field that appears to be a text field, but is in fact a textarea field. This variation is used due to the limited space on mobile devices—the field is only as large as is needed at any one time, and expands to additional lines as more text is entered. See Figure 11-4.

Interaction details

Text Input Areas may only be used while in focus. Focus may be granted by scrolling with Directional Entry keys or by direct selection with a pen or finger. Scroll-and-select devices may require explicit selection of the field by selecting the "OK/Enter" button. This allows the form to be scrolled through quickly, and only when a field needs to be entered will full focus be granted.

While a field is in focus, pressing the "OK/Enter" key (when available) will perform different functions, depending on the context of use. Web forms will work as they usually do for the desktop web. For fields in mobile applications, "OK/Enter" will generally commit the field and will transfer focus to the next field in the form—in a mode ready for entry—or the next item in the list. If no other item is available, the entire form will often be submitted instead.

Whenever content is known to the system, make sure it is prepopulated so that the user does not have to enter it. Prepopulated content is interacted with just as user-entered text, and may be edited, added to, or removed. When large amounts are prepopulated a Clear Entry item should be provided.

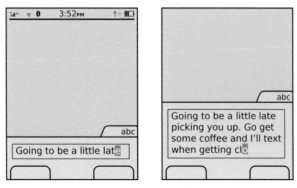

Figure 11-4. *Textarea fields are essentially identical to text fields, except for the height. Convertible fields, as shown here, only occupy space as needed.*

The focus of a field must always be very clearly delineated, with border, background, or other effects. Cursors must always be used when fields are editable, to denote the state change and the position of the text insertion point. See the Focus & Cursors pattern for additional details.

You should always use an Input Method Indicator of some sort to denote the available text entry modes.

Though some fields are clear based on context (message composition fields in SMS applications), you will generally need to label each field. Labels adjacent to the field should be to the left or above the field, and must be close enough to make the relationship clear. Use regular alignment and designed grids to ensure clarity.

If space is short, some abbreviations or icons may also be used to save space. Figure 11-3 shows several space-saving options.

To save even more space you can place labels inside the form field itself. You can also use this method when labels outside the field are provided, for *hint text* to give information on the type or limits of information to be entered in the field. Label or hint text in the field must be clearly differentiated from actual content (whether user-entered or prepopulated). Typically it will be gray, and it may be italicized or a different size to make the difference clearer under all conditions. Content specifics, such as trailing ellipses, may also serve to make this distinction clearer.

Labels or hints in the field are not like prepopulated text, and will disappear when focus is granted. When all content is removed from the field (or none has been entered) and focus is removed, the label or hint text will appear again. The loss of the label or hint text, once the user types, can be a problem, so you should consider if you can live with this interaction.

If hint text is required and the field is not available (due to the label being in the field or the likelihood of prepopulated information), hint text should appear adjacent to the field, either below or to the right.

Whenever possible, validate forms at the field level as they are typed or when the field loses focus (indicating the user has completed entry). Successful validation can be indicated adjacent to the field, near any hint text. Errors in validation, indicating improper entry, should appear in the same location. The relevant hint text, such as field constraints, can be highlighted if space allows.

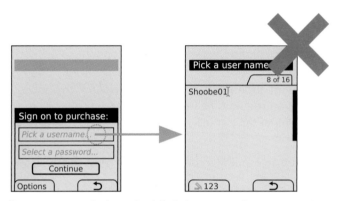

Figure 11-5. *Full-screen entry methods are the default for J2ME and some entire OSes. Aside from simplicity of development, they were originally developed to offer all entry options, counters, and other features in a small amount of space. Today, they are just confusing, as the user is removed from the context entirely. This example typifies one key issue: a large field is exposed for entry that cannot exceed 16 characters.*

Do not use full-screen input panels, as shown in Figure 11-5. In several OSes or environments, selection of an input area will display a common full-screen panel by default. This removes the user from any context, makes hints and other text invisible, and makes implied limits (such as the difference between a text field and a textarea field) invisible. Even if you are working in such a platform, there are workarounds, and often fairly easy ones. Be sure to avoid this behavior whenever possible and allow editing of text in the page context.

Form fields, and especially text Input Areas, are among the most prone to frustration and error due to free-form entry. Design carefully to avoid confusion, and use supporting patterns such as Autocomplete & Prediction and the principles of Cancel Protection to make entry more efficient and speedy.

Form Selections

Problem

You must support making single or multiple selections from preloaded lists of options.

Form Selections, like radio and pull-down lists, are universally supported in a fairly consistent manner.

Solution

 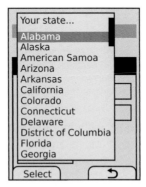

Figure 11-6. *A typical select or "pull-down" element, alongside other form elements. When selected, a pop-up scrolling list appears, from which one item can be selected.*

Mobile devices, even more than other computing devices, can use forms to good effect for selection from already-provided information. Careful design choices can reduce user input, improve accuracy, and greatly reduce frustration.

Extending the ease of entry, mobiles have several regularized variations and subtypes of single and multiple selections. Consider the entire range of variants within this pattern before completing the design of any selection mechanism.

For mobile web use of standard forms, see existing standards from organizations such as the W3C for the presentation, interaction, and design of these form fields. When making mobile-specific websites, feel free to also consider use of the other variations outlined in this pattern.

Remember to use sensors, already-entered information, and any other data to prepopulate or make informed assumptions whenever possible. Do not make users enter information they already have, or which the system should know.

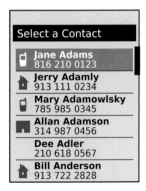
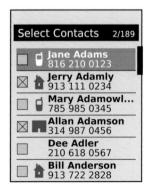

Figure 11-7. *Lists are heavily used in mobile devices, either as single-select, where only one click selects and submits, or for multiple selections. Long lists should usually appear as scrolling lists, using the Vertical List pattern, but shorter ones such as the Power key options on the right should more clearly express the actions, by using lists of buttons instead.*

Several variations exist, including some that do not appear to be form elements based on conventional web form understanding. These can be categorized into three basic types:

Single-select

When this is used in conjunction with other form elements, a single item may be selected from a provided list. When all form selections are made a Button or other submission element (such as a soft key) is used to submit the form.

Select

The drop-down or pull-down list as seen on desktop and web forms can be placed within a mobile application. When selected, it opens as an anchored layer, allowing scrolling and selection. See Figure 11-6.

Radio

Each item in a list is preceded by a line-item selection mechanism, just like a radio button list in desktop or web forms. These work best for vertical text lists. In other cases they may be unrecognized or not easily associated with a particular item. Compare this with a checkbox in Figure 11-8.

Multiple-select

When this is used either in conjunction with other form elements or as a list on its own, one, several, or all items from a provided list may be selected. Items already selected (or preselected by the system) may be deselected. When all form selections are made a Button or other submission element (such as a soft key) is used to submit the form. Figure 11-7 shows some of these.

Checkbox

Each item in a list is preceded by a line-item selection mechanism, just like a checkbox list in desktop or web forms. Used in conjunction with other form elements, these work best for vertical text lists but can be made to work in multiple columns and for other types of selections. For complexly displayed multiple selections, try the next variation.

Multiple-select list

This is a variation of the Select List pattern, where each item displayed has an associated checkbox, allowing selection and deselection.

Single-click

When this is used when there is no form, or if this is the only element, a single item may be selected from a provided list. This does not mean it may be the only interactive item on the page. No button is required to submit the form, as the single selection is committed immediately.

Button list

This is used not for radio buttons, but for actual form-submission types of buttons. A list of these can be provided to choose from. This is only used for short lists, where no scrolling is required and the action is committed immediately.

Select list

When a longer list is needed, the setting does not take effect immediately, or confirmation of the setting should take place inline (by showing the current state in the same display, refreshed) a Select List, usually a simple text Vertical List, will serve the purpose.

Interaction details

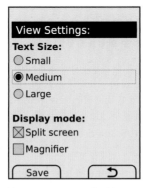

Figure 11-8. *Radio buttons and checkboxes are most valuable in conventional forms, such as this mixed form, with conventional submit buttons. Some handsets use the soft keys as their default buttons.*

Any of the types of Form Selections may be viewed in their passive state at any time, or while not in focus. They may be scrolled through and selected (or deselected, if permissible) while the selection mechanism is in focus.

The definition of focus here may not be entirely clear, but generally it does not concern the end user. For the sake of better understanding, the following (somewhat circular) definition will suffice: any time the list can be scrolled through, it is in focus. Examples may help.

- A drop-down list is not in focus when it appears as a single line. When the field is in focus, it will open to reveal what is functionally a scrolling, vertical Select List. Whenever the field is in focus, for a scroll-and-select device one item within the list will be in focus, and using the "OK/Enter" button will commit the field selection and close the drop down. For touch or pen devices, no items are in focus and pointing at an item in the opened list will select and close the list.

- Any list used inline in the context of a page, such as radio button list, can be considered to be in focus at all times. The individual items are selected in the same manner as for drop-down lists, though the list itself does not change, just the selection mechanism.

- Select lists are mostly used full-page, so the list is much like a drop down in that it scrolls as a complete list. When it is a multiselect list, this works like an inline list and selection only changes the indicator. When it is a single-select list, it is more like the drop-down list and closes the list or continues to the next state.

Whenever content is known to the system or it has been entered by the user previously, it should be preselected in the field. Prepopulated content is interacted with just as a user-entered selection, and may be deselected or another selection made.

Presentation details

Focus of a field and selection item must always be very clearly delineated, with border, background, or other effects. See the Focus & Cursors pattern for additional details.

Selection mechanisms should always appear as their intended function. Use current web standards as a heuristic evaluation for the design of such mechanisms; even if OS standard form widgets are used, ensure that they are understandable to the typical user. Do not mix functions with different appearances, such as allowing multiple selections with selectors that look like radio buttons.

Labels should accompany the field whenever space provides, and may be abbreviated or iconic to save space. Labels adjacent to the field should be to the left or above the field, and must be close enough to make the relationship clear. Use regular alignment and designed grids to ensure clarity.

Labels may be placed inside the form field itself, when there is space inside, such as for a pull-down list. You can also use this method when labels outside the field are provided, for hint text to give information on the type or limits of information to be entered in the field. Label or hint text in the field must be clearly differentiated from actual content (whether user-entered or prepopulated). This is usually by language or formatting; "Select a state..." is clearly different at a glance from "Kansas."

Labels in the field are generally selection text, but they will be invalid selections. Do not allow them to be submitted (or not have an assigned value, just a label). If hint text is required, such as reasons for entering the information or clarification about what the information means, it should appear adjacent to the field, below or to the right.

Whenever possible, validate forms at the field level, as they are selected, or when the field loses focus (indicating the user has completed entry). Successful validation can be indicated adjacent to the field, near any hint text. Errors in validation, indicating improper entry, should appear in the same location. The relevant hint text, such as field constraints, can be highlighted if space allows.

Validation may also allow automatic selection of items. For example, on an address form, if the user enters a zip code (postal code), the corresponding region information may be filled in as well, before the user can scroll down to fill in the information.

Antipatterns

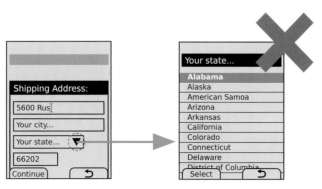

Figure 11-9. *Full-screen entry methods are the default for J2ME and some entire OSes. Especially for selection lists, they are just confusing, as the user is removed from the context entirely. In this case, the state is a part of a larger address, and the context (shipping) may not be the default case the user considers. Staying in context can prevent errors, mistakes, and confusion.*

Do not use full-screen input panels, as shown in Figure 11-9, as the selection mechanism for small components such as pull downs. In several OSes or environments, selection of any input or select field will display the choices as a full-screen panel. This removes the user from any context and makes hints and other text invisible.

Use caution when designing forms if selection mechanisms require odd display methods. Some OSes will not display a pull down as attached to the initiating area, but as a free-standing Pop-Up. Some even do the same for radio button lists. Although there are often workarounds, sometimes there are not, or they would break too far from the OS standards. If this is the case, consider using a different design, such as only selecting one piece of information at a time. A settings page could, instead of a long form, be a list of items, each of which opens a single settings panel.

Never overuse form-style input methods; if a single Select List or Button list is needed, simply allow selection to commit the actions. Do not also make the user select a submit button to commit the action.

Use the right type of selector for the information to be entered. Do not attempt to provide hints on how to multiselect from a pull down, but use a multiselect mechanism (a checkbox or Select List) instead.

Mechanical Style Controls

Problem

You must provide more obvious methods for users to make changes to a setting, level, or value.

Mechanical Style Controls are entirely a matter of design and implementation. Though they really require touch or (possibly) pen input hardware, there are few other absolute requirements.

Solution

Figure 11-10. *Spinners and tapes are form elements, and they can be mixed with others, as shown here. Be sure to use the best type of element for each piece of data, but also ensure that the styles interact adequately, so it feels like a single form.*

Although some machine-era inputs and outputs forced restrictions based on systematic limits, many were well conceived and grounded in good ergonomic and human factors principles.

Some are simply helpful to use because the mapped interface is very well recognized, or there is no other way to "directly" control the output. Audio output, for example, has no obvious way to simply evoke more sound waves, so a continuously moving volume control is the best available solution.

These can be repurposed for digital input, especially on "direct entry" touch or pen interfaces. You may find they can serve as more compact, graphically oriented variants of certain single-select Form Selections, as shown in Figure 11-10. For their use with other forms, see that pattern.

These types of controls have value, especially over other types of selectors, in three key attributes:

- They are graphically oriented, so they tend to provide information at a glance.

- They provide direct control when using touch or pen interfaces, or very close coupling of the input to the available display methods for scroll-and-select devices, with an implicit scale. Compare this to a simple list of values, which the user must map to a model of the system.

- The manner in which they are controlled is apparent (if well designed), so no instruction and minimal previous knowledge of computers is required. Even for scroll-and-select devices, as long as the user can control the device to this point, he is likely to be able to use the control.

All of these are in opposition to many typical computing forms, which are usable only to those who have become accustomed to them in other contexts. Mobile devices are so pervasive, however, that they are many people's first exposure to robust computing systems, so more obvious controls are necessary.

Variations

Although several types of Mechanical Style Controls have been simulated in digital environments, only two distinct classes have really been successful. Each of these, with their subvariations, is discussed in detail here.

Tapes or *sliders* are vertically or horizontally oriented indication and control mechanisms (Figure 11-12). They are used to control levels, such as for volume or zoom level. Zoom & Scale is in fact a specialized subset of these, with the results displayed live. Display Brightness Controls and volume controls (see Chapter 13) are other common implementations.

Continuous

A single strip is shown, and the level can be set anywhere on the tape. Generally this just means a large number of increments (e.g., 100), but the setting is smooth and continuous.

Incremental

The strip is divided into visible increments, and when the indicator is moved, it will jump to (or if smoothly dragged, will only lock at) specific intervals. If there are only eight volume levels, for example, an incremental tape is the right solution.

Sliding switch

When an incremental tape has only two positions, it is a sliding switch instead. Expressing a toggle as a switch with two positions instead of a form element such as a pair of radio buttons or a single checkbox can be very helpful for many direct-control interfaces.

The term used here is derived from linear "tape" gauges, as for indication purposes they reflect this closely. The closest machine-era term would be *sliding potentiometer*, but when shortened this can be confused with other interface elements, and it discusses technology too closely.

Spinners, as shown in Figure 11-11, simulate, often with significant visual fidelity, a small-diameter wheel on a horizontal axis. Only the current setting is visible through a port in the screen, and moving it rotates the wheel in the vertical direction to reveal other settings.

Click wheel

Like the mechanical thumb wheel, gestures (with finger or pen) up and down are used to move the number directly. This can be combined with the other methods below for all touch interfaces.

Hardware buttons

When in focus, using up/down buttons or some other hardware control can increment and decrement the value.

Increment button

A single on-screen button is provided, which increments the value with each button push. This is good for items that are mostly incremented, and for small ranges of values so that mistakes can be fixed by going around. An example is a snooze control on an alarm. Each time the user pushes the button, 5-, 10-, or 30-minute increments are provided.

Increment and decrement buttons

Instead of one, two buttons are provided, above and below the value indicator. The top one increments and the bottom one decrements.

Direct entry

When the field is in focus, entry can be typed directly. This is most useful for numeric values, but you can make it work as the live jumping variant of the Search Within pattern for text items.

The term *spinner* is already somewhat in use. The most generic name for a machine-era device like this would be a *thumbwheel*, but this implies only the click-wheel style of interaction. There is historically no single, noncumbersome name for such controls.

Rotary dials and other types of continuous, glanceable controls or indicators are not used enough on mobile devices for patterns to have emerged. Many of these are simply unsuitable, due to the interfaces currently available. Rotary controls, for example, rely on a fixed axis of rotation, which is simply not present in touchscreens. It is possible that advanced Haptic Output methods will, in the near future, allow these to work by simulating grasping of an object.

Attempting to use some of these other control types may not be obvious or easy for the end user. Pushing the interface concept can lead to confusion and mistakes in input.

Figure 11-11. *Two variations of a full-featured spinner, with increment and decrement buttons. On a touch device, these could both support gestural control as well, using the clicky thumbwheel variation. Note that on even the low-fidelity version, the numbers are within boxes that are clearly differentiated from the rest of the interface. The high-fidelity version appears to be split by individual characters, but this is purely for visual display and style purposes; the minutes and hours are changed as a whole set.*

Interaction details

The aforementioned variants are classes, and they provide guidelines, but there are numerous spaces in the middle where new interfaces, perhaps inspired by machine-era antecedents, may be used. Consider the standards used by the OS and the user's needs to enter data quickly and to protect entry, when selecting one of these methods or designing another one.

Pick the most suitable entry method, even if it induces design inconsistency. A spinner is a terrible way to select from two values, such as a.m./p.m. when setting time. A sliding-switch style of tape would be the obvious replacement, and it has similar direct interaction, so you can make it match the overall style of the interface.

Values do not have to be at consistent intervals. As in the snooze button example, they should be sensible and predictable, but they do not have to be regularly spaced.

Ideally, the entire display should be shown as though it is an actual mechanical object, though it may be styled in a shiny, perfect, digital manner. It can also be made to work with simple type and flat colors. Just keep the principles in mind: the background of the value should still be a different color, a border should exist, and the entire element to include increment buttons must be in a single container.

Spinners should snap to displayed values, and never be able to stop with partial values in the window. Slight misalignment at lockup may be tolerated if it is deliberately built and it follows the design ethic of being a quirky, actual machine. But this is not typical, so use it carefully.

Tapes should also be designed as though they are physical items, with a moving selector handle/indicator. Always offer a handle that can be easily grabbed for touch or pen devices. Additional communication of the value can be offered by filling in the background of the strip. The edge of the filled space indicates the position.

Tapes with incremental control must display the increments. You can use any style that is clear, including gaps, ticks, or other types of indicators within or immediately adjacent to the displayed bar. You can also provide a scale, or values for each level, but you will often want to suppress them, as they may be arbitrary. You can also avoid displaying the value of each tick, if the current value of the setting is displayed, attached to a moving indicator, or at the end of the tape.

Use nonvisual responses to inputs whenever possible, and especially for gestural inputs. Clicks should be felt (or if needed, heard) at each increment. For nonincremental values, you can provide a subtler response (very light clicks or a resistant, almost scraping feedback) to indicate that the control is moving and to encourage the perception that these are physical controls being moved. See the Haptic Output pattern for additional details.

Antipatterns

All Mechanical Style Controls must perform smoothly and respond without delay to user input. Delays will cause the user to attempt a new entry, and built-in input buffers may result in the user accidentally making the wrong selection.

Avoid using variants such as the click wheel, where the display is obscured by the user's finger, without haptic or audio feedback. Either do not obscure the input, or always allow some other feedback of incrementing.

Other dedicated level-changing keys the user may have become accustomed to, such as the volume control, may also be useful for changing the level of spinners and tapes. However, this advice may conflict with other interface expectations or the OS standards. If the volume keys are reserved for volume only, or are used in some cases to scroll the page, do not use them to change other levels.

Clear Entry

You must provide a control that allows users to remove contents from fields, or sometimes whole forms, without undue effort and with a low risk of accidental activation.

Clear Entry functions are built into some form elements in some platforms, but in others (such as some web browsers) they may be difficult to implement due to restrictions on refreshing data without replacing the entire contents of the page.

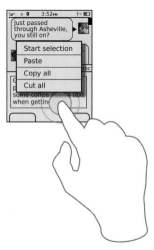

Figure 11-13. Option menus, such as this contextual menu on a touch device, offer options for dealing with text already. Adding a Clear function may not be natural, but a dual-purpose function with automatic recovery, such as this "Cut all" function, can be a very convenient way to include it. The user is additionally led to develop a mental model that includes this, from her more frequent use of typical editing functions.

Just as text entry on mobiles is difficult, removing already-filled-in values can be as difficult, and even more tedious. Long URLs, or other old, prepopulated or prefilled data may be inaccurate, and it can be very slow to delete all this content. Providing an explicit method to remove the content can sometimes be helpful.

Clear Entry functions should only rarely be used, and can be dangerous, as they may accidentally encourage deletion. But they should be considered for some types of fields and forms. This pattern is in no way meant to condone the common but poor practice of the "Reset" Form button, or to encourage the use of clear functions on every field.

Additionally, the "Reset" Form button does not allow selection of individual fields. Even on touch devices, it can be difficult to use conventional desktop paradigms of selection and deletion, so without additional assistance users have to resort to repeated keystrokes and other time-consuming actions. To make matters worse, certain actions we take as designers to prevent loss of data can make it hard to get around; discarding a text message, for example, often saves the message as a draft and there is sometimes no way to get around this.

A series of functions and tactics should be considered to assist the user with this task, without conflicting with primary tasks and principles of preservation of data.

Variations

Most Clear Entry functions for mobiles are at the field level, and these almost always empty text Input Areas:

- Explicit functions are controls that must be activated and then, in one step, clear or reset an entire field. The reset function will revert to any defaults, and also means it can be used on any type of input field or function.

- Assistive functions reuse, repurpose, or exaggerate existing features of the interface to assist with field clearing, such as that shown in Figure 11-13.

Form-level clear and reset buttons are discouraged, but they can be useful in some processes. Additionally, do not overlook the possibility that your form may have essentially one field in it. An SMS message composition page can use a "Clear" button at the form level, which functionally will only clear the one field.

Interaction details

You should not automatically include Clear Entry in every page, or even every project, but you should always consider it. Think of all form design—as you think of all interactive design—in the context of the user. If repeated use of a field is likely, or there is no other way to abandon the information, you may need to add a method to clear the input.

If you are implementing this as an explicit function, simply add buttons adjacent to the field (or within it), or as options available in a menu. These may overlap with the intent of the implicit function by simply offering a "Cut all" function; this has other purposes and in fact loads the clipboard with the data, but also serves to clear the field. Make sure this function is visible enough. Many edit controls are only available via Press-and-Hold or other, low-affordance functions.

One classic implicit function is to simply overdrive the key repeat. The typical user will, when faced with a large delete task, simply enter the field and begin pressing the Delete key. Many will be aware of key repeat, so they will keep the key pressed down if the whole field must be cleared. Key repeat usually has a relatively low maximum speed, to allow for stopping at a specific point. But if this is unlikely, such as in the URL field example or after a longer-than-usual time, the repeat may accelerate, or simply scroll across the entire field.

This may be a very marked acceleration, even to the point where the entire field is suddenly deleted within a fraction of a second. This must always be coupled with behavior that stops key presses at the end of the field, and does not move to the next field in line without the user stopping the action, even if the Back key is used to cycle to the previous in-focus area otherwise. This behavior can be applied to arrow keys when used for selection as well, but is not applied to other keys. This acceleration method should work for both selection and deletion functions.

These variants and examples should not rule out other methods being developed or used. Gestural clear functions are becoming common on Pen Input fields; however, these have not yet developed into a regularized enough set to be considered a single pattern.

For all of these actions, use the principles of Cancel Protection. Explicit cancel guards are not usually needed if the control is well designed, but saved states, or other paths to retrieve the information, should often be developed. URLs can usually be revisited with the existing history functions, and the "Cut all" menu option described earlier does save the cut data to the clipboard; pasting will recover it.

Presentation details

Label all buttons, menus, and other functions clearly. Rarely can a reset function be a compact button, labeled only with a graphic, due to the potentially catastrophic loss of information. Use text labels when possible.

Be sure to label clearly. "Clear," "Reset," and "Cut" are very different actions, with different implications. Do not confuse terms or assume the user base is not savvy enough to understand subtle semantic differences.

Live functions such as high-speed key repeat, or gestures to select or delete functions, must show the action happening in real time. If the screen cannot update at the speed of the actual behavior, do not use the function. Even at maximum speed, the user must be able to release the key and expect to have the cursor stop where he sees it at that moment.

Always use conventional display methods for exaggerated behaviors. Selection at high speeds should use the same highlighting as normal-speed selection, for example.

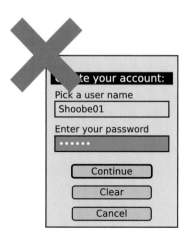

Figure 11-14. *Even on the desktop web, the Clear button is massively overused, and causes errors more than it fulfills real use cases. This very small form has no need for such a button, and it is also undifferentiated from the other two, and far too close. Even if identified, it can easily be accidentally activated.*

Form-level Clear buttons, if used, must be very carefully placed to avoid accidental activation and should still often use explicit Cancel Protection methods. Do not use the poor convention of the Clear button immediately next to the Submit button at the bottom of every form, as shown in Figure 11-14.

Chapter 12

Audio and Vibration

The Big Tooter

High atop Mount Oread, in the picturesque city of Lawrence, Kansas, stands a whistle—a whistle with quite a history composed of tradition, controversy, and headache that started about a hundred years ago. The whistle is known as the "Big Tooter."

March 25, 1912, 9:50 a.m.: a deafening shrill begins. For five earsplitting seconds the power plant steam whistle at the University of Kansas sounds. The sound is so loud it can be heard from one side of the city to the other. It's the first time the whistle is used to signal the end of each hour of class time.

According to the student newspaper *The Daily Kansan*, the whistle was not only used to replace the untimely and inconsistent ringing of bells with a standard schedule of marked time, but was also used to remind professors to end their lectures immediately. Prior to the whistle, too often professors would keep students past the 55 minutes of class time, causing them to be late for their next class. With the new sound system in place, even the chancellor had something to say.

"If the instructor isn't through when the whistle blows," said KU Chancellor Frank Strong to the student body, "get up and go."

The Big Tooter Today

For the past 100 years, the Big Tooter has been the deafening reminder to faculty and students about punctuality and when to cover their ears. I can say I, too, was one of those students who would purposely alter my walk to class to avoid that sound at its loudest range. But despite the fact that the steam whistle was excruciatingly loud, it served its purpose as an audio alert. It was so unique that it was never misunderstood, and is so reliable that it is always trusted.

The Importance of Audition

From the preceding example we see that people can benefit from specific sounds that are associated with contextual meaning. Using audition in the mobile space can take advantage of this very important concept, for the following reasons:

- Our mobile devices may be placed and used anywhere. In these constantly changing environmental contexts, users are surrounded by external stimuli that are constantly fighting for their attention.

- The device may be out of our field of view or range of vision, but not our auditory sensitivity levels.

- Using audition together with other sensory cues can help to reinforce and strengthen users' understanding of the interactive context.

- The user may have impaired vision—either due to a physiological deficit or from transient environmental or behavioral conditions—thus requiring additional auditory feedback to assist him in his needs.

- The user may require auditory cues to refocus his attention on something needing immediate action.

Figure 12-1. Control of volume level is usually still performed by hardware keys due to its importance. This makes it an interesting control as it is always pseudomodal. The user enters the mode by using the keys, which loads a screen, layer, or other widget to indicate the volume change. Related controls, silence and vibrate, are associated with this mode as well.

Auditory Classifications

Audible sounds and notifications have become so commonplace today that we have learned to understand their meaning and quickly decide whether we need to attend to them in a particular context.

Warnings

Audible warnings indicate both a presence of danger and the fact that action is required to ensure one's safety. These sounds have loud decibels (up to 130 dB), and some use dual frequencies to quickly distinguish themselves from other external noise that may be occurring at that time. Most often these warning are used with visual outputs as well. Examples of warning sounds include:

- Railroad train crossings
- Emergency vehicle sirens
- Fire alarms
- Tornado sirens
- Emergency broadcast interrupts
- Fog horns
- Vehicle horns

Alerts and Notifications

Alerts are not used to signal immediate action due to safety. Instead, they are used to capture your attention to indicate that an action may be required or to let you know an action has completed. Alerts can be a single sound, one that is repetitive over a period of time and can change in frequency.

Mobile alerts are quite common. They must be distinguishable and never occur at the same time as others. When appropriate, use visual indicators to reinforce their meanings. Use a limited number of alert sounds; otherwise, the user will not retain their contextual meaning. Examples of alert and notification sounds include:

- The beep of a metal detector when it detects something under the sand at the beach
- The chime of a doorbell
- The ring of an elevator upon arrival
- The beep of a crosswalk indicator
- The sound of a device being turned on or off
- The sound of a voicemail being received on your mobile device
- A low-battery-level notification
- A sound to indicate a meeting reminder

Error Tones

Error tones are a form of immediate or slightly delayed feedback based on user input. These errors must occur in the current context. Mobile error tones are often buzzers to indicate:

- The wrong choice or key was entered during input.
- A loading or synching process failed.

Voice Notifications

Voice notifications can be used as reminders when you are not holding your device, as well as notifications of incorrect and undetectable input through voice, touch, or keypad. Use syntax that makes it clear what is being communicated. Keep the voice notification messages short and simple. Examples of voice notifications include:

- A reminder to take your medication
- Turn-by-turn directions
- A request to repeat your last input because the system didn't understand it

Feedback Tones

Feedback tones occur immediately after you press a key or button such as a dialer. They confirm that an action has been completed. These may appear as clicks or single tones. Feedback tones can occur when you:

- Enter phone numbers on a dialer
- Enter characters on a keyboard
- Hold down a key for an extended time to access an application such as voicemail
- Press a button to submit user-generated data
- Select incremental data on a tape or slider selector

Audio Guidelines in the Mobile Space

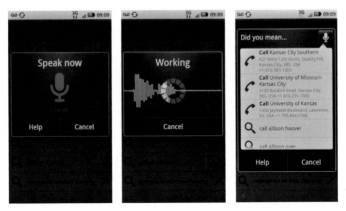

Figure 12-2. *Voice inputs can help users in situations where visibility is limited or nonexistent, but they have to be part of a complete voice UI. Don't just perform one or two steps of a process and then require the user to read and provide feedback on the screen.*

Signal-to-Noise Ratio Guidelines

People will use their mobile devices in any environment and context. In many situations, they will rely on the device's voice input and output functions during use. See Figure 12-2. But, whether the person is inside or outside, her ability to hear certain speech decibels apart from other external noises can be quite challenging. Here are some guidelines to follow when designing mobile devices that rely on speech output and input:

- Signal-to-noise ratio (S/N) is calculated by subtracting the noise decibels from the speech decibels.

- To successfully communicate voice messages in background noise, the speech level should exceed the noise level by at least 6 decibels (dB) (Bailey 1996).

- A user's audio recall is enhanced when grammatical pauses are inserted in synthetic speech (Nooteboom 1983).

- Synthetic speech is less intelligible in the presence of background noise at a 10 dB S/N.

- When the noise level is +12 dB to the signal level, the consonants *m, n, d, g, b, v*, and *z* are confused with one another.

- When the noise level is +18 dB to the signal level, all consonants are confused with one another (Kryter 1972).

Speech Recognition Guidelines

In addition to the signal-to-noise guidelines in the preceding section, you must understand how users recognize speech. The following guidelines will assist you when designing Voice Notifications:

- Words in context are recognized more when they are used in a sentence than when they are isolated, especially in environments with background noise.

- Word recognition increases when the words are common and familiar to the user.

- Word recognition increases if the user is given prior knowledge of the sentence topic.

Audio Accessibility in the Mobile Space

When designing for mobile, as with any device, always consider your users, their needs, and their abilities. Many people who use mobile devices experience visual impairments. We need to create an enriching experience for them as well.

Recently companies have been addressing accessibility needs as standard functions in mobile devices. Before this, visually impaired users were forced to purchase supplemental screen reader software that worked on only a few compatible devices and browsers. These are quite expensive, starting at around $200 to $500.

Accessibility Resources

Here are useful resources on audio accessibility. Included is information on types of assistance technologies companies are using in mobile devices today.

- Apple has integrated VoiceOver, a screen access technology, into its iPhone, iPod, and iPad devices. For more information on Apple's accessibility commitment, visit *http://www.apple.com/accessibility*.

- Companies such as Code Factory have created Mobile Speak, a screen reader for multi-OS devices. See Code Factory's site at *http://www.codefactory.es/en*.

- For additional information on accessibility and technology assistance for the visually impaired, we recommend viewing the American Foundation for the Blind's website, *http://www.afb.org*.

- We also recommend the following websites geared toward accessibility of mobile devices for all:

    - *http://www.accesswireless.org/Home.aspx*

    - *http://www.mobileaccessibility.info/*

    - *http://www.nokiaaccessibility.com/*

The Importance of Vibration

Figure 12-3. *Vibrate on most devices is coarse, and is provided by a simple motor with an off-center weight. Here, it is the silver cylinder between the camera and the external screen; the motor is mostly covered by a ribbon cable. It is mounted into a rubber casing, to avoid vibrating the phone to pieces, but this also reduces the fidelity of specific vibrate patterns, if you were to try to use it for that purpose. The figure on the right shows the motor assembly on its own.*

Depending on our users' needs, their sensory limitations, and the environment in which mobile is used, vibration feedback can provide another powerful sensation to communicate meaning.

Since the largest organ in our body is our skin, which responds to pressure, we can sense vibrations anywhere on our body. Whether we are holding our device in our hands or carrying it in our pocket, we can feel the haptic output our devices produce.

When designing mobile devices that incorporate haptics, be familiar with the following information:

- Haptic sense can provide support when the visual and auditory channels are overloaded.
- The touch sense can respond to stimuli just as quickly as the auditory sense, and can respond even faster than the visual sense (Bailey 1996).
- In high-noise-level areas or where visual and auditory detection is limited, haptics can provide an advantage.

Common Haptic Outputs on Mobile Devices

Many mobile devices today use haptics to communicate a direct response to an action:

- A localized vibration on key entry or button push

- A ring tone set to vibrate

- A device vibration to indicate an in-application response, such as playing an interactive game (i.e., the phone might vibrate when a fish is caught, or when the car you're steering accidentally crashes)

Haptic Concerns

Using haptics appropriately is a great way to provide users with additional sensory feedback. However, you still need to be aware that:

- Using haptics can quicken the process of draining battery life. Provide the option to turn haptic feedback on and off.

- Use haptics when appropriate. Too much might reduce the user's attention to the stimuli and ignore the response.

Patterns for Audio and Vibration

Using audio and vibration control appropriately provides users with methods to engage with the device other than relying on their visual sense. These controls can be very effective when users may be at a distance from their device, or are unable to directly look at the display but require alerts, feedback, or notifications. In other situations, a visually impaired user may require these controls because they provide accessibility. We will discuss the following patterns in this chapter:

Tones
Nonverbal auditory tones must be used to provide feedback or alert users to conditions or events, but must not become confusing, lost in the background, or so frequent that critical alerts are disregarded. See Figure 12-1.

Voice Input
A method must be provided to control some or all of the functions of the mobile device, or provide text input, without handling the device. See Figure 12-2.

Voice Readback
Mobile devices must be able to read text displayed on the screen, so it can be accessed and understood by users who cannot use or read the screen.

Voice Notifications
Mobile devices must provide users with conditions, alarms, alerts, and other contextually relevant or time-bound content without reading the device screen.

Haptic Output

Vibrating alerts and tactile feedback should be provided to help ensure perception and emphasize the nature of UI mechanisms. See Figure 12-3.

Tones

You must use nonverbal auditory tones to provide feedback or alert users to conditions or events, without them becoming confusing, lost in the background, or so frequent that critical alerts are disregarded.

Practically all mobile devices have audio output of some sort, and it can be accessed by almost every application or website. There can be strict limits, such as devices that only output over headsets, or those which only send phone call audio over Bluetooth, which can limit the use of some tones.

Figure 12-4. Feedback tones provide confirmation that the target was contacted correctly and that the device accepted the input. This is especially helpful for touch and pen devices, which have little or no tactile feedback, unlike hardware keys that move and (generally) click as part of the key action. Feedback has developed over time as a result of keys being obscured by the user's finger or hand, so visual feedback is of limited value.

Interaction design, like much design work, focuses heavily on visual components. However, users of all devices have a multitude of senses, all of which can and should be engaged to provide a more complete experience.

Mobiles are carried all the time, often out of sight, and must use audio alerting and Haptic Output to get the user's attention, or to communicate information more rapidly and clearly. Tones should be used not just for alerting or for problems, but as a secondary channel to emphasize successful input, such as on-screen and hardware button presses, scrolls, and other interactions.

Alerts are used to regain user focus on the device. Additional communication can then be made to the user, usually visually, with on-screen display of Notifications or other features. However, this may also involve Voice Readback or initiating voice reminders.

Whenever you are implementing this pattern, be aware of the danger of the "better safe than sorry" approach to auditory Tones. Too much sound can be as bad as too little in that it can add confusion or induce *alarm fatigue*. At its worst, this can be akin to the well-known phenomenon of *banner blindness* with some or all device tones simply edited out by the user's brain.

Variations

Feedback tones are used to provide direct response to a user action. You can serve up a tone as alternate and immediate feedback to indicate that the system has accepted the user action. This is usually a direct action such as a button press or gesture. This style of feedback developed in the machine era, as a result of the user's fingers and hands obscuring the input devices; even when a button moved or was illuminated, more efficient input was ensured by the inclusion of auditory and Haptic Output. See Figure 12-4.

Alerts serve to notify the user of a condition which occurs independent of an action, such as a calendar appointment or new SMS. These also date back far before electronic systems. Due to the multitude of alarms, control systems such as those used for aircraft and power plants are more useful antecedents to study than simple items such as alarm clocks. See Figure 12-5.

Delayed response alerts exist in a difficult space between these two—for example, when an action is submitted and a remote system responds with an error several seconds later. The response is in context, so it is similar to a feedback tone, but it needs to inform the user to look at the screen again, much as an alert.

Interaction details

Phenomena such as the *McGurk effect*—in which speech comprehension is related to the visual component—appear to exist for other types of perception. Tones should support interaction that is directly related to the visual portions of the interface.

Tones must always correspond to at least one visual or tactile component. If an alert sounds, you should always make sure there is a clear, actionable on-screen element. These should always appear in the expected manner for the type of event. Feedback tones correspond to key presses or other actions the user takes, and alert tones should display in the conventional method the device uses for Notifications.

Displayed notifications must not disappear without deliberate user input. Users may not be able to see the screen or act on a particular alert for some time; the visual display to explain the alert must remain in place.

Use caution with reminder alerts, or repeatedly announcing the same alert again. You should generally only use these for conditions that are time-sensitive and require the user to perform a real-world action, such as an alarm clock. There is generally no need to repeatedly sound an audible alert for normal-priority items such as email or SMS messages.

Alert silencing—removing the auditory portion or the audible and LED blinking portion of an alert—should be unrelated to any other notifications. It should be possible to silence an alert without dismissing the entire condition. Often, this can be done by simply opening or unlocking the device; if the device was previously in a locked mode, the user can be assumed to have seen and acknowledged all conditions. You can stop sending tones, but until the user has explicitly dismissed the message, the alert condition is marked as unread.

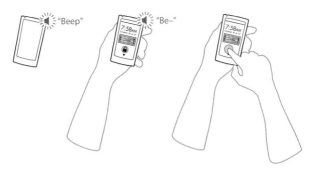

Figure 12-5. *A sample process by which alerts are displayed, with audio as a component. While the device is idle and locked, the alert condition occurs. The device LED illuminates and the alert tone plays. The user retrieves the device and taps the power button to illuminate the lock screen; this user action silences the audio alarm immediately, but does not cancel the alert otherwise, so the LED and notifications remain in place. Once the user unlocks the device, the notification will appear in the conventional location on the Idle screen, and can be interacted with as usual.*

Presentation details

Any Tones you send must correspond to the class and type of action, and should imply something about the action by the type of tone. Clicks denote key presses, whereas buzzers denote errors, for example.

Sound design is an entire field of study. If sounds must be created from scratch, a sound designer should be hired to develop the tones based on existing principles. Be sure to denote the meaning—and emotional characteristics—for each sound.

Vibration and audio are closely coupled, so you should consider them hand in hand. If Tones are used in concert with Haptic Output, ensure that they do not conflict with each other. For devices without haptic or vibration hardware, audio can serve this need in a limited manner. Short, sharp tones and especially those at very low frequencies provided through the device speaker can provide a tactile response greater than their audio response. Test on the actual hardware to be used, to make sure the device has appropriate responsiveness and that low-frequency tones do not induce unwanted buzzing.

The "illusory continuity" of tones can be exploited in a manner similar to how "persistence of vision" is for video playback. Simple tones can be output with gaps or in a rapid stairstep and be perceived to be a continuous tone. You can use this to simulate higher-fidelity audio than can be achieved on devices with constrained processing hardware, or when the rest of your application is using so much of the processor power that you need to make components such as the audio output more efficient.

Do not allow alerts or other Tones to override voice communications. Many other types of audio output, such as video or music playback, may also demand some tones be greatly reduced or entirely eliminated.

Generally, alerts can be grouped. If three calendar appointments occur at the same time, only one alert tone needs to be played, not three in sequence. Never play more than one tone at a time. You only need to get the user's attention once for her to glance at the screen and see the details.

Use caution with the display of multiple alerts that closely follow each other. If the user has not acknowledged an incoming SMS and a new voicemail arrives moments later, there may be no need to play the second alert tone. The delay time required to play a second alert tone depends on the user, the type of alert, and the context of use.

Carefully consider *audibility*, or how any sound can be perceived as separate from the background. You can use automatic adjustments by opening the microphone and analyzing ambient noise. However, do not simply make sounds louder to compensate for this. Adjust the pitch or select prebuilt variations more appropriate to the environment.

Be aware of the relationship between pitch and sound pressure, and adjust based on output volume. Consider using auditory patterns (or tricks) such as *Sheppard-Risset tones* to simulate rising or falling tones, which can be more easily picked out of the background or located in space.

Antipatterns

Alarm fatigue is the greatest risk when designing notification or response Tones. If your audio output is too generic, too common, or too similar to other tones in the environment (whether similar to natural sounds or electronic tones from other devices), it can be discounted by the user. Use as few different sounds as possible. This may be deliberate, where the user consciously discards "yet another alert" for a time, or unconscious as the user's auditory center eventually filters out repeated tones as noise. This can even result in users sleeping through quite loud alarms because their brains consider the tone to be noise.

Do not go too far in the other direction and have no auditory feedback, especially for errors. Delayed response conditions are especially prone to having no audible component. This induces a great risk in users who may believe they have sent a message or set a condition to be alerted later, and may not look at the device for some time.

It is even possible for accidental entry, or deliberate eyes-off entry, to clear a visible error so that users are not informed of alerts or delayed responses at all. Be sure to use notification Tones whenever the condition calls for it.

Keep in mind that users may be wearing headsets. Using tones to emulate Haptic Output will not work in this case, and can be very annoying. Consider changing the output when headsets are in use. In addition, some headsets will not accept all output, so if your application relies on tones, make sure there is a way for the device to still make these sounds.

Voice Input

Problem

You must provide a method to control some or all of the functions of the mobile device, or provide text input, without handling the device.

Most mobiles are centered on audio communications, so they can accept Voice Input in at least some ways. Some application frameworks and most web interfaces cannot receive inbound audio, but there are sometimes workarounds, using the voice network. Devices not built around voice communications may not have the appropriate hardware and cannot use this pattern at all.

Solution

Voice Input has for many decades promised to relieve users of all sorts of systems from executing complex commands in unnatural or distracting ways. Some specialized, expensive products have met these goals in fields such as aviation, while desktop computing continues to not meet promises, and not gain wide acceptance.

Mobile is, however, uniquely positioned to exploit voice as an input and control mechanism, and has uniquely higher demand for such a feature. The ubiquity of the device means many users with low vision or poor motor function (and thus poor entry) demand alternative methods of input. Near universal use and contextual requirements such as safety—for example, to use navigation devices while operating a vehicle—demand eyes-off and hands-off control methods.

And lastly, many mobile devices are (or are based on) mobile handsets, so they already have speakers, microphones designed for voice-quality communications, and voice processing embedded into the device chipset.

Since most mobile devices are now connected, or only are useful when connected to the network, an increasingly useful option is for a remote server to perform all the speech recognition functions. This can even be used for fairly core functions, such as dialing the handset, as long as a network connection is required for the function to be performed anyway. For mobile handsets, the use of the voice channel is especially advantageous, as no special effort must be made to gather or encode the input audio.

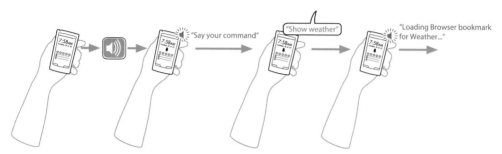

Figure 12-6. Voice input must be considered as an entire replacement UI, with Voice Readback as the output channel. Use the same principles of organizing information and communicating clearly but concisely when designing voice overlays to a conventional UI. State changes must be clearly communicated, as the user cannot necessarily glance at the interface to see what is happening. Here, the initiation of the voice interface and the acceptance of the command are stated. Note how the user input was read back not as stated, but as interpreted by the system.

Variations

Voice command uses voice to input a limited number of preset commands. The commands must be spoken exactly as the device expects, as the device cannot interpret arbitrary commands. These can be considered akin to Accesskeys, as they are sort of shortcuts to control the device. The command set is generally very large, offering the entire control domain. Often, this is enabled at the OS level and the entire handset can be used without any button presses. Any other domain can also be used; dialing a mobile handset is also a common use, in which case the entire address book becomes part of the limited command set as well. See Figure 12-6.

Speech-to-text or speech recognition enables the user to type arbitrary text by talking to the mobile device. Though methods vary widely, and there are limits, generally the user can speak any word, phrase, or character and expect it to be recognized with reasonable accuracy.

Note that voice recognition implies user-dependent input, meaning the user must set up his voice profile before working. User-independent systems are strongly preferred for general use, as users can use them without having to set them up. Only build user voice profiles when this would be acceptable to the user, such as when a setup process already exists and is expected.

A detailed discussion of the methods used for recognizing speech is beyond the scope of this book, and is covered in detail in a number of other sources.

Constantly listening for Voice Input can be a drain on power and can lead to accidental activations. Generally, you will need to provide an activation key press which either enables the mode (Press-and-Hold to input commands or text) momentarily, or switches to the Voice Input mode. A common activation sequence is a long press of the speakerphone key.

You will need to find other keys or methods when there are no suitable hardware keys. Devices whose primary entry method is the screen (touch or pen) may have trouble activating Voice Input without undue effort by those who cannot see the screen.

Consider any active Voice Input session to be entirely centered on eyes-off use, and to require hands-off use whenever possible. Couple the input method with Tones and Voice Readback as the output methods, for a complete audio UI.

When input has been completed, a Voice Readback of the command as interpreted should be stated before it is committed. Offer a brief delay after the readback, so the user may correct or cancel the command. Without further input, the command will be executed.

For speech-to-text, this readback phase should happen once the entire field is completed, or when the user requests readback in order to confirm or correct the entry.

You must carefully design the switch between voice command and speech-to-text. Special phrases can be issued to switch to voice command during a speech-to-text session. Key presses or gestures may also be used, but may violate key use cases behind the need for a Voice Input system.

Correction generally requires complete reentry, though it is possible to use IVR-style lists, as examples or to allow choosing from list items.

You should consider the use of voice command in the same way as the use of a five-way pad for a touchscreen device. Include as much interactivity as possible, to offer a complete voice UI. When controlling the device OS, you must enable all the basic functions to be performed by offering controls such as Directional Entry and the ability to activate menus. This also may mean a complete scroll-and-select-style focus assignment system is required, even for devices that otherwise rely purely on touch or pen input.

Provide an easy method to abandon the Voice Input function and return to keyboard or touchscreen entry, without abandoning the entire current process. The best method for this will reverse the command used to enter the mode, such as the Press-and-Hold speakerphone key.

Figure 12-7. Whenever there is space, such as in this dedicated search box, selecting voice control or entry should communicate the mode switch with icons and labels, and describe the actions to be taken, or the action the system is taking, very clearly. Additional hints, and links to get even more information, can also be provided.

Whenever the Voice Input mode becomes active, you should make it announce this in the audio channel with a brief tone. When this is rarely used, for first-time users or whenever it may be confusing, a Voice Readback of the condition (e.g., "Say a command") may be used instead. After this announcement is completed, the system accepts input.

All Voice Input should have a visual component, as in Figure 12-7, to support glancing at the device, or completing the task by switching to a hands-on/eyes-on mode. When in the Voice Input mode, this should be clearly denoted at all times. The indicator may be very similar to an Input Method Indicator, or may actually reside in the same location, especially when in speech-to-text modes.

Hints should be provided on-screen to activate voice command or speech-to-text modes. Use common shorthand icons, such as a microphone, when possible. For speech-to-text especially, this may be provided as a mode switch or adjacent to the input mode switch.

When space provides—such as a search, which provides text input into a single field—you should display on-screen instructions so that first-time users may become accustomed to the functionality of the speech-to-text system. Additional help documentation, including lists of command phrases for voice command, should also be made available, either in text on-screen or as a Voice Readback.

You cannot rely on audio systems and processing to be *full duplex*, or to be heard at full fidelity while speaking. Give decent pauses, so tones and readback output don't get in the way of user inputs. Consider extra cognitive loads of users in environments where

Voice Input is most needed; they may require much more time to make decisions than in focused environments, so pauses must be longer than for visual interfaces. These timings may not emerge from typical lab studies even, and will require additional insights.

Voice Readback

You must allow certain classes of users or any user in certain contexts to consume content without reading the screen.

Practically all mobile devices have audio output of some sort, and it can be accessed by almost every application or website. There can be strict limits, such as devices which only output over headsets, or those which only send phone call audio over Bluetooth, which can limit the use of some tones.

Mobile devices must be able to read text displayed on the screen for the user, so it can be accessed and understood by those who cannot use the screen.

Due to mobiles being contextually employed, there are numerous instances in which the user may not be able to, not be allowed to, or choose not to read the screen.

The user may well choose to use Voice Readback so that she can use her hands and eyes for other purposes. While working, or performing hobbies which do not require excessive cognitive load themselves, such as most driving, radio and other audio output is used to gather information or provide entertainment. Video is generally more entertaining, but is totally unsuitable for these situations.

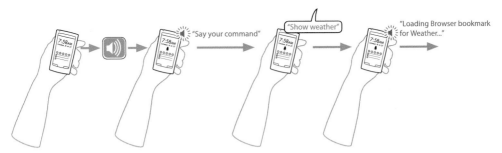

Figure 12-8. Voice Readback can form an integral part of a complete voice UI for mobile devices. Readback is used to prompt for commands, and then confirms the user input or declares how the system has interpreted the command. It will also read on-screen displays and options, to allow the user to select appropriate items without looking at the screen.

Voice Readback always works in broadly the same way, but what is being read varies:

Universal

> The entire interface is read, to allow the device to be used without any view of the display. This is usually combined with the Voice Input pattern to create a complete voice UI, as an alternative to the conventional button (or touch) and screen UI native to the device. Even if used for only one section, action, or phrase, this same method is used for any readback of voice commands. See Figure 12-8.

Elemental

> An entire document, such as a PDF, email, or web page, is read until the user cancels the action or the entire document is read. See Figure 12-10.

Selected

> A selection the user has specified within any context—for example, by highlighting text in a web page—is read in its entirety. See Figure 12-9.

Voice output that is presented based on conditions, such as position or time, is discussed under the Voice Notifications pattern.

Figure 12-9. *Selections made by the user can be read. The play control is usually contextual and is related to the selection such as the Pop-Up menu shown here.*

Voice Readback can be turned on as a setting for the entire OS, or on an application basis. It will then be used automatically, whenever a change in the application is initiated.

Other input methods, such as keypads and directional controls, will still function. Buttons will generally be needed to unlock or refresh a screen so that the current condition is read aloud.

Readback can also be initiated, for the elemental and selection variations, within the application or as a contextual control such as a Pop-Up, menu, Annotation, or other control.

Readback for single-use cases of UI control is the result of Voice Input. Initiation is discussed under that pattern.

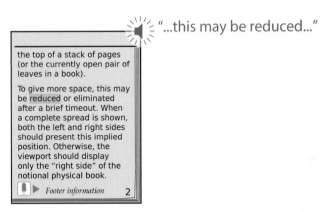

Figure 12-10. *When entire documents, long passages, or even just marquee text is being read, the document will scroll to always have the current reading selection visible in the viewport. A cursor or highlight, as shown here, should be displayed to correspond to the word currently being read. An indicator of audio playback should be on the screen at all times, and a control should be provided to immediately mute or pause the audio.*

Audio should be played by default through the external speaker or speakerphone. The last-set in-call volume (or equivalent playback volume for nonphones) should be used. Whenever possible, detect the ambient noise level and adjust the volume accordingly, in order to make it audible.

When a headset is attached (either physically or by a link such as Bluetooth), the playback should default to this device, and use the last-set in-call volume for this device.

Content read must be identical to that printed on the screen. The condition that resulted in the user employing Voice Readback may be temporary and transient. Allow the user to switch between the screen and audio channels. The user may even wish to read along with the voice output. Even for users with a vision deficit, others may be accompanying them, who may also wish to use the device.

Also be sure that content scrolls as the audio gets to that part so that the item being read back is in the viewport.

There may be delays between phrases, or before the start of the audio readback. To inform the user that audio is about to commence and to prepare the user for the volume level, play a subtle tone immediately beforehand.

Use a similar tone when Voice Readback has stopped for a significant time, or to communicate a selected volume setting, to confirm this condition to the user.

Antipatterns

Avoid mixing readback of commands and text. When the two must be used together, use delays, tones, changes in voice, and clear syntax (such as "You said...") to indicate the difference.

The voice you select must be as understandable as possible. Text-to-voice translation of names especially can be difficult to understand or improperly pronounced. If quality is too low with the available hardware and software, do not implement the solution.

Keep in mind that users may be wearing headsets. Some headsets will not accept all output, so if your application relies on Voice Readback, make sure the targeted devices support your application or service sending audio to all attached headsets or other audio devices.

Voice Notifications

Problem

You must inform certain classes of users—or any user in certain contexts—of conditions, alarms, alerts, and other contextually relevant or time-bound content without requiring them to read the device screen.

Practically all mobile devices have audio output of some sort, and it can be accessed by almost every application or website. There can be strict limits, such as devices that only output over headsets, or those that only send phone call audio over Bluetooth, which can limit the use of some tones.

Solution

Much as Notifications appear on the device screen (and sound alarms and blink the LED) when messages arrive or other such alerts activate, Voice Notifications read the content of a notification so that the information can be used when the device is not in the hand or cannot be viewed.

These can be especially useful for users or contexts in which reading would be difficult. A common case is turn-by-turn directions. The reminders are based on telemetry information instead of time (like alarms) or remote information (like messages), but the principle is identical: a message is read, without specific user input, at the relevant time.

Voice Notifications also allow the screen to remain available for the display of other content. It may be able to be glanced at to retrieve certain types of data (such as a map, in the navigation example), but not to the level or type of detail of the read-aloud text.

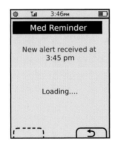

"Time to take your 3:45 pill... Take one blue and yellow, 500 milligram Abelcet, with water."

Figure 12-11. *The best way to notify is to use technologies such as push messaging, supported by a surprising number of devices, on most networks. Remote messages can launch the application, then not only read the alert but also present a correspondingly well-formatted page to look at. Note that the audio only plays when the entire message has loaded, so the user may glance at the device to get other information and follow along. The audio must always be able to be paused or stopped, in case it is disrupting an event or the user has privacy concerns.*

Variations

There are only two basic variations:

In context

> The device is operating in the mode or application from which the alert is launched. Often, the alert is integral to the application, and on-screen display accompanies the alert in a critical manner. Turn-by-turn directions are the key example of this variation. See Figure 12-12.

Change of context

> The device is in an arbitrary mode, on the Idle screen, or locked when the alert is read from an application that is not in focus. A simple example is any typical notification, such as an appointment or incoming SMS. See Figure 12-11.

If the user requests voice output, or voice output is used as a part of generally navigating the device UI, this is a Voice Readback pattern instead.

Interaction details

Voice Notifications occur based on time, position, remote messaging, or other actions outside the user's direct control.

Treat reminders currently being read the same way as other alarms. Make sure they appear in the Notifications area, or as Pop-Up dialogs, as the rest of the OS standards dictate. You must also enable the same actions; they may be muted, snoozed, or canceled. Do not snooze alarms that are irrelevant at a later time, such as position-related messages.

If the voice component is muted, this is the same as silencing an alarm of Tones. The notification itself will remain in the Notifications area, and may be manually selected and viewed (or listened to), or be cleared.

A privacy mode may only read a generic version of the message. When the user interacts with the device, a more detailed message can be seen, or read aloud. This will rapidly become the Voice Readback pattern. See that for specifics of the design and interaction.

Consider informing users of the risks of reading private data aloud when setting up Voice Notifications, and provide methods to change or cancel them easily.

Presentation details

"Turn coming up... Turn right onto Missouri Highway Y - Y in 500 yards... 6 miles to next turn."

Figure 12-12. In-context Voice Notifications are most often used for features such as turn-by-turn navigation. The display is mostly reserved for pictorial displays of information, which are more glanceable and allow multiple types of information to be presented at once. A summary of the audio message should be presented on-screen, even if you expect it to be unreadable; the user may change contexts (e.g., stop the car), another individual may be able to assist the user, and the mere presence of the information box reinforces that the message is originating from this device and relates to this task.

Make sure all alerts are presented visually, as well as via audio. See the patterns under Notifications for display of notifications. Even when in context, the alert should be noted in some manner, though this may be more specific, such as an Annotation over a map. Only use Pop-Up dialogs or other intrusive measures when absolutely required.

To inform the user that audio is about to commence and to prepare the user for the volume level, play a subtle tone or unimportant audio message immediately beforehand. This may also be the normal alert tone to emphasize that the device is communicating an alert. Make sure the volume of this tone is similar to (or lower than) the audio to be read aloud, as the intention is not just to consciously alert the user, but to prime the user's auditory system for receiving messages.

Use syntax that makes it clear what is being communicated and attempt to give users a moment to acclimate to the voice and the instruction format, even after any notification Tones. For example, state the type of message or action the user must take, then the details:

- "Turn coming up . . . turn right in 500 yards."

- "Time to take your medications . . . take one Ritalin now."

Repetition generally is not perceived in a negative way in spoken phrases as it is in print, due to the immediacy and the way audio versus visual perception works.

Use caution when choosing what sort of content will be read—and to what level of detail. Alerts will generally be sent through the speaker, so they have minimal privacy. Since anyone nearby can hear, messages must be formatted to be general or have no explicitly secret information. For example, instead of stating the medicine and dosage as noted earlier, the message could just be, "Time to take your medication. . . . Be sure to take it with a meal."

Use a consistent voice and tone of voice for all notifications. This way, the user can become accustomed to the reminder, and be able to comprehend the intent without listening as closely to opening phrases.

Antipatterns

Avoid "blurting out" alerts or instructions, especially with important information in the beginning of the phrase. Directions such as "Turn right in 500 yards" are often not useful, as the user is still acclimating to the voice speaking through the direction statement; the user may only hear "...in 500 yards."

Voices provide strong emotional responses, and when the user is not accustomed to them or expecting them, they may be startling. Avoid using Voice Notifications when users are asleep, and be sure to precede all notifications with Tones to make it clearer that this is a machine, not a scary intruder sneaking up from behind.

The voice you use must be as understandable as possible. Even within a single language, be sure to use the most generic accents possible, and be aware of regional idioms.

Text-to-voice translation of names especially can be difficult to understand or improperly pronounced. If quality is too low with the available hardware and software, do not implement the solution.

Haptic Output

Problem

You should use vibrating alerts and tactile feedback to help ensure perception and emphasize the nature of UI mechanisms.

Haptic vibration is available on many devices, and can be activated by most any application. The use of sound to simulate vibration is a viable alternative. Future technology will improve many aspects, and may broaden the base of devices with this feature.

Haptics refers to receiving information by touch. Practically, with current, generally available technology, this refers to the use of vibration to communicate with the user. In much the same way that mobiles have evolved to contain a variety of sensors, they also have a broad range of output devices; most mobiles have some sort of vibration, or an external speaker, which can be used for at least basic Haptic Output.

These vibrations are generally propagated through the entire device, but they can be localized, or placed only in specific components, such as the pen in a Pen Input device.

Most vibration is very coarse and is accomplished by a simple off-center weighted motor. Control is only by intensity and switching the motor on and off. Transducer-based vibration is rarer but also available, using tools typically employed for audio output to make more nuanced vibrations. This may simply involve the bone conduction transducer (BCT) included for use of hearing-impaired users being repurposed for general Haptic Output. Use of the BCT for general audio output is not part of this pattern, and is otherwise beyond the scope of this book.

Additional methods of haptic output are being developed in the laboratory, and may appear in production soon. These include the ability to sense objects that are not physically present, enabling tactile virtual keyboards, for example.

Figure 12-13. *When the user performs a function on the device, haptic response can make this more obvious, can confirm that it happened, and confirms that the device perceived and is reacting to the action. For touch devices, the haptic response takes the place of physical response, and allows more natural and assured use of the interface.*

For mobile devices, Haptic Output is currently used in two ways, both of which are dynamic output methods:

Response

When the user makes an action, the device confirms that it has received and understood the action, and makes it seem more real by adding a response. Most typical is a "click" when virtual buttons are used, but any action may use this, including scrolling, gestures (the system indicates when a complete gesture is accepted), and even the use of physical buttons. See Figure 12-13.

Notification

Alert lights and tones, used to notify the user of events when out of context, may be emphasized or replaced with haptics. Vibrate has been heavily used for this for so long that it has become a well-understood pattern by the general population. Vibration also helps users detect alerts when the sound is off or muffled, and can help to localize the device when in a pocket or purse. For additional details, especially on display in other channels, refer to the Notifications and Tones patterns. See Figure 12-14.

As discussed earlier, haptics may soon also be used for other types of tactile output. Especially useful will be the ability to detect shape and texture—to allow users to feel virtual on-screen elements. This will form a third variation of output perceived by the user to be passive and representing static, physical spaces and objects. It may even enable new interactions, where users may turn dials around a resisting axis, or be able to pinch and grab UI elements as though they are physical objects.

Interaction details

Use Haptic Output to emphasize that an action happened or to communicate in another channel, when users may not notice audio or visual cues. You should never use haptics alone, and must only use vibration to reinforce actions by the user, reactions on-screen, or displays on-screen, by LED or by audio.

Phenomena such as the McGurk effect—where speech comprehension is related to the visual component—appear to exist for other types of perception. Vibration should support interaction with the visual portions of the interface, by being related to the physical area interacted with whenever possible and by having a corresponding visual or physical component.

If an alert is expressed as Haptic Output, you should always have a clear, actionable on-screen element. These should always appear in the expected manner for the type of event, as described in the Notifications pattern.

Haptics resulting from key presses or other actions the user takes should be the only response aside from the intended action (e.g., typing the character). You can also use them to support a Wait Indicator, or replace them for very short time frames (less than one-quarter of a second), when indicating that the device has received and understands the input.

Consider Haptic Output for Notifications, including ring tones, to be contiguous with the volume control system. Vibrate is the setting between off and the lowest level, or it can be enabled as a switch so that it is on (or off) at all volume levels. While being used, vibration will be silenced in the same manner and at the same moment as Tones or Voice Notifications are. The same mute, cancel, or silence function will be presented when only vibration is enabled, and there is no audio output.

Figure 12-14. Notifications can use vibrating output to help users locate the device, or to provide a (generally) more silent "ring tone" than audio output. Typically, vibrate is activated alongside tones. Be sure to alternate audio and vibrate, to avoid the two conflicting with each other.

When using Haptic Output as response items, you should not use haptic vibration generically but have the signal carry meaning in much the same way audio does. Clicks should feel like clicks, errors should be harsh as though being rejected, and so on.

When emulating real objects, the tactile characteristics of those objects can be measured, and codified for the simulation. Tricks of audio playback sometimes work for vibration; this is in fact why vibration can be used to simulate tactile perception. However, touch is different from audio, so different guidelines must be used and specialists in the design of vibration should be employed.

For devices without haptic or vibration hardware, audio can serve this need in a limited manner. Short, sharp tones, or very low-frequency tones, output through the device speaker can provide a tactile response greater than their audio response.

Very low bass tones can provide some of the best response, and most speakers can actually generate tones below the threshold of human hearing. However, devices are generally not designed in this manner, so the amplifier hardware may not go this low, the software to access the hardware may not accept output this low, or the tones can generate distractingly improper resonance in the case, ruining the effect. Test on targeted devices before use.

Also use care when generating tones at the edge of human hearing with the intent of using them purely as haptics. Younger populations, and certain individuals, have wider ranges of hearing and may be able to detect it audibly. This may be an acceptable side effect, but ensure that it is not distracting or interferes with the use of audio in other frequencies for those users.

Vibration and sound are closely coupled behaviors and can influence each other in undesirable ways. When both are needed, they generally must be alternating to avoid conflicting with or modifying each other. Vibration devices can cause unintended buzzing or other distorting resonance in audio devices. Alerts, for example, generally vibrate briefly, then sound a brief tone, and repeat this until the notification alert time expires or the user cancels or mutes the output.

Antipatterns

Do not overuse Haptic Output without a good understanding of all effects of the device hardware. Besides the audio conflicts described earlier, if traditional motor-driven vibration is used it may consume the battery excessively quickly.

Do not play multiple vibrations at one time. Even if the device supports this, they cannot generally be perceived accurately by the user, and may end up as noise, even if in multiple localized areas.

Make sure haptic alerts send vibrations repeatedly. Phantom ringer behaviors are known to exist for vibration as well as audio; single vibrations may be written off as an accident of perception by the user, and the alert will be missed.

When employing an audio output device for haptic vibration (either the device speaker or the BCT), audio has priority, especially when used for playback of media or for ongoing communications. Never interrupt the voice channel of a call to play a haptic alert.

Haptic output can easily induce alarm fatigue in the same manner as for alert Tones. If the alert method is too generic, too common, or too similar to other tones in the environment (whether similar to natural sounds or electronic tones from other devices) the user can discount the alert. Use as few different sounds as possible, but avoid using simple "vibrate" for every alert. Note that disregarding alerts is not always a deliberately conscious action; the tone or vibration is eventually considered noise by the user's brain, such that it will not longer be noticed.

Chapter 13

Screens, Lights, and Sensors

The Relationship

Let's travel down memory lane. Take a moment to think back to your first date. Stop! Just kidding; that's outside the scope of this book. But do take a moment to remember when you bought your first cell phone. Even though I bought mine some time ago, I clearly remember the exact year, model, and reason I bought it in the first place.

The year: 1997, while in college.

The model: Motorola StarTAC, 2G GSM; 4 × 15 character, monochrome graphic display (see Figure 13-1).

The reason: Cool factor! A flip phone and the smallest cell phone available. I could send and receive calls, SMS, and store up to 100 contacts. And I could show it off ever so smoothly when I clamped it onto my belt. Back then, that was all the functionality I could ever want! It was love at first sight!

The Breakup

Well, I'm sad to say that that phone is no longer with me. I've had to keep up with the times and the technology. Since that StarTAC, I've gone through an extensive number of mobile phones. I believe the count total now is eight. That number seems reasonable, considering the rate at which people upgrade their mobile phones today—once every 18 months, according to Gizmodo.com. Today, my mobile requirements consist of greater interactive control and highly visible functionality on a powerfully crisp and color display. That original 4 × 15 monochrome graphic display, if used today, would be quite limiting and unsatisfying.

Figure 13-1. *Over time, display technology has improved dramatically. Now color depth and resolution are high enough that traditional pixel-based display concerns are beginning to change. However, there is still a trade-off in cost and power consumption. Devices will always have a variety of screen quality levels. By the way, on the left is a StarTAC, just like my first phone.*

I'm Not "Everyman"

Not everyone needs what I need in a mobile phone. Mobile design is *never* about you and me. It's about all the other people who are using a range of multiple devices, with varying needs in limitless contexts.

Here are the things we must consider in terms of screens, lights, and sensors in order to create an enriching user experience while considering everyone:

- Context of use
- Display size
- Display resolution
- Display technology

Context of Use

One of the mobile design principles centered on in this book is that mobile devices must work in all contexts. They must function properly and behave appropriately in a variety of environments. Let's consider the following contexts that can affect the way we interact with a mobile display.

Outdoors

The outdoors is the most complicated environment to design around. It's highly unpredictable, constantly dynamic, and uncontrollable. External stimuli such as bright sunlight, cloudy days, moonlight, darkness, and street lights aren't controlled by the user. We can't just switch on and off the sun or blow the clouds away. All of these stimuli can make it more difficult to view the screen, thereby affecting the way our users interact with the device.

Indoors

The indoor environment may be more predictable, less dynamic, and more controllable than the outdoors, but it is still highly complicated. External stimuli present may include natural light from windows, doors, or skylights, or generated light from bulbs, fluorescent light, incandescent light, LEDs, halogens, and high-pressure sodium lamps.

Both

Mobile users are constantly transient, which changes their environment. They may use their device outside in daylight to read the news, and then walk inside a public building with dim lighting.

This change affects the amount of time it takes our rod receptors in the back of our retina to respond to light differences. The greater the difference between the two environments, the longer it will take our eyes to appropriately adjust. This affects our ability to quickly detect and identify details in that time—such as text, small images, and even controls.

Imagine this scenario: a user takes his mobile smartphone into a movie theater. The lights inside are very dim. He wakes his phone to check the time. The screen display is excessively bright, causing him to look away. This is not just a screen display technology issue; it's a UI design problem.

Here are two solutions:

1. Have the automatic brightness sensor pick up the ambient light around the user and the phone (not the user's face, which is what typically occurs). The sensor recognizes the low level of light and automatically dims the screen to an appropriate level to minimize eye strain and maximize viewability.

But not all users like automatic brightness on, partly as it can rapidly drain the device's battery.

2. Provide immediate access to brightness controls. Rather than have them buried in a system setting, consider using the physical keys (e.g., volume) that can open a menu to control the display settings. These physical controls have eyes-off functionality and the user can interact with them without even looking at the device.

Displays

Mobile device displays can range in size, resolution, and pixel density (ppi). As a mobile designer and developer, it's helpful to become familiar with these differences so that you can make appropriate decisions throughout the design process. Depending on the requirements of the project, you may be designing for one particular device or for multiple devices with varying displays.

Screen Sizes

Display sizes vary with the type of mobile device. They are measured diagonally from the display's corners. Typically, smartphones have larger screen sizes than feature phones. A common misconception with mobile devices is that the screen size limits the user's ability to see the screen.

However, people automatically adjust the visual angle between the device and their eyes. Entirely aside from phones being handheld, they get moved to the right distance to see things correctly.

Broadly speaking, video display ends up occupying the same field of view regardless of the device size. If we watch a movie on a 46-inch HDTV, we will distance ourselves enough so that we are able to see the entire TV in our visual field. That visual angle can be similar when we view a movie on a phone. This is why people actually do watch video on phones, and will continue to do so.

Other factors that affect how we automatically adjust our visual angle to see the screen include:

- Vision impairments a person might have
- The size of the display content, such as images, text, buttons, and indicators
- The amount of light the device is giving off (illuminance)
- The amount of light reflected from the light's surface (luminance)

Screen Resolutions

The screen resolution is determined by the total number of physical pixels on a screen. Note that the actual size of a pixel varies across devices as well as the type of technology being used in the display. So it can be very misleading when people say to design an object with a minimum touch target size of 44 pixels. Across devices with different screen resolutions, this touch target size will not be consistent. Therefore, when designing for appropriate touch target sizes, use unit measures that aren't variable. For more information on proper use of touch target size, refer to the section "General Touch Interaction Guidelines" in Appendix D.

Here are common screen resolutions (pixels) across mobile devices, including feature phones, smartphones, and tablets:

Small: 128, 176, 208, 220
Medium: 240, 320
Large: 320, 360, 480+
Tablet: 600, 800, 768, 1,024

Pixel Density

Pixel density (ppi) is based on the screen resolution. It equals the number of pixels within an area, such as a square inch (units do vary, so refer to numbers you find very carefully).

A screen with a lower pixel density has fewer available pixels spread across the screen width and height. A screen with a higher density has more pixels spread across the same area.

When designing for mobile displays, it's important to be aware of the targeted device's pixel density. Images and components that are designed in a particular height and width in screen pixels appear larger on the lower-density screen and smaller on the higher-density screen. Objects that become too small will affect legibility, readability, and detail detection.

Mobile Display Technology

Mobile devices use a range of display technology. Some devices may use multiple types of hardware within a single device. Each technology can serve a unique purpose in functionality—backlight, primary display, and flashing indicators.

As a designer and developer, it's important to understand that each of these technologies has limitations in terms of the device's battery life, the device's lifespan, creation of glare, or restriction of orientation modes (see Figure 13-2). Therefore, we need to create UIs that can maximize the user experience around these limitations. Here are descriptions of some common types of display technology:

LED

Light-emitting diodes are a superconductor light source. When the diode is switched on, the electrons move within the device and recombine with electron holes. This causes a release of photons.

- – Used in annunciator illumination.

- – Benefits include low power consumption, small size, and efficient and fast operation.

- – Limitations include the ability to communicate only one piece of information at a time; also, they are single-channel, single-bit devices.

OLED

Organic light-emitting diodes contain an organic semiconductor film that emits light in response to an electrical current.

– Used in the primary display.

– Benefits include the ability to function without a backlight, thin width, the ability to achieve a high contrast ratio, and when inactive does not produce light or use power.

– Limitations include the fact that it uses a lot of power to display an image (such as black text) with a white background, and may develop screen burns over time.

AMOLED

Active-matrix OLED is a technology that stacks cathode, organic, and anode layers on top of a thin-film transistor. This allows for line-scanning control of pixels.

– Used in the primary display.

– Benefits include the ability to function without a backlight and the ability to be used for large-screen displays.

– Limitations include known problems with viewability in glare and direct sunlight.

ePaper

ePaper generally relies on reflected light, not emitted light, by suspending particles in a liquid; a charge causes the dark particles to rise and be visible, or bicolor particles to rotate from light to dark (technologies vary) and display dark areas on a lighter surface, much like ink on paper.

– Used in the primary display.

– Benefits include low power consumption, reduced glare, and high contrast ratio.

– Limitations include slow refresh rates, and the fact that color technologies are just emerging.

OLED and AMOLED displays are lit-pixel displays (without a backlight). White text on black will, unlike a backlit display, use much less power than black text on white. This may be a key design consideration for those devices.

Retroreflectivity

Retroreflective displays include a reflective layer behind the backlight layer. Ambient light coming into the display is dispersed across the surface and reflected back to provide passive, power-free illumination of the display content. This can be used to save power, to provide some view of the display when the device is in sleep mode, or to boost the performance of the display in bright light; instead of fighting the sun with high-power backlight, it is simply reflected back.

In some displays, this is the primary or only method to illuminate the display, and there is no backlight. In this case, the layer is located immediately behind the backplane.

Transflective layers are similar in concept, but use a transmissive/reflective layer that can be placed in front of the backlight. The key problem with these two technologies is that high-density color screens with touch overlays limit the amount of light passing through. Placing the reflective layer as far forward as possible increases the efficiency.

This works only for pixel masking displays, such as LCDs. Lit-pixel displays, such as OLED, cannot use it as they have no backlight—all pixels are opaque and must provide their own illumination. Although ePaper uses this concept, some technologies also cannot use it directly, and the reflectivity is integral to the "white" component of the display.

Display illumination on devices with reflective layers must take this into account to provide the most suitable illumination, and to extend battery life when not needed. In sleep states, the lock screen must continue to display useful information since it can be read at any time.

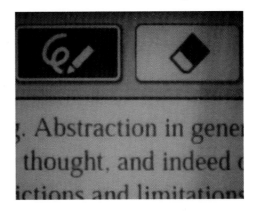

Figure 13-2. *Low-power screens, or low-power modes, must make trade-offs in other display characteristics. ePaper technologies may evolve, but currently they have grayish blacks and very unwhite whites, as well as slow refresh times. The variable power of OLED allows the primary display to stay illuminated while sleeping, but only a few items can be illuminated, and contrast masks the fact that the white pixels are actually not at full intensity either.*

Input Overlays

If the display includes a touch or other input function, this is generally attached as a separate layer on top of the display screen. This is a problem in two ways. First, every layer over the display plane adds depth, which increases the parallax.

Second, nothing is truly transparent. Stacking several layers, with different indexes of refraction as well, can significantly reduce the brightness the user is able to perceive. Any touchscreen will be dimmer than the same hardware without a touchscreen.

Touch and other input layers are frequently sold as an integrated package by component manufacturers. Therefore, solutions such as Super AMOLED are available that allow the sensing layer to be thinner and more optically integrated than separately developed components.

Sensors

Another important concept that we will discuss later in this chapter is location-based services used by operators and GPS satellites. These services identify tracking sensors in your mobile device to determine your location with varying degrees of precision and accuracy. Location services can provide an enriching user experience for a multitude of reasons.

They can work tangentially with your device's OS and existing applications to trigger notifications or create appropriate device state changes without compulsory user interaction.

For example, say you are driving to the airport to catch a flight. Your mobile "Itinerary travel application" has you scheduled for a 3:00 p.m. departure at ORD.

As you get within a particular range of the airport, location services identify your current location and proximity to the airport. After you send that information to your Itinerary application, a notification pops up mentioning the latest departure time, and any occurring gate changes.

Figure 13-3. *Automatic brightness will probably remain imperfect for a long time, so brightness controls should be a function of other settings or a home-page widget so that the user can get to them easily.*

Patterns for Screens, Lights, and Sensors

The patterns in this chapter describe how you can use screens, lights, and sensors to increase visual and device performance, provide additional notifications, and incorporate location-based services with native applications. We will discuss the following patterns in this chapter:

LED
> Notice of status should have a visual component that does not require use of the primary display.

Display Brightness Controls
> Displays are the most critical output on smartphones. A method to control the display settings must be easy to access, be easy to use, and allow for manual adjustments. See Figure 13-3.

Orientation
> Devices must allow for orientation changes to display content in the manner most readable and comfortable to the user.

Location
> Availability of a location-based service, and the actual location of the handset, must be easily and accurately communicated.

LED

Problem

You should ensure that status notices have a visual component that does not require use of the primary display.

Every device has at least an annunciator LED, used for power and other key status messaging. This light can be used by many applications, but not usually by web pages.

Solution

Figure 13-4. *LEDs are most often encountered above the screen, usually to the left, and generally facing the direction of the screen, though some are placed so that they can be seen at other angles. On-screen indicators must communicate what the LED is intending to signal. Here, the red LED is reflected in the red battery meter level in the annunciator. Critical alerts will often demand more strident notifications, such as Pop-Up dialogs.*

Annunciator illumination has been used for status and alert purposes since before the invention of electric light. Mobile devices took on the legacy from larger electronics, and machine-era devices with only minor changes, to denote power, connectivity, and alerts, among other things.

In all electronics for the past decade at least, the small, efficient, light-emitting diode is used for annunciator lighting. The term *LED* has come to have this general meaning, even in the now-rare event that another type of lighting technology is used.

Mobiles use LEDs more than other devices due largely to power consumption. The display is the largest consumer of power, so it is only illuminated when needed. The LED uses very little power, especially when blinking, so it can be used to notify the user with essentially no effect on battery life. LEDs may even be used to inform the user that the battery is entirely dead, when no other systems have sufficient power to operate.

The LED has become so heavily used that it is always designed in, but it may not be needed in the future. OLED-based displays, for example, use power only for the lit pixels and so can perform many of these display actions without the concern for power. LEDs may be used less often, or differently, in the future.

Variations

Even in the machine era, multiple types of annunciator lights were used for different purposes. Each of these has its own useful variation in mobile use:

Fixed
> A single light is placed in an obvious, expected location, usually on the front face of the device, above and to the left of the screen. One light is provided for each screen on flip phones and the like, or is positioned so that it is visible when either screen is in use. See Figure 13-4.

Relational
> This is a simple light, like a fixed variation, but is provided adjacent to a physical item or area of the screen for the notification it refers to. A secondary light adjacent to the charge port indicates that it is operating; a light along the top edge of the screen indicates a Notification that the user can see by activating the screen and looking at this area. See Figure 13-5.

Ambient
> Illumination of a major component provides not just visibility, but also signaling. The keypad backlight may change from white to red to indicate low battery, or the charge port may be illuminated internally. These may be relational (such as the charge port example) or abstract. The keypad has no direct relation to the battery, but is visible and obvious.

These are not exclusive choices, and may be combined in one device.

Figure 13-5. *Relational LEDs send a signal not just by abstracting to colored lights, but by their placement. Here, the light adjacent to the USB port is indicating that the device is charging. No light, when the cable is plugged in, means no power is being provided. No color or blink codes must be interpreted, or another part of the device consulted.*

The LED is always a display item, with no inherent interaction. However, it must be considered as a part of the overall interaction of the device, so it is coupled with other functions and behaviors.

Never use the LED as the only notification of a status change. Even relational items should always display the status on the device screen as well, generally as a Notification or in the Annunciator Row. This does not have to be visible immediately; if the device is asleep, the user can be expected to wake or unlock it first.

You should usually make audio Tones and Haptic Output accompany any LED notifications. The light serves to confirm that the mobile device made the sound, and to help the user locate the device if it is not visible or if the environment is dark.

Relational lighting uses the same principles as Focus & Cursors, where position is most crucial. For example, when an LED is provided for camera flash (Xenon flash has more technical limits), you may also use the same light at very low power to indicate to the subject that the photo has been taken. There is no need to signal in any other way, as the meme of "camera flash" means the photo has been taken, regardless of the position or intensity.

LEDs are single-channel, single-bit devices. They can only communicate one piece of simple information at a time. Generally, multiple types of messages can be sent over the same LED, but not easily at the same time. One of three tactics must be used:

Alternate
Each displays in turn. This can lead to confusion and some signals may be missed

Priority
Only the highest-priority message is displayed. When cleared, the next-highest-priority message is displayed.

Alert

A general notifier is presented for all cases, and the user is expected to view the Notifications panel to see what alerts are present. When all alerts are cleared, the LED will no longer illuminate.

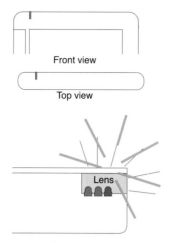

Figure 13-6. *Lenses obscure the operation of the LED, allowing users to see the final, combined color, and to see even illumination from multiple angles.*

Signaling by lights falls into three categories:

- Position
- Color
- Speed

Position is partly discussed earlier in this pattern, but it also has to do with simply being attached to the device. Always consider the contextual nature of the device, and as much as possible signal only useful messages to the user. When multiple lights are available on the device, use the one most relevant to the information being communicated.

Color has cultural meaning, though for electronics, the US standards have largely taken hold where red means "bad" and green means "good." These are not universal, but should generally be, or users may be confused. Sometimes branding will supersede these colors; if the brand of the device involves much red, and the default backlight is red, a red LED may lose the "bad" meaning; another color will be needed to express urgency, such as yellow.

Speed refers to blinking (at speeds visible to the user). These should be used very rarely and carefully on mobile devices. As these devices are frequently glanced at, often for only a fraction of a second, blink rates that cycle slower than one-quarter of a second are unlikely to be seen reliably. High-speed blinking can be used to express urgency. If movement is needed, throbbing and pulsing, where the light is never entirely extinguished, are better choices.

LEDs are inherently directional and must have a lens or it would be blinding at one angle, and almost invisible from others. An "LED," as perceived by the user and communicated in the device user guide, is actually the lens, often fed by multiple LEDs to provide different colors, as described in Figure 13-6. Generally, no more than three are used, as at this point (if red, green, and blue are used), combinations of all of them can be used to make essentially any color.

Brightness should be changed based on ambient light level. The LED should never disappear in bright sun, or blind the user (or light up the whole room) in the dark. See the Display Brightness Controls pattern for details.

Antipatterns

LEDs like are not Tones or Haptic Output devices. You should only use them for alerts and hardware status changes, and not generally as feedback for key entry or otherwise as a regular feedback mechanism.

Avoid excessively technical signaling using the LED. If the message would not be popped up on the screen, do not display it as a light.

Be aware of how LEDs work before employing them in a specialized manner. For example, LEDs do not have useful low-power modes, and actually blink at very high rates to simulate reduced output. This blink rate can be detected if the device is moving relative to the user, or can cause interference with other blinking when used for multicolor output.

Avoid blinking to indicate any important state change. Users may glance repeatedly at the device, but it always seems to be during the off cycle, and they will then never see the signal.

Never use blink ratios, even for errors or diagnostic purposes, where the LED is off more than twice as much as it is on, and avoid using differences in timing to encode signaling. Users are poor at perceiving rates or time without a reference.

Codes that are more complex than a simple blink should be avoided whenever possible, and always for general use. There is no mnemonic to remember that three short blinks and then one long one means the battery is full. Users will have to refer to documentation, or will simply disregard the message.

Display Brightness Controls

Automatic light sensors and user-controlled display settings are difficult to find, are difficult to use, and often cannot be made to yield an appropriate light level without significant user input. You must make display controls easy to access, and devise schemes to ensure that automatic controls are effective and useful.

Display controls are built into the device OS, but some platforms allow adding to the standard control sets.

Figure 13-7. Brightness controls are routinely located inside a settings panel with various other related controls. The level must be displayed numerically as well as graphically, and the automatic brightness switch should be immediately adjacent to the level control.

Displays are the most critical output on modern smartphones, as well as a highly touted feature. Even minor degradations in performance can render the display uncomfortable to use or unreadable, or can reduce the perception of the quality of the display.

This pattern focuses on higher-quality color displays, with light sensors suitable for use in setting automatic brightness. Other display types may use many of the components of the patterns, but may not be able to meet all of them. This also presumes backlit screens; lit-pixel screens such as OLED (and AMOLED) screens may have to modify the technical information to achieve the intended result.

Users must be able to get to display settings easily, preferably without much on-screen interaction, in case the screen is hard to read and they need to adjust it. The primary display setting is brightness (white point). Contrast (black point) and other controls may also be provided, but will not be well understood by the typical user; good median values must be selected by default for all other display adjustments.

The user must be able to make manual adjustments even if an automatic setting is provided, and should be able to provide user input to the automatic settings, as well as provide user calibrations to the light sensor.

As a general rule, keyboard and keypad backlight should follow the same illumination guidelines as the display, though manual control may provide for separate settings. Key backlight color changes or blinking as a signaling method is discussed under the LED pattern.

Variations

Manual is the most common mode, available on almost all devices with a screen. A simple one-axis user control is provided on a settings panel as in Figure 13-7.

Modern mobile devices with high-quality screens mostly come set to automatic mode by default. While in automatic, a light sensor detects the user's face and surmises the ambient light level to determine optimal output. The user may switch to manual mode, which disables the automatic settings.

An automatic mode with manual input should also be provided, to allow user modifications to any failure of the automatic system to detect the optimal output level.

A learning mode would be a useful fourth variation, using the principles of learning user predictive text, voice, or gestural input. The device will monitor user overrides to automatic input and/or accept multiple user calibrations in order to adjust how automatic behaviors work, with the intent of an entirely satisfactory automatic behavior, eventually.

Interaction details

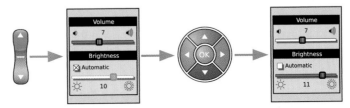

Figure 13-8. *Easy access to brightness should be provided when it can be. The example Pop-Up appears when the volume control is used. Using the left and right directional controls changes focus to the brightness. Automatic brightness will be disabled when manual changes are made. Changes should always be immediately implemented, without explicit saving.*

Whenever possible, make access to the brightness settings immediately accessible from any screen. A well-used option is to have an existing function (such as volume or power) make a general settings panel appear, with brightness, volume, synch, and power-off/restart functions (see Figure 13-8). Once it appears, volume keys may change the volume, and (for example) the Left and Right Directional Entry keys may change brightness. Other solutions may be more suitable to your specific interface or operating system.

Eyes-off brightness control is necessary in case the brightness is set so low that the screen cannot be seen; the user may remember the interface, and be able to activate the panel and increase brightness, without viewing the screen.

The primary brightness adjustment control is a tape, as described in Mechanical Style Controls, or similar control with implied analog input. In fact, this will control input in steps based on the software controlling the display backlight. Allow direct numeric input of these values.

Make sure the brightness control allows the user to set true minimum and maximum output. Mobile devices can be used in complete darkness, where users' eyes may be fully adjusted or bright screens may distract. Do not select an arbitrary minimum brightness above the adjustment range value of 1.

A simple setting on the brightness control panel must allow the user to enable and disable the automatic brightness control. This may be set from other places as well.

Provide a function to add user adjustment to auto-brightness. This will not disable automatic control; it will just add the user control as a new temporary zeroing point. Ideally, the device will remember this user setting and factor it in over time in a learning mode.

Sensors will continue to be most useful when placed on the front of the device (among other things, it is probably best for checking light where the user is). Calibrate sensors from the factory not to neutral gray, but to people's faces at typical distances. Provide a function to allow users to recalibrate the sensor to their face and typical viewing angle and distance. This may also be implemented as an explicit learning mode, with the user performing this for several typical environments in which she uses the device.

Presentation details

Whenever room is available in the Annunciator Row, in a Notifications area, or in similar areas where current settings are displayed, the current brightness settings should be displayed. This should indicate automatic or manual, and the level in the conventional scale of measure used on the adjustment panel itself.

Within the brightness panel, it must be very clear when automatic brightness is enabled. When enabled, display the actual brightness selected by the automatic controls in a manner similar to the manual setting.

When the user is adjusting brightness, make sure the screen reflects the adjustments in real time, so the user can tell how bright he wants it. This also may mean the brightness setting dialog may need to either have sufficient information or not gray out the parent window, so there is a chance for the user to see what typical screens look like.

It is highly suggested that adjustments be simulated when the brightness panel is displayed. This is so that the adjustment panel itself can be seen in all conditions. Set the screen backlight to 100% brightness, with the adjustment panel as a modal Pop-Up with the controls and instructions simulated (via the pixel color) to 50% brightness, with very high contrast, such as white letters and outlines on a black background. The parent window behind the Pop-Up will simulate (with color changes to the pixels themselves or opacity to a black overlay...) the adjusted-to value.

Within the display adjustment panel, present the brightness numerically as well as via any graphical control indicators. Some users will understand numerical values better, and most users will be able to recall numerical values that work in certain environments, so they may be able to adjust to those by memory if automatic systems do not work.

All hardware keys must be visible in darkness or marginal lighting conditions. Certain devices or keys may be able to use phosphorescent ("glow in the dark") keys, but active backlight is generally required. All functions of the keys must be visible in all lighting conditions.

Avoid brightness-changing functions that are not discoverable or learnable.

Do not overly simplify the brightness control software (either automatic or manual) by limiting it to a small number of large steps. With even 16 steps, users can detect the difference between individual steps. They may not find a suitable brightness, or may find the display jumps settings (exhibits "seeking" behavior) when in automatic mode. Using the full complement of steps (e.g., 100 or 256) will provide sufficient levels and smoothness that even rapid adjustments will be perceived as smooth changes in brightness.

Do not illuminate only some portions or functions of a key. The key labels must be clearly visible in both reflected light as well as emitted light, when only visible under backlighting. For example, "function" modifiers may be yellow and correspond to a yellow-painted key on an otherwise white label set. When illuminated, and key labels are generally white, the function labels for each key must remain yellow, and the function key should be entirely yellow, to match the reflected color scheme and style.

On light-colored keys, keyboard backlight can conspire with ambient lighting to make the key labels entirely invisible to the user. White backlight, at low light levels, can easily have the same luminance as a white or silver key. Carefully test and adjust backlight settings or select contrasting keys or backlight colors.

Orientation

You must provide a way for any device orientation changes to display content in the manner most readable and comfortable to the user.

Many devices have accelerometers, dedicated rotation switches, or both. This hardware and any other OS-level orientation changes should be respected by all applications and other content. When not available at the OS level, orientation can often be changed anyway, if it would be more helpful to the user.

Solution

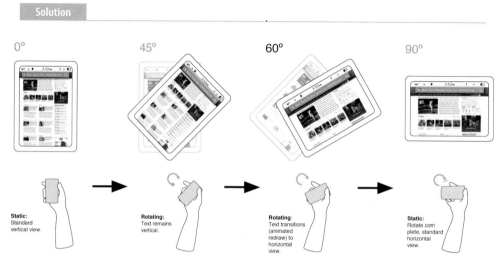

Figure 13-9. *An illustration of how device orientation should not take effect immediately, but requires significant commitment. Note how additional information is visible in some cases (wider title bar, more page width available allowing for larger text and images to be used) but less information is visible in others (two-thirds of the page does not fit in the viewport). This trade-off is why users must be allowed to control their viewing orientation.*

Mobile devices are small and portable, and unlike desktop computers they can be manipulated and viewed in any manner. Naturally, users will rapidly face the screen in the correct direction, but after this they should be allowed to choose their preferred viewing Orientation. Content must be presented in a useful format in whichever orientation is chosen, modified to fit the screen, but without changing context or modifying existing user entry.

This change must be performed either automatically or with an obvious or deeply integrated physical mode switch.

A related function is for devices with screens on both sides, or other form factors; flipping, sliding, or otherwise switching screens can follow this pattern, but is not discussed explicitly due to the small number of such devices.

This pattern exclusively addresses screen orientation or activation, and does not discuss use of the orientation sensors as a gestural control. For generalized discussion of position and orientation, see the Kinesthetic Gestures pattern. For some methods of employing subtler orientation changes, see the pattern on Simulated 3D Effects.

Variations in Orientation generally follow the device form factor and consider the effectiveness of the overall device, including input methods.

Fixed form factors, and especially "all touch" devices, generally should use *automatic sensing* to switch from portrait (vertical) to landscape (horizontal) mode (see Figure 13-9). Sensors have generally been accelerometers, to sense the device orientation relative to the ground. The device camera may also be used to sense the user's head position and adjust relative to this plane. Level sensors can sometimes give poor results (when content is being shared or when the user is in an unusual position, such as lying down); the *machine vision* solution attempts to overcome this.

Devices with any sort of variable form factor generally also change the use of its display. This *physical mode switch* serves as a switch to the screen as well. This can be considered to include the activation of an external screen when a clamshell or flip phone is closed. Typical cases for rotation are the now-common messaging-oriented phones, with Directional Entry keys (or sometimes even a numeric pad) when closed, but a fold- or slide-out keyboard (see Figure 13-10). The switch in screen orientation corresponds to the orientation of the hardware entry method.

Figure 13-10. *Changing orientation does not just mean zooming content or reflowing text. Here, modules for a portal landing page switch from being simply stacked in the portrait view to being arranged in two columns in the landscape orientation. This device is typical of a large class that switches orientation on the physical mode change of opening the keyboard.*

For automatic sensing devices, much leeway must be granted before a rotation is committed to. Users will not hold devices perfectly upright, and may use them in unusual orientations.

The angle of commitment must be a large value, such as 60 degrees from the vertical. Rotating to the opposite orientation (180 degrees, such as from one landscape orientation to the other) should likewise not take place as soon as the device passes level, but only after exceeding about 20 degrees of the opposite angle.

Once a switch has occurred, use a new baseline to measure from. The angle (e.g., 60 degrees) is measured from the current ideal vertical. The angle at which the orientation switches back must not be the same one for both, or it will flip back and forth repeatedly. This is a key reason 45 degrees cannot be used as the switching angle.

You can provide additional leeway based on context. Consider a device placed on a table. When picked up, it may be accidentally over-rotated, away from the user. The device should probably not immediately rotate 180 degrees on the expectation that someone else has picked it up from the opposite side of the table.

Here, the sensors can determine that the device has been set down and has not been used for some time, and can reset the baseline to be "from the tabletop." Then something like the 60-degree guideline can be used. Only when rotated backward that far will the screen rotate 180 degrees to be usable at the opposite viewing angle.

The user's selections or inputs on the screen must in no way be changed during an Orientation switch. Items in focus must remain so through the change, with selections to the character or pixel when applicable.

No uncommanded actions must take place. Do not clear entry, submit forms, scroll, or perform any actions that would not have been performed had the orientation change not taken place.

Presentation details

Whenever possible, you should animate the rotation. Actually show the starting orientation rapidly rotating to the new position. The entire change does not have to animate, reflowing to the new shape; but the initial animation helps to explain the change and confirm to the user that the rotation action is occurring, and not some other, possibly uncommanded, behavior.

Avoid blanking the screen, as this can be perceived as an error or bug, and the discontinuity requires the user to reorient herself to the screen.

Make content reflow to fit to the screen. Never allow items to fall off the screen, or for gaps or margins to appear to the side as a way to cheat the rotation.

This will necessitate reflowing of text (changing where line breaks occur), or even switching to an entirely different display template, such as placing modules side by side, instead of stacked in a list. However, every effort must be made to have all content from the old orientation visible in the new one.

When this is not possible, the focus area from the initial orientation must always remain in the viewport in the new orientation. For example, if a list view is shown in portrait, and the bottom item on the screen is in focus, when rotated it may be off the screen. Instead, the page must scroll so that the item remains in the viewport in the horizontal screen, even though it means other items have been scrolled up off the screen.

At no point should the in-focus item be off the viewport, so use caution when designing an animation to depict this. For example, do not rotate and then scroll down to the in-focus item. The time when the in-focus item is out of view may be disorienting.

Antipatterns

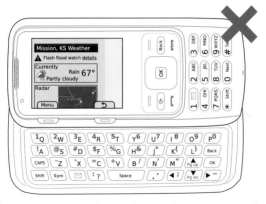

Figure 13-11. *Never change orientation without redrawing the content to use the space more effectively. In this example, extra space is unused on the screen, but other cases exist where content flows off the screen. Don't consider user controls that allow zooming as an excuse for this behavior—always deliver the best possible experience automatically, based on the current context.*

Always redraw items on the screen to take advantage of the new orientation. Never leave blank space, as in Figure 13-11, allow content to fall off the screen, or otherwise take shortcuts that use the new orientation in a less-than-optimal manner.

A *manual override* is often provided for devices using automatic sensing. This may take the form of a manual selector or a lock (to prevent rotation from the current position). These are generally arbitrary buttons or on-screen actions, so they must be learned. In general, they should be considered a half-measure, and everything should be done to make automatic sensors and their algorithms as good as possible.

Use caution when including both physical mode switches and automatic sensing in a single device. Consider a perfectly usable "all touch" device, with automatic sensing for orientation changes. When the hidden hardware keyboard is slid out, it makes no sense to rotate, as the keyboard would face the wrong way, but the user may have become accustomed to having the ability to rotate orientation.

Location

You must clearly communicate the availability of location services, and use these location services to make services more personalized and relevant.

Aside from the many dedicated navigation devices, GPS is ubiquitous in many other devices, and other location technologies can extend this capability to almost any connected device. Access is sometimes restricted for websites and some application types, but there are workarounds.

Figure 13-12. *Location can present more that just a pinpoint on a map. Direction can be indicated organically, with the pointer and by rotating the map to match the direction of travel, as well as with compasses. Speed and other attributes (e.g., height, relative position) can also be displayed.*

Mobile devices have numerous methods of retrieving location information. These include:

- Cell
- Sector
- Triangulation
- GPS telemetry
- WAAS

- AGPS

- WLANs and PANs

Note that only one method is the use of GPS. You should try to understand at least the basics of each technology, and their capabilities and limitations. Each is detailed in Appendix A, "Mobile Radiotelephony."

One more key method of retrieving information is, of course, to ask. Users often know where they are. If they cannot get any location, or there's a reason they might want to override it (even one as simple as that the data is bad), let your users enter a location. If they do this, allow the device to accept lots of methods. Only accepting zip codes or postal codes is not useful for travelers, who do not generally know such details.

It is also important that you understand the difference between precision and accuracy. In brief, *precision* is the number of decimal places you measure something to; *accuracy* is how correct it is. The less accurate you think your measurement is, the less precisely you should report it.

Privacy and security concerns are beyond the scope of this book, but they must be considered when designing location-based services. In many countries, there are legislative or regulatory restrictions on enabling and using location, so notification of background tracking is a legal requirement, not just a best practice.

Location is not the same as Orientation or other types of position information. When these must be communicated—as for augmented reality—they are generally deeply integral to the visualization and interactive design of the application. No distinct patterns yet exist for these behaviors, but best practices from aviation may serve to guide designs.

Variations

Location service is often used as a background service, as a way to enable smarter ambient computing. When directly referred to, it may only be momentarily switched to while other tasks occupy the remaining time. Therefore, both *explicit* and *background* indicator cases may be present, not just in the same system, but also at the same time:

Explicit
> Location services are currently mostly used for explicit location applications, such as mapping, directions, local information, and augmented reality. These explicit representations of location are discussed in detail shortly.

Background
> Location services can also be used as a trigger to initiate notifications or to change the behavior of the device. Geofencing, location-driven advertising, and location-based profile changes (e.g., silent when in a meeting room) have all been tested, but are not widely adopted.

Settings, including installation of applications using location services, must also take into account these principles, especially those of privacy and awareness.

Though this pattern discusses such behaviors and features on a mobile device, they may also be used for remote devices in much the same manner. "Child trackers" and similar functions may be used from another mobile device or from desktop computers to track a remote mobile device. In this case, both the viewing terminal and the mobile device must consider the display and behavior attributes of this pattern.

Figure 13-13. *When information is not available, such as speed and direction of travel, it should generally not be shown at all, instead of simply being zeroed out. Precision must always be shown with a true-to-scale error scale circle, and numerical precision whenever possible.*

You should present indicators of state only in the Annunciator Row so that the user cannot interact with it. Make a system setting available to control the use of GPS, WiFi, and other controls, as well as to set privacy controls for sharing and how much automatic (versus manual) control is allowed over these systems.

Provide an easy method for the user to gain access to this control, without drilling into multiple submenus. Whenever possible, add this control shortcut to any application that uses location services. Another good method, when technically feasible, is to add the control as an interactive Icon on the Idle screen. In this way, the user can simply pop back to the home screen and change the setting before using any application.

Whenever possible, applications or services that work best with location should automatically enable the required hardware. OS-level restrictions or user settings may interfere with this behavior, and require user intervention each time. If so, an interstitial Pop-Up, despite the intrusive nature, will end up being the best way to request this access.

When the Pop-Up itself is restricted from offering a function to enable the GPS (or other hardware), use it sparingly, as a reminder to the user instead, and provide a link to the appropriate settings panel. When settings are changed, always return the user to the originally requested application.

Only use the precision required for the task at hand. Weather, for example, rarely requires more than city-level precision. The GPS is not required for basic conditions and forecast information, so if other location services are providing sufficient detail, do not enable the always power-hungry GPS.

On the other hand, always present the most relevant details for each view. If the user chooses to view a local radar map, the user should not be required to know her location but should be presented it systematically. This example weather application may have to turn on and off various location services as the application is used, and should not enable all services when it is opened, or get by with only the minimal set and present less-than-optimal information.

Presentation details

Figure 13-14. *Icons to denote location services are not just optional variations, but can imply different meanings. The crosshair is a standard used to denote location-enabling as well as the fact that the mobile handset sends its location to the operator, for e911 and other purposes. The satellite dish and satellite denote GPS specifically, and may be used to denote other conditions. For example, the dish may imply the device requests service, and the satellite only appears when a link is established.*

When location service is enabled, be sure to display an icon (see Figure 13-14) in the Annunciator Row. The location icon (crosshair) has come to mean GPS is enabled, so there may be challenges in communicating other types of location services. GPS specifically can use other types of icons, such as a satellite dish. GPS requires a short time to find position, can lose position due to interference, clutter, or other conditions, and must display the current status. Generally, animating the satellite dish (implying "scanning for service") works well.

Avoid duplicating indicators. If the indicator is in the Annunciator Row, there is generally no need to also include such indicators within an application. This is another good reason to preserve the Annunciator Row bar, instead of loading applications full-screen.

For explicit graphical display of location, such as on a map, use a cursor to denote the current position. Note that this is actually the centroid of the "circular error of probability" (CEP). A probability threshold is programmatically established; based on the technology used to determine location, the device is almost certainly located within this circle. (Note that it is impossible to absolutely determine location [hence the "probability" phrasing], but this is not terribly important for general discussions.)

Display this CEP circle with the cursor at the center, as a visual method of communicating the precision of the location technology (see Figure 13-13). Unless additional precision is needed but is currently unavailable, the default view should hide the CEP circle from the user. For a weather map, for example, the default zoom level should be large enough that a 100 m CEP is smaller than the cursor, so it disappears under it. Good selection of map zoom levels can help to communicate the degree of precision available.

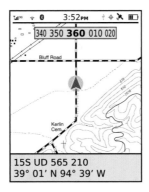

Figure 13-15. *Coordinates and other location information, when relevant and useful, should be presented on-screen in an easy-to-read format. Multiple formats may be presented at once, as shown here. Many digits have been removed in this example, in the interest of using only the degree of precision available.*

When the display is zoomed such that the CEP circle is at least 10 times the size of the cursor, or the edge is partly or entirely outside the viewport, print an explicit display of precision adjacent to the cursor.

An even greater pitfall in the display of precision information is with printed location, especially in coordinate systems (see Figure 13-15). Whenever possible, use standard, well-understood nomenclature. Avoid printing the location when it is not useful; lat/long is difficult to interpret and very few general users will understand it.

Account for precision in the display of coordinates and other printed locations. For this example, the MGRS coordinate system will be employed as it is based on simple measurements (meters from a regional baseline). This is unlikely to be employed in general consumer applications, but is convenient as an example. A complete location to 1 meter precision is listed as:

15S UD 12345 67890

(The 10 numbers at the end show the position in meters vertically from the equator, and horizontally from a point too complex to explain here.)

However, most location technologies, in most instances, cannot give such precision. To correct for this, remove digits until only the correct level of precision is shown. For example, if using cell sector, with precision in the 100 m range, only display:

15S UD 123 678

The last digit will always be of less precision than the others, and even when displayed digitally can be used like that on a dial or scale. Simply restrict the display to only the 5 or 0 digit.

Precision should be explicitly displayed when coordinates are printed. Use the existing settings for units of measure, but scale them appropriately. In the following examples, the user has set his navigation tool to use miles:

- "Accuracy 19 ft"—Correct. Appropriate scale (versus yards or miles) for the size, and in the same system.

- "Accuracy 6 m"—Incorrect. In the wrong system of measure.

- "Accuracy 0.004 mi"—Incorrect. Too large a measure with many decimal places and not easily understood.

Note that the term *precision* is not well understood, so it is often replaced with *accuracy* in general communications, as in the preceding examples, although it is not strictly true.

Addresses, likewise, should only be displayed when that level of precision is available. Otherwise, use existing best practices for general location:

- 5600 Russell Ave., Mission, Kansas 66202

    - 56th St. & Russell Ave., Mission, Kansas 66202—Street or Intersection

    - West Crossland, Mission, Kansas 66202—Neighborhood

    - Mission, Kansas 66202—City

    - 10 miles west of downtown Kansas City, MO—Relative Location

    - Near Kansas City, MO—Area

Only display the larger-scale information such as state and country when needed. When moving between two areas only a few miles apart, the default location display should probably just be the street address portion. The same filtering logic should be used for coordinate systems, when applicable. Although Lat/Long is a global system, UTM and MGRS are divided into regions; the "15S UD" portion in the preceding example can be made much less prominent, as most precision navigation is not also global navigation, so it will not be used as much as the remainder of the displayed coordinates.

Only display the device's direction of travel when available (see Figure 13-12), and just like the location, only display it to the degree known. Generally, the cursor can simply become a pointer. For free-standing displays, when users will get value from it display the current bearing in degrees and by named directions (northeast), but degrade to cardinal directions as the system's understanding becomes poorer.

Cardinal directions (e.g., "north") can imply precision if simply printed on the screen; use graphical displays (either dials or circular tapes) to communicate the degree of precision available, make small changes in bearing angle visible to the user, and make the display more glanceable.

When direction of travel cannot be determined (there is no compass and the device is not moving), display no direction of travel indicator, or an icon indicating it is unknown. The cursor may change to a circle, or otherwise become blunted to reduce or remove the implication that direction is known. Do not display the last direction, as this is likely to be spurious data arising from loss of signal, or coming to a stop.

When direction of travel is available, orient the map so that direction of travel is forward, or up. If there is a specific reason not to do this by default, provide a function so that the user can switch to this mode.

Antipatterns

Never equate "location" with "GPS," especially to the degree that a location-aware service or application may only launch when the GPS is enabled.

Never display or imply precision that is not available. Avoid the use of crosshairs and other implied attributes, even if a CEP circle is displayed as well.

Summary

Wrap-Up

People interact with their devices in unique ways that are most comfortable and natural to them. Some prefer using the keyboard, while others who are less familiar with text input may use the pen. Those who enjoy using natural gestures will use devices with touchscreens and accelerometers.

The environment and context also influence how users interact with these devices. Some people prefer to work outside no matter the weather condition. They might be bundled up in a coat and wearing gloves, and they still expect to input data. Those who use their device indoors while studying may require the use of lights and haptics as notifiers so that they aren't interrupted. People who are constantly traveling may require an integration of location-based services with their current application use.

With all of these variables affecting how people interact within the mobile landscape, it's important to carefully consider the mobile design principles when designing mobile interfaces:

- Respect user data.
- Mobiles are personal.
- Lives take precedence.
- Mobiles must work in all contexts.

- Use your sensors and your smarts.

- User tasks usually take precedence.

- Ensure consistency.

- Respect information.

Pattern Reference Charts

The pattern reference charts in the following subsections list all the patterns found within each chapter described in this part of the book. Each pattern has a general description of how it can apply to a design problem while offering a broad solution.

Cross-referencing patterns are common throughout this book. Design patterns often have variations in which other patterns can be used due to the common principles and guidelines they share. These cross-referenced patterns are listed in the following charts.

Chapter 9, "Text and Character Input"

Despite the existence of more efficient ways to input text, people still may choose to use what they are most comfortable with. Some people are comfortable with handwriting, others with keyboard input. Some may prefer to use a pencil, pen, or stylus. Always default to the most common method they can be expected to be familiar with, and provide options. Text and numeric entry methods must be simple, easy, visible, and so predictable in behavior that they may be performed by any likely user with little or no instruction.

Pattern	Design problem	Solution	Other patterns to reference
Keyboards & Keypads	Text and numeric entry must be simple, easy, and so predictable in behavior it may be performed by any likely user with little or no instruction.	Three options exit: hardware/virtual, keyboard/keypad, and direct/multitap. Consider the constraints held by the device, as well as cultural norms, and adhere to basic keyboard principles.	Input Method Indicator Dialer Autocomplete & Prediction Directional Entry Button Remote Gestures Exit Guard
Pen Input	A method must be provided to input text that is simple, is natural, requires little training, and can be used in any environment.	Pen interfaces can be an alternative to keypads or touchscreens when touch is unavailable due to environmental conditions (gloves, rain), when the device is used while on the move, or just to provide additional precision and obscure less of the screen.	Keyboards & Keypads Autocomplete & Prediction Wait Indicator Cancel Protection
Mode Switches	Provide access to additional and alternative controls without taking up more hardware or screen space through the use of mode switching.	Mode switches provide extra capabilities of a keyboard or keypad by including access to variations of characters and symbols.	Tabs Keyboards & Keypads Pen Input Dialer Input Method Indicator

Pattern	Design problem	Solution	Other patterns to reference
Input Method Indicator	The user must be made aware of the current mode of the selected input method, and any limits on selecting modes for a particular entry field.	An indicator should be placed on the screen to explicitly indicate the current input mode or method.	Pen Input Mode Switches Annunciator Row Notifications Accesskeys Dialer
Autocomplete & Prediction	Whenever possible, use assistive technology to reduce text entry effort, and to reduce errors.	Autocomplete and predictive entry have proven especially valuable due to the relative difficulty and reduced speed of text entry, and especially for complex, technical character entry such as URLs.	Infinite List Pen Input Tooltip Cancel Protection

Chapter 10, "General Interactive Controls"

In addition to the keyboard and keypad, users expect to interact with the UI using many other methods of control. These controls may be hardware keys located on the device, touchscreens that allow for finger input, or sensors that allow for kinesthetic and remote gestural interaction. The type of control used depends on the context, the user's preference and comfort level, and the technology available on the device. The control's behavior must be well understood, provide immediate feedback, and use constraints to limit error.

Pattern	Design problem	Solution	Other patterns to reference
Directional Entry	To select and otherwise interact with items on the screen, a regular, predictable method of input must be made available.	Two separate solutions work hand in hand to meet these needs: hardware input devices (buttons and controls) and the behaviors (axis and planar) that govern them.	Focus & Cursors Scroll Wait Indicator Mode Switches
Press-and-Hold	In certain contexts, an alternative cursor-initiated function should be made available. A simple, universal method of applying this mode switch should be provided.	A press-and-hold behavior may present a contextual menu of options, or other contextually relevant optional selections.	Keyboards & Keypads Mode Switches Revealable Menu Fixed Menu Annotation
Focus & Cursors	The position of input behaviors must be clearly communicated to the user. Within the screen, inputs may often occur at any number of locations, and especially for text entry, the current insertion point must be clearly communicated at all times.	Areas and items for which immediate interaction is possible are considered "in focus." When an exact point is in use within this area, this is indicated by a cursor.	Directional Entry LED Input Method Indicator Mode Switches Wait Indicator

Pattern	Design problem	Solution	Other patterns to reference
Other Hardware Keys	Functions on the device, and in the interface, are controlled by a series of keys arrayed around the periphery of the device. Users must be able to understand, learn, and control their behavior.	All mobile devices have numerous additional keys. Regardless of their function, all must comply with some basic behavioral standards in order to be useful, usable, and valuable to the user.	Keyboards & Keypads Directional Entry Annunciator Row Pop-Up Mode Switches Exit Guard
Accesskeys	Provide single-click access to an arbitrary set of key functions for each page or view.	Accesskeys provide one-click access to functions and features of the handset, application, or site for any device with a hardware keyboard or keypad.	Mode Switches Dialer
Dialer	Most mobile devices are still centered on voice networks connected to the PSTN. Access to a phone dialer must be provided.	Common numeric entry methods of operation that users are accustomed to must be followed to provide easy and accurate access to the voice network.	Home & Idle Screens Dialer Tooltip Autocomplete & Prediction Keyboards & Keypads
On-screen Gestures	Instead of physical buttons and other input devices mapped to interactions, allow the user to directly interact with on-screen objects and controls.	Items in the screen are assumed to be physical objects that can be "directly" manipulated in realistic (if sometimes practically impossible) ways.	Directional Entry Vertical List Film Strip Infinite Area Pop-Up
Kinesthetic Gestures	Mobile devices should react to user behaviors, movements, and the relationship of the device to the user in a natural and understandable manner.	The mobile device uses sensors to detect and react to proximity, action, and orientation.	Orientation Location Jump Remote Gestures Tones Voice Notifications Haptic Output LED Tooltip Annotation
Remote Gestures	A handheld remote device—or the user alone—is the best, only, or most immediate method to communicate with another, nearby device with a display.	Ready availability of accelerometers, machine vision cameras, and other sensors now allows the use of gesture to control or provide ambient input.	Kinesthetic Gestures Simulated 3D Effects Directional Entry On-screen Gestures Focus & Cursors Tooltip Annotation

Chapter 11, "Input and Selection"

In today's mobile landscape, filling out fields and forms is a trite task in any mobile context. But such a common task continues to have a common problem: user input errors. Whether it's from incorrect character input or accidental clearing, users can become quickly aggravated with input and selection. So it is essential to design these functions with the user in mind. This chapter discussed the input and selection methods that allow users to quickly and easily enter and remove text and other character-based information without restriction.

Pattern	Design problem	Solution	Other patterns to reference
Input Areas	A method must be provided for users to enter text and other character-based information without restriction.	Text and textarea input fields are heavily used to accept user-generated text input in all types of computing.	Dialer Directional Entry Clear Entry Focus & Cursors Input Method Indicator Autocomplete & Prediction Cancel Protection
Form Selections	A method must be provided for users to easily make single or multiple selections from preloaded lists of options.	Mobile devices, even more than other computing devices, can use forms to good effect for selection from already-provided information. Careful design choices can reduce user input, improve accuracy, and greatly reduce frustration.	Button Select List Vertical List Focus & Cursors
Mechanical Style Controls	A simple, space-efficient method must be provided for users to easily make changes to a setting level or value.	Mechanical Style Controls are really just compact, graphically oriented variants of the single-select Form Selections.	Form Selections Vertical List Zoom & Scale Search Within
Clear Entry	Users must be able to remove contents from fields, or sometimes whole forms, without undue effort and with a low risk of accidental activation.	Clear control can be used to clear or reset an entire text input field.	Input Areas Cancel Protection

Chapter 12, "Audio and Vibration"

Using audio and vibration can be an effective way to communicate alerts and notifications. Tones, haptics, and voice notifications are used to get the user's attention, communicate information more rapidly and clearly, or read the content of a notification when the device is not in the hand, cannot be viewed, or is chosen to not to be seen.

Pattern	Design problem	Solution	Other patterns to reference
Tones	Nonverbal auditory tones must be used to provide feedback or alert users to conditions or events, but must not become confusing, lost in the background, or so frequent that critical alerts are disregarded.	Tones should be used for alerting, and as a channel to emphasize successful input, such as on-screen and hardware button presses, scrolls, and other interactions.	Haptic Output Notifications Voice Readback Voice Input LED
Voice Input	A method must be provided to control some or all of the functions of the mobile device, or provide text input, without handling the device.	Voice Input allows voice to be used as a control and input mechanism to relieve users of eyes-on, hands-on control methods.	Accesskeys Voice Readback Directional Entry Input Method Indicator Tones Mode Switches
Voice Readback	Certain classes of users, or any user in certain contexts, must be able to consume content without reading the screen.	Voice Readback allows the device to read text displayed on the screen, so it can be accessed and understood by users who cannot use the screen.	Voice Notifications Directional Entry Annotation Pop-Up Voice Input
Voice Notifications	Certain classes of users, or any user in certain contexts, must be informed of conditions, alarms, alerts, and other contextually relevant or time-bound content without reading the device screen.	Voice Notifications read the content of a notification so that they can be used when the device is not in the hand, or cannot be viewed.	Notifications LED Annotation
Haptic Output	Vibrating alerts and tactile feedback should be provided to help ensure perception and emphasize the nature of UI mechanisms.	Most mobiles have some sort of vibration, or an external speaker, that can be used for at least basic haptic output.	Pen Input Tones Notifications LED Wait Indicator Voice Notifications

Chapter 13, "Screens, Lights, and Sensors"

Mobile devices today are equipped with a range of technologies meant to improve our interactive experiences. These devices may be equipped with advanced display technology to improve viewability while offering better battery life, and incorporate location-based services integrated within other applications.

Pattern	Design problem	Solution	Other patterns to reference
LED	Notice of status should have a visual component that does not require use of the primary display.	LED, a low-power-consuming diode, is used to indicate states and status of power, connectivity, and alerts on mobile devices.	Notifications Annunciator Row Tones Haptic Output Focus & Cursors Display Brightness Controls
Display Brightness Controls	Displays are the most critical output on modern smartphones. Even minor degradations in performance can render the display uncomfortable to use or unreadable, or can reduce the perception of the quality of the display.	Although most modern mobile devices come set to automatic mode by default, the user needs manual control to modify the current screen brightness settings.	LED Pop-Up Mechanical Style Controls Directional Entry Other Hardware Controls
Orientation	Devices must allow for orientation changes to display content in the manner most readable and comfortable to the user.	Mobile devices may use automatic sensing or respond to a physical mode switch to change the orientation of the display.	Kinesthetic Gestures Simulated 3D Effects Directional Entry
Location	Availability of location-based service, and the actual location of the handset, must be easily and accurately communicated.	Mobile devices have numerous methods of retrieving location information, which can be used for explicit location applications, and as a trigger to initiate notifications or change behaviors of the device.	Orientation Annunciator Row Icon Pop-Up

Additional Reading Material

If you would like to further explore the topics discussed in this part of the book, check out the following appendixes:

The section "Human Factors and Physiology" in Appendix D
> This appendix provides additional information on our human sensation, visual perception, and information processing abilities.

The section "Hearing" in Appendix D
> Here you can become familiar with how our auditory sense works and how sound is measured.

The section "Brightness, Luminance, and Contrast" in Appendix D
> This appendix discusses the differences between perceived brightness and emitted luminance as well as appropriate levels needed for visual acuity.

The section "General Touch Interaction Guidelines" in Appendix D
> This appendix provides valuable information on appropriate sizes for visual targets and touch sizes for interactive displays.

Appendixes

To keep the patterns focused on design and implementation, we have pulled all kinds of supporting information out of them. However, a lot of it is still very interesting. And there's no good way for a designer or developer to get a summary of this sort of information.

So we have included it here in the form of appendixes, ordered so that you can just pretty much read it from one end to the other.

You'll find that a few of the appendixes are actually just lists of resources. And in this day and age, resources are links to websites—which, of course, will go out of date soon. Luckily, we keep this up to date on the 4ourth Mobile Design wiki (*http://4ourth.com/wiki/*).

Visit any time to get the latest updates, or just to avoid typing in long links from a piece of paper. And please add your own information, or update old or changed links.

Mobile Radiotelephony

An Introduction to Mobile Radiotelephony

Aside from simply understanding your place in history, the underpinnings of mobile communications can be crucial to ensuring that your particular application, site, or service works correctly. Decisions made long ago, whether technical or regulatory, influence how mobile telephony evolved, and are still felt today.

An example might help. My favorite is that SMS (text messaging) isn't data. It looks like data, because it's typed; email and IM are data, right? But SMS is in the *paging channel*, or the part that is used to ring the phone and send caller ID data. What does that mean for pricing, availability, and traffic management? Well, too much to go into here, but importantly different things than managing data services.

I have gone out of my way to take actual RF engineering classes. It's pretty arduous, and I no longer remember how to calculate Walsh codes by hand, for example. But as it turns out, almost no one does. I was in class with guys who had EE degrees and had been working as radio techs for mobile operators for years, and they still didn't know the history or how parts of the system outside their domain work.

Hence, I feel pretty good about boiling several thousand pages of lecture slides and books into this short appendix. Just understanding the basics can matter a lot to your everyday work.

The Electromagnetic Spectrum

We'll start with junior high physics, to make sure everyone is on the same page. Everything from light to radio to x-rays (and much more) is part of the electromagnetic spectrum. These various parts of it are named, and discussed as separate elements based on the way the radiation interacts with physical matter. Visible light excites electrons at a frequency convenient to biochemical processes, so there are rods and cones in our eyes that detect it. Radio collectively oscillates materials, like all the electrons in an antenna at the same time.

Think of the entire electromagnetic spectrum in the same way as the spectrum of visible light; there is a clear area that is "red" and a clear area that is "orange," but also a space in between, parts of which could be considered "red-orange," or red, or orange. Though the individual components are discussed as though they are separate components, they are also part of a continuum and certain interactions overlap quite strongly.

Everything on the electromagnetic spectrum has a frequency, wavelength, and power. Radio especially is very commonly discussed in these terms, as it influences range and the speed and capacity of the information carried.

Radio is generally considered to be the frequencies between 3 kHz and 300 GHz. That "collective oscillation" mentioned earlier means you can feed an RF frequency electrical signal into an antenna (of the right materials, size, and shape) and it will send waves through the air, radiating much like ripples in a pond. When another antenna (again, of the right configuration) gets struck by these waves, they vibrate the electrons in the antenna in such a way that it generates electrical signals, and the receiving electronics do their job. See Figure A-1.

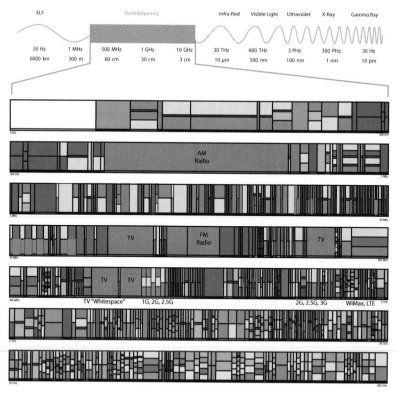

Figure A-1. *The electromagnetic spectrum, with a detail of the radiofrequency spectrum, including the allocations of various services, within the United States as of 2003. As you can see, a lot of different services are vying for a limited amount of space. Key mobile frequencies and a few others are labeled. All others have been removed for clarity.*

Frequencies are measured in Hertz, where 1 Hz is one "cycle" per second. Electrical power in the United States is delivered at 60 Hz. Mobile phones are much higher frequency, measured in megahertz (MHz) or gigahertz (GHz).

In general, longer wavelengths travel farther, and may go through and around objects. Very low frequencies have global ranges. High frequencies carry much more information, but cannot penetrate objects.

Certain frequencies are unavailable due to physical phenomena, such as cosmic rays and the sun. Frequencies are managed by national governments (in the United States, by the FCC), generally through international agreements. The spectrum is quite full of traffic now, such that adding any service requires disabling another one. The switch to digital television, for example, freed up quite a bit of useful bandwidth, just now coming into service for some next-generation mobile networks.

History

I always start presentations about mobile telephony by getting a baseline of the audience. I ask when mobile telephony was first instituted, and where. Occasionally someone will refer to the DynaTac, and correctly insist it all started in the United States. Much to the derision of their youthful cohorts, who are sure it came about no earlier than 1985, in Japan or, for the very clever, Finland.

No one guesses that it emerged directly from experience with miniaturization of radios in World War II, and was first placed into service by the Bell system right after the war, in 1946. See Figure A-2.

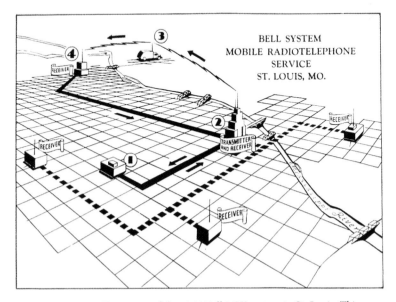

Figure A-2. *A contemporary illustration of the 1946 Bell MTS system in St. Louis. This was a quite complex installation for MTS. Due to the size of the surrounding areas, multiple receive antennas were installed, something not common with the first systems.*

These first systems were based in relatively small areas, covering a single city, for example. A single radio antenna on a tall building downtown sent signals to a large radio in your car, and received signals from it. To dial a call, you had to get an operator to physically connect the call by dialing it on a switchboard, then plugging a patch cable between the mobile network and the wireline phone network.

These Mobile Telephone Systems were replaced starting in 1963 with an improved version, which added antennas for larger areas such as the St. Louis region shown in Figure A-2, and had some capacity for handoff between them, allowing larger range and lower power. Devices became smaller (though still installed, they were the size of a briefcase, not a trunk). The customer could even direct-dial calls straight from his car.

These were mostly replaced by 1995, but at least one IMTS network was operating in Canada until 2002. The long range of the individual towers (up to 25 miles) was an advantage in the specific locale.

But these were still not cellular devices.

Legal and regulatory

Much of the behavior of these systems, and even how they work today, is tied up in regulatory requirements. The FCC allocates frequencies in the United States; other countries have their own bodies, but radio waves do not respect international borders, so they generally coordinate through an organization called the ITU.

The delay moving to IMTS, for example, was due entirely to the allocation of bandwidth to the mobile operators. The FCC denied anyone wanted or needed mobile phones, so it sat on the requests for more than 10 years.

Despite the name, mobile phones are not really phones. The wireline phone network is more precisely called the PSTN, or Public Switched Telephone Network. It is a single system that connects to (more or less) the entire globe. It also must generally provide a certain level of service, offering subsidies to the poor or disabled, for example. As usual, specifics vary by country.

Related to this is Quality of Service. This is not quality in the sense of resolution or clarity, but degree of service. For example, if the PSTN is totally jammed up—perhaps due to some disaster—when the fire department picks up their phone, someone else (you or I) is forcibly dropped to provide room. Gradations between this public safety override and general access, where you may pay for better access for your business, also exist.

Traditionally, there is no such equivalent practice in any private network, such as satellite or mobile telephony. In some countries, the devices must dial emergency services, and there is a small trend, starting in Scandinavia, to make mobile service and complete coverage a right of every citizen.

Additional regulations are as or more crucial to understanding how and why services work in a particular manner, or what you must do to comply with them. In the United States, location must be made available to emergency operators by offering GPS telemetry (this under the e911 mandates), but cannot be used for any other purpose for individuals under the age of 13 without quite specific consent rules. How do you enforce this, without overly burdensome rules?

Likewise, there are restrictions on how your personal data can be used to market to you, and how it can be sold. In the United States these are all consolidated under a concept of Customer Proprietary Network Information, which has more recently been used to ensure the security of your data as well. If you get phone service in the United States, you get a mailing at least once a year outlining your rights under CPNI, much like the privacy brochure you are given at the doctor's office.

This all means that, aside from technical and moral limits, there are significant and generally well-enforced restrictions to what can be done with personal information, and is much the reason that it is difficult to get customer information from mobile operators. Understand local technology, markets, laws, and regulations.

Early cellular

Although it seems no one could have foreseen the growth in mobile we are continuing to experience, there was a great deal of pent-up demand for more, more portable, and more flexible mobile telephony. After numerous delays, mostly due to regulatory approval, a new service was launched starting in 1983. The day this service launched, there were some 10,000 people on waiting lists for an IMTS device in New York City alone.

The Advanced Mobile Phone System, what we all call the "old analog system" today— once again, first launched in the United States—was a massive success, and was used in one guise or another throughout the world.

This was the first true cellular mobile network because it shared two key characteristics, both designed for efficiency at multiple levels:

- Handoff between base stations
- Frequency reuse

These allow much more traffic in the network, by a very simple frequency division multiplexing scheme. Although signals are still fixed to a channel, that channel uses a dedicated frequency range, and your call uses the whole channel, it is much narrower than in MTS, and is not fixed for the duration of the call. When you switch to another tower, the handoff may change the channel to suit the available space.

Aside from some cleverness with handoffs, and being over radio instead of wires, this was still very similar to the classic, wired PSTN. At any one moment there is a dedicated voice circuit. Replacing the analog voice carrier with a digital signal allows even more capacity, as well as reductions in power. This is what D-AMPS (D for digital) sought to do in an effort to extend their network life in the face of the next-generation networks.

D-AMPS encoded the voice (by adding a *vocoder*, which turns it into data), then chunks the data stream into "frames," which are then sent in intervals, with a specific time slot for each user on the network in that cell.

This time-based system is called Time Division Multiple Access, or TDMA. Digital signals allow compression of the voice signal. Combined with very rapid time slice switching, the same frequency can be used for several users at once.

Cells and backhaul

There are two fundamental components of a cellular network. The first is the most common, the *handset* (or aircard, or tablet, or eReader, or home modem, etc.). Operators also call this a terminal, or customer terminal, or use the old CPE terminology, for Customer Premises Equipment.

The other side of the equation is the *cell site*, which is actually a small complex of items. Although we all use the term as a shorthand, the tower is actually just the mast to which antennas are mounted. Wires run down this to a shed which houses transmitters, and often backup power such as generators. GPS antenna and other equipment are also included on the tower or in the shelter. The whole arrangement is more technically called a BTS, or Base Transceiver Station.

The antennas themselves are arranged around the top of the tower, in a flat ring of three (or sometimes four), grouped with two or three antennas pointing in each direction. The multiple antennas work together for power management and position finding, using math not worth getting into.

Whenever you see more than one stack of antennas on a tower, this is just because they lease to their competitors. This is a very profitable business alone, so competing operators rarely restrict one another from mounting antennas to their tower, at least for the right money.

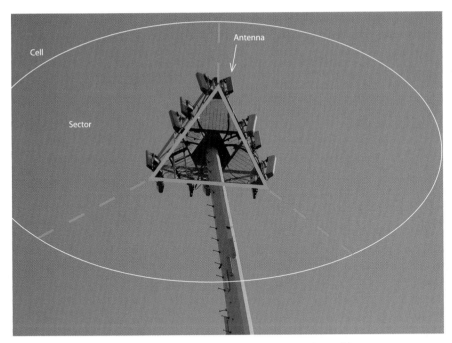

Figure A-3. *The arrangement of the antennas on a tower is very specific, and not just to cover more space, but integral to the way cellular mobile telephony works.*

Each group of antennas pointing in a specific direction constitutes a *sector* (see Figure A-3). When you are on a call or using data services and you're moving through the network, your handset stays connected by negotiating with the BTS to "hand off" between cells and between sectors. When you run out of available coverage in your home network, you can even be handed off to a roaming network.

There are differences in the various types of handoff, but they are fairly unimportant for our purposes. Networks are generally voice-optimized still, so they spend a lot of time making sure calls are not interrupted during handoff. Data services often are interrupted, and this can have implications in terms of network-based services. Websites in particular can become confused and suddenly reset when they find you coming from a new IP range.

Handoff does require continuous coverage, so cells and sectors overlap. Your handset is, as you now might expect, communicating not just with the tower it's getting voice and data from, but with all adjacent towers. They all work together to decide when to hand off to ensure maximum efficiency and prevent drops in coverage.

As discussed earlier, the mobile network must attach to the PSTN at some point. The first step is to connect the BTS to all other nearby BTSes in that operator's network. Then they connect to a central office, and at various points interchange with the PSTN. The whole scheme of connecting the towers to one another and to the central office is called *backhaul*. This can severely limit the speed or capacity of the network, especially when not considered in the design of new radio technologies.

Last year I was on vacation in Door County, Iowa. It is a relatively remote peninsula, but we were able to get mostly decent mobile coverage (high signal strength, low noise). However, the data speeds were awful and it even sometimes took forever to connect a call. It turned out that the backhaul was all done by a microwave link, one tower to the next in a line down the peninsula. It was not able to handle the traffic load of tourist season, even though the individual towers easily could.

Backhaul is generally microwave, via yet another antenna on the tower, or cables in the ground. Some other radio technologies are used, and some have backups and so provide both. Central offices (which may be collocated with a BTS) sometimes have satellite links as a last-resort backup even, to offer service in the event the local PSTN fails.

The 2G networks

Quite soon after AMPS took over the world, it became clear that mobile telephony was a big deal, and eventually the current networks simply wouldn't be able to handle the load. Several schemes were put forward, and these all-new digital networks took over the mobile landscape by the late 1990s.

Since these were considered an upgrade to AMPS (et al.), or the second generation, they came to be called *2G networks*. In the United States, the FCC codified the system under the moniker of Personal Communications Service, or PCS, which you could see appended to the name of several operators.

It also became clear that mobile telephony was a huge industry, and would only grow. This ended up changing the industry landscape, as it suddenly became very expensive to be a mobile operator. The reason is not strictly one of scale or technology, but of competition. As discussed already, national governments control the bandwidth in their borders. When a new type of network launches, it generally appears on an all-new band, so it can coexist with the older network and we can all take most of a decade to switch over.

To decide how to allocate this bandwidth to the various operators, there are auctions. Other portions are involved (you have to prove there's a valid business plan and a market for the service, and agree to actually offer service), but basically the one with the most money wins. Many of the small or local operators simply could not compete in the face of big telecoms and consortia of smaller ones. In some countries, of course, the nationalized telecom simply assigned the rights to itself.

The 2G networks led to much more reliable, wide-reaching, and eventually data-driven mobile use. However, despite being able to be grouped as a class, they are not a single service like AMPS became. Instead, two competing network technologies emerged, and are still dominant in their follow-on versions.

GSM

Although pretty much the entire United States and many other regions implemented AMPS with fairly good consistency, Europe suffered severe fragmentation, both by carrier and across international borders. This just extended the muddled, expensive system already in place, with toll calls (on a complex scale) being the norm even for much local service. For both technical and market pressures, by the mid 1980s it became clear that a common standard would be needed, and it should be built into a next-generation system.

GSM, or its follow-on variations, did meet much of the promise and is a global standard that a majority of handsets use. Many implementations are even seamlessly interoperable, so you can enjoy relatively global coverage with a single handset. Relatively, of course, is not "complete," and Japan, for example, has no GSM coverage at all.

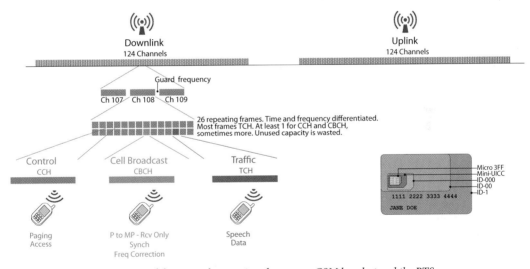

Figure A-4. *A summary of the network operations between a GSM handset and the BTS*

A key feature of GSM is the Subscriber Identity Module, or SIM card (shown on the right in Figure A-4). The SIM is a smart card that carries the user's subscription information and a phone book. Theoretically, the user can then just swap the SIM freely between handsets. Practically, there are a few different sizes, so some phones aren't compatible with the card you own, the SIM is always older technology, so address book data is more limited than you want, and many operators just don't let you swap cards. A practice called *SIM locking* means you cannot get the SIM to work in another device. This is for business reasons, when the operator underwrites the price of the handset, and makes you sign a contract. The freedom of the SIM violates this principle, so the operator might not get its money back. In some countries, consumer freedom laws prevent SIM locking.

GSM is an entirely digital standard, encoding the voice as a stream of data. This is then assigned to *frames*, which are sent out over specific timeslots, mixed in with all the other traffic on the network. At the other end, it is extracted from the frame, bolted to the rest of the stream, and then turned back into continuous voice. The time slices make this a TDMA or Time Division Multiple Access system. On each of the 124 frequency channels, 26 different conversations can be had. This is possible partly due to the efficiency of digital encoding; the coded traffic takes less time to encode, send, and decode than it does to say or hear, so it can be broken up into bits without you noticing the lag.

Additional parts of the network are dedicated to various traffic control functions, and to power management. Among the traffic control functions is a message section, originally just for network technicians to send short text phrases between the BTSes, then repurposed to let the handsets replace pagers. As you might have guessed, this is where SMS text messaging came from.

Despite all the voice traffic being digital, this does not mean early GSM supported data. The concept was still of delivering circuit-switched voice to the PSTN, so data essentially operated like a dial-up modem, and was inefficient.

CDMA

The first thing most people notice about CDMA, when comparing it to GSM, is that it has no SIM. There is actually a similar part of the device that has the subscriber information, but it is not removable. When comparing to SIM-locked devices, though, this is an unimportant distinction.

Another key difference is that CDMA is less of a standard. There are several implementations of it, and even the standard ones detailed here (Figure A-5) can be made so that they are not interoperable with others. You generally cannot use your CDMA phone in any other country, and often it will not even work on a competing CDMA operator's network in the same country.

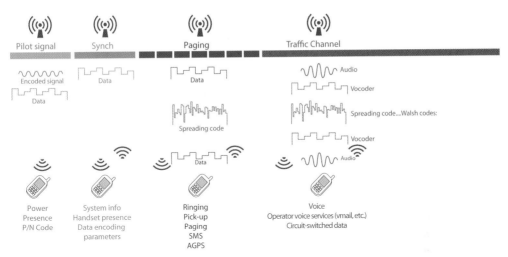

Figure A-5. *A summary of the network operations between a CDMA (IS-95) handset and the BTS.*

Another interesting feature is that the underlying technology is covered by a handful of very specific patents, and the result has been that a company called QUALCOMM effectively owns the technology. Instead of (for the most part) literally licensing, it simply retains the right to provide essentially all the chipsets that run these devices. It has parlayed this into being among the largest mobile chipset suppliers for all types of networks, worldwide. A surprising amount of what mobile telephony can do, and will do in the future, is at the whim of a single company.

CDMA uses the name of the underlying technology, Code Division Multiple Access. Voice is again encoded digitally, but this time the encoding itself is part of the message scheme. The voice signal is encoded in such a way that when sent over the air it is mixed willy-nilly with the bits from every other conversation or bit of data on the same frequency. At the other end, a code is used to extract each conversation individually from the pool of random-looking data.

This scheme is much more secure, and ends up being faster and more efficient. Although the original implementation did not directly support data, the underpinnings were better able to support it, so CDMA-based networks have gained some dominance with the move to data networks in 2.5 and 3G.

As with GSM, other portions of the signal are used for traffic control measures, power control, and signaling. Although not originally implemented in the same way, a paging-like one-way text sending system evolved to be 100% compatible with SMS text messaging.

Other Networks and Concepts

These are the two largest basic network technologies, but others do exist. PDC is a network used almost exclusively in Japan, and nowhere else. Many others who create taxonomies of mobile history consider D-AMPS to be a 2G network, due to it being a digital service (I don't, because it seemed like a half-measure at the time and in retrospect more so, as it was a dead end).

Both of the ubiquitous mobile technologies have benefits and risks or limitations. Many of the downsides are not inherent, and get fixed in later implementations.

The key features that mobile networks are evaluated on tend to be:

Cost
> This includes the density of the networks, and the ability to reuse existing towers, antennas, switchgear, and backhaul.

Efficiency
> This is mostly about compression and reuse of assets. You never want unused frames or channels.

Capacity
> As many handsets as possible must be able to be connected to a particular tower. Capacity regularly doubles, or more, when a new generation is released. There seems to be no end to network tricks.

Reliability
> If the network gets to its rated capacity, it had better be able to continue functioning, and have some overhead where it can function at degraded call quality. Drops and blocks must be kept to a minimum, and the tower must never, ever simply crash so that it is off the network entirely.

Interference
> This has largely been cracked, but it is still waiting in the sidelines. Interference from other radios and from obstacles must be tolerated or worked around, and the mobile signal must not interfere with other radios.

Power
> This includes both total radiated power and power management. Modern mobiles use single-digit milliwatts of power when connected correctly, and this continues to go down all the time. The less power, the less the tower needs, and the more that any battery on the handset needs to use for radios (and more for big screens and other features). Poor power management means the handset sends out higher-power signals trying to get a good signal.

Much of this is about the costs associated with the network. It costs billions just to own the rights to the network (those auctions), and billions more to build a network. A national-scale operator can spend a billion US dollars every year on network maintenance and improvements.

2.5 and 3G

Without getting into too much detail, third-generation networks (or 3G) are directly descended from 2G. They retain the basic underpinnings of the technology (TDMA versus CDMA) and much of the general layout of control and network configuration, but can also naturally handle data, by turning some traffic channels into packet data channels.

Naturally, a change to the network architecture also means other parts were changed, and the networks use even less power, manage power and interference better, and offer markedly increased capacity.

The 3G networks had to meet certain requirements by international certifying bodies, especially those of data transmission speeds and capacity. The "2.5G" networks are simply steppingstones, which were so different from 2G that they were worth mentioning, but not yet good enough to be "3G." This is, however, an insider's term. Consumers didn't mostly get told all this, and thus began the marketing of network specs, and a set of lies. Many 2.5G networks were marketed at the time as 3G. The same is happening with 4G today; although good WiMax and LTE networks are active, none are really 4G-certified yet.

Future networks

To a certain degree, the future is already here. It is just, to borrow a phrase, unevenly distributed. And for that largest of mass markets, mobile telephony, that means everything can still change. LTE and WiMax networks certainly point the way, but it is still hard to tell which will be most dominant, if either of them, and what features will actually make it to devices when they are more broadly adopted.

As far as networks in general, one clear trend is to data-only operation. This is far past simply encoding voice traffic for efficiency. The expectation of more and more pure data services, the possibility of using any traffic channel for any type of traffic, and other efficiencies mean that VoIP will replace dedicated voice channels, and data networks will rule in the future. This requires all-new networks, not just updates, partly so that things such as data handoff happen seamlessly. Without this, voice service (still by orders of magnitude the largest communications method on mobile networks) will suffer a severe degradation, and operators will not be able to sell their customers on the changeover.

This theoretically means any data connection could be used as well, such as ad hoc WiFi networks. People who use services such as Skype can tell you there is no overwhelming technical impediment to this today. However, they also can tell you it isn't really business class, or grandmother-proof. Aside from this reliability, the whole model of operating a mobile network makes this unlikely in the foreseeable future; operators want to retain more complete control, for revenue, perceptions of trust, and customer support purposes.

However, if there is anything I have learned from my time in the industry, it is not to predict anything. No specific network technology is clearly going to win out. Whatever does emerge will probably be used in a way that is somewhat different from how it was conceived, and by the time it is in widespread service, the replacement will be well on its way to market.

An Introduction to Location Technologies

Appropriate use of sensors is one of the key methods to make mobile devices behave in a useful, intelligent manner. Most of these are fairly easy to comprehend. A camera is a camera, and even accelerometers are understandable in a general sense with only minimal exposure.

Location services, however, are very complex due to their scale, the opacity of the operation, and poor choices in marketing communications. The most important thing for understanding location technology on mobile devices is to never, ever, ever conflate "GPS" with "location." We discussed the numerous technologies briefly in the Location pattern. Longer descriptions of the same are provided in this appendix.

The following are key guidelines to keep in mind when designing location services on mobile devices:

- Every phone is location-enabled. GPS or not, working or not, there's some location available as long as it is attached to the network. If your service offers up a weather report or local news, a few thousand yards is practically pinpoint precision.

- All the available location technologies must be addressed when designing your application or service. Don't disable the whole service, even if it's about something that seems to require precision such as mapping, just because the GPS is off.

- Precision and accuracy must be understood by designers, and correctly exploited by the product. When the GPS is off, or doing badly, show the CEP circle so that it's clear that we don't know "exactly" where you are. Admit the same lack of precision in text; say "Turn right in 'about' one-quarter of a mile," and so on.

- If you work for a carrier, exploiting the network like this should be a snap. If not, your devices or software may or may not be able to talk to the phone enough, and you might need to negotiate with the carrier or find a third-party provider. There are providers of location services that have already set up agreements with the operators.

- Consider what to store locally. Does the handset really need to be on the network, or should you give some functionality when the connection is bad or unavailable?

- Look useful without acting creepy. And don't just follow the letter of the law. Be careful with your user's data, share only what is needed, and store only what you really have to. Often, these last two are answered as "nothing." If that is the decision, make sure it's implemented as such, and no one gets lazy with the development or doesn't get the memo. Don't keep location information that is not relevant anymore, and be sure to encode information that has to be stored. Presence is considered personal information, so it will be perceived as a breach of confidence, even if not a breach of law.

Location Technologies

As we have already mentioned, GPS is not the only location technology available. A whole series of them are employed, and work together to give a sense of location to any device. Whenever possible, these technologies are listed from least to most precise.

Cell

Mobile telephony is based on communications and handoff between cells, each of which is (generally) controlled by a single base station, generally on a tower with multiple antennas. The entire assemblage of tower, antennas, transmitters, shed to house it in, and so on is called a Base Transceiver Station or BTS.

The tower location is well known, and its ID is attached to the call data record (used for billing and network control purposes), so it can be used by certain systems.

This is excellent as a fallback, when other things are not available. Better technologies are generally not available when roaming, and the data is not or cannot (due to technology or politics) be shared at more precision with your home network or because you are from another operator.

Precision will vary widely by the size of the cell, and there is no (good, universally accepted) method to determine and communicate precision and accuracy to the handset or any systems that use it. A good rule of thumb—and something similar is generally used by devices—is to assume the device falls within a circle 3,000 yards (m) across, though this varies by the technology used.

Sector

Recall that each tower has multiple antennas. Those three antennas (or four, depending on the technology and carrier implementation) point in different directions to cover an entire circle. Each of those is an independent mobile radio sector. Just like signals are handed off between cells, they are handed off between sectors, and the data is generally known to the handset and can sometimes be shared with services or providers. However, since the direction the sector faces is really only known to the company that installed it, and the network operator, they are the only ones that can use the data well.

Precision will vary widely by the size of the cell, the antennas used, the number on the cell, the down-angle, and so on. There are some semi-useful ways to determine precision and accuracy for a single read, but they are not universally implemented and there is no way to send the data to the information providers (e.g., the map program). A good rule of thumb—and something similar is generally used by devices—is a circle 1,200 yards (m) across, collocated with the radio centroid of the sector (which can be determined from maps and some educated guesswork; others have already done this). Yes, sectors are not circles, but it's also hard to communicate shape and orientation, so this works.

Triangulation

Generally, triangulation is the practice of determining the location of a point by measuring the angle and/or distance from several known locations. More points give more precision. It can get more complex when you try to find points in 3D space, but all mobile triangulation systems mostly assume the Earth is locally perfectly flat, and simply find the location on the geoid. Methods of triangulating get really complex and vary by network type and equipment available; multilateration and signal interpolation are some of those. The math can get pretty harrowing.

Precision varies mostly by the number of sectors being used, the distance between them, and other variables of the network. There may be a method of communicating to services employing it an approximation of the precision available. In any remotely built-up area, where it most often has enough sectors to work, a circle 50 yards (m) across is typical; 12 yard precision is possible, with good accuracy.

GPS telemetry

The Global Positioning System is a GNSS (Global Navigation Satellite System) consisting of a ground-based control system, a series of satellites, and any number of anonymous receivers. So, reading that alone dismisses many misconceptions; it's one-way, it's a system, and the signal is from satellites and has nothing to do with mobile networks.

Telemetry means data sent from a remote (usually mobile) site such as a rocket, back to base. Here it means the device figures out where it is by listening to the satellites, then tells the network or website or whatever where it is.

GPS can yield precision to fractions of an inch at least. But those are large, complex devices used for special surveying and scientific tasks, and require the antenna to be fixed to the ground for a period of time, and then rotated for reasons not worth getting into here. Portable, handheld receivers like those in mobile handsets can generally be assumed to give precision of 20 to 50 feet, though antenna and signal processing improves constantly and precision can be in single-digit feet.

Selective availability is the capability of the US government to partly or entirely disable the commercial GPS network so that enemy forces cannot use the system. Military receivers are functionally identical, but use a special, coded signal instead. In practice, this is unlikely to ever take place outside of a global catastrophe in which you won't much care about GPS anyway.

However, it exemplifies the theoretical risk of a major system being held by a single country. Therefore, there are other national and regional GNSSes, including Galileo, GLONASS, and COMPASS as well. None of these are really fully active, and there are few or no receivers in mobiles as yet. In theory, eventually, using multiple unrelated systems will give better accuracy yet, and give better coverage when in touchy situations like in dense cities.

WAAS

GPS uses radios, and the atmosphere and Earth are imperfect and can displace or distort the signals. WAAS is a satellite-based augmentation system based in North America (others exist in the rest of the world, but are less well established) that sends additional signals via another set of satellites to allow GPS receivers to adjust for those inaccuracies. I know of no mobile phones that have a built-in WAAS receiver, but it's a location technology that does exist, is almost universally used on every dedicated GPS device, and is employed on BTSes using AGPS.

AGPS

Assisted GPS requires the use of mobile networks, so it only works when in data communication range. The BTS has its own GPS receiver (which many already have in order to get precise time signals), and usually a WAAS receiver. AGPS provides a bit more precision, assures it of more accuracy, but mostly helps speed up cold or warm starts.

Briefly, for a GPS to work, the receiver has to download data about all the satellites before it can calculate the position. AGPS caches this info for the phone, and sends over just the relevant bits, cutting minutes to a second or two.

This is a reason you may not see WAAS, and might see some benefits from another GNSS (say, GLONASS) before the receivers are added to the handsets. AGPS generally gives precision, even in dense areas and on the move, of 10 feet or better. Two-foot indicated precision has been observed.

AGPS is not generally selectable as a separate service; when the user chooses to turn on or off "GPS" and has a network connection (and an AGPS-compliant phone, etc.), he gets the advantages of AGPS if it is available on the network and BTS. You can often see this at work, as your location will be almost perfect when in a home network, but suddenly be worse when traveling. Check, and you will probably be roaming, and therefore incapable of communicating with the AGPS features of the roaming network BTS.

WLANs and PANs

Local networks such as WiFi and Bluetooth are not exactly more precise, but are at the bottom of this list because they are local, and can add another layer of location, so they could increase precision. They can also be used instead of other technologies when there is no GPS signal or a bad mobile signal in which triangulation cannot be used.

Both of these currently use the "cell" (or network identifier) and triangulation methods described earlier, but with their transmitters. There are no sectors, and no universally adopted AGPS to work with device telemetry. Naturally, a handset getting good GPS data can send that telemetry to anyone over any network, including WiFi, but that's not the same thing.

They are both relatively rarer, and often are supported by third-party software or individual services (which may have specific capabilities), and are not universally implemented on the handset. There are no reliable, independently verified numbers on the real-world precision, but tests of WiFi triangulation alone have gone down to inches. On the other hand, poor network identification has led to placing users on the wrong continent. There's room to improve here.

Ask

People often know where they are. If you cannot get any location, or there's a reason your users might want to override it (even as simple as the data is bad), let them enter a location. Never override user-entered data—at least not without providing a warning.

If you do this, allow lots of methods and think about what the user knows, and is comfortable with. Do not require a street address unless it is really needed. Do not use simple entries such as a zip (or postal) code, if the user is unlikely to know it, such as when traveling.

Key Topics in Implementing Location Services

Location is not just GPS

All of the above is pretty well known by network engineers, chip makers, and various low-level handset software designers. But it is often missed a bit by various software implementations (and certainly by marketing). All too often GPS "is" navigation, such that turning it off kills all location services. Certain systems are becoming aware of this, and (for example) just give poorer precision when you turn the GPS off, but they do allow location services to work.

Vertical accuracy

Vertical location is not known with many systems, and is much worse than the horizontal accuracy even with GPS/AGPS. Consider this in your design if users in buildings or even just multilevel roadways are likely to be encountered. Technology may not reveal the vertical position with sufficient fidelity for the system to respond usefully. But if you can derive vertical location, start thinking about using it. A person 120 feet off the ground does not want a coupon for the Starbucks across the street; he's probably in a meeting or at his desk in a building. Eventually, map data should account for this, and even simple, obvious things like multilevel roadway navigation will also become easier.

Precision versus accuracy

Precision and accuracy are often confused. A quick primer: precision is the number of decimal places you measure something to; accuracy is how correct it is. The less accurate a measurement is, the less precisely it should be reported. Practically every location system gives location to the same degree of precision (e.g., 1 meter, using full UTM precision) at all times, and presumes the accuracy measure will be considered as well.

Visual representations are similar. The moving arrow (or car, or person icon) is the centroid of the current location precision circle. Even when that accuracy circle is displayed, the centroid marker implies much higher precision than is actually being achieved. Consider better ways to represent location to avoid user confusion.

Appropriateness of information

Display of information services should be related to the current accuracy. For example, turn-by-turn directions always assume perfect location precision (using the centroid of the current location accuracy circle), and say "You have arrived" instead of "It's the big, white building on the north side of the street." What if the location is vague and the target or turn is 30 yards behind you, or 50 yards in front of you? The user is confused. Close roads (in dense cities), multitier highways, and other situations can exacerbate the effect of this behavior.

Design to accommodate imperfect location instead.

Data integrity and reliability

If your map might be wrong, or imprecise or inaccurate (and they all might be), you should think about this when planning the interaction. Don't assume your data is perfect, and give contextually helpful information. Let users note problems, and do something with the problem reports.

Local versus network storage

Handheld dedicated GPS units and most in-car units store map data locally, so when a GPS fix is found, a map can be displayed immediately. Mobiles generally assume functionality only when on the network, even when a GPS is included. If your service is usable when out of network range (e.g., it's just for navigation, not for social media or communications), consider storing or caching enough information locally for the device to work when out of network coverage.

Design Templates and UI Guidelines

Drawing Tools and Templates

For many years, there was no particular assistance provided to those seeking to design for mobile handsets. Over time, books and other documents and supporting information has emerged. Now, major software vendors such as Adobe are embedding mobile-centric technology into their products.

New ones are constantly being added, or replaced, so please help us keep this up to date. Visit the wiki at *http://www.4ourth.com/wiki* or contact us with updates you may encounter.

Templates and Stencils

Templates and stencils are graphic items you can use with various drawing programs to create concepts, mockups, process diagrams, comps, or graphics for final designs.

First listed are organizations with either general templates or a number of templates for different devices and OSes. This is followed by a list by platform, as recently most designers and agencies have become concentrated on one or two devices or platforms, leading to a surge in such templates.

Note that this is just those we've found, or found useful. Many more may exist, or be included as part of the various manufacturer and OS developer links in the "UI Guidelines" section. Also be aware that by no means are all of these reviewed for quality, and are just checked to make sure that (at the time of this writing) they were valid links to real files, and are in the right category.

Design Organizations

The following organizations provide valuable resources for drawing tools and templates that you can use:

4ourth Mobile

The following is a list of the templates we use every day for designing mobile device interfaces. The list includes a number of mobile components as well as additional information, such as typography, for general mobile design. It also includes gesture guidelines, and some handset/OS-specific items. The illustrations and diagrams in this book were created with these components.

 - Adobe InDesign CS4 (*.indd*) format, compressed ZIP file, 15.7 MB; *http://www.4ourth.com/downloads/MobileDesignElements-2011april15.indd.zip*

 - PDF for viewing, or opening in, Adobe Illustrator, et al., 1.4 MB; *http://www.4ourth.com/downloads/MobileDesignElements-2011april15.pdf*

 If you are still working in Macromedia Freehand, the old (September 2009) version is still available. It has Little Springs branding, as that's who Steven worked for when he made it.

 - Macromedia Freehand MX 2004 (*.fh11*) format, compressed, 1.1 MB; *http://www.littlespringsdesign.com/downloads/mobileDesignElements/mobileDesignElements21sept2009.zip*

Graffletopia

Mobile UI Stencils (Omnigraffle); *http://www.graffletopia.com/search/mobile*

Mobility

Mobility, a free set of mobile UI design elements (Photoshop); *http://www.fullcreative.com/2010/10/mobility-a-free-set-of-mobile-ui-design-elements/*

Punchcut

Punchcut Toolset for Managing Screen Resolutions, a collection of devices of various sizes, in PSD format, to assist with developing content or design at multiple resolutions; *http://punchcut.com/perspectives/expanding-universe-toolset-managing-screen-resolutions*

Sqetch

Wireframe Sketching Kit, with tools for Browser, iPad upright, iPad landscape, Smartphone, GUI-Elements and Form-Elements. Not sure what some mean, but I guess "Smartphone" means they are trying to not just be iOS, so they get a category (Illustrator); *http://www.eleqtriq.com/2010/08/sqetch-wireframe-toolkit/*

Yahoo!

Yahoo! Stencil Kit has many nonmobile parts, but also useful general and iPhone bits and pieces; *http://developer.yahoo.com/ypatterns/about/stencils/*

Specific Operating Systems

Android OS

(All versions, all platforms, all operator overlays)

- Library of Android UI Elements: from *Smashing Magazine*. Photoshop library of Android (default) elements only (Photoshop); *http://www.smashingmagazine.com/2009/08/18/android-gui-psd-vector-kit/*.

- Android 2.2 Elements: based on the *Smashing Magazine* work (Photoshop); *http://thiago-silva.deviantart.com/art/Android-2-2-GUI-171047600*.

- Android Wireframe Tools (OmniGraffle); *http://www.marlinmobile.com/blog/?page_id=285*.

- Android GUI Set: has a lot of UI widgets, and includes the typography needed to make it all work (Photoshop); *http://www.webdesignshock.com/freebies/free-photoshop-android-interface-gui/*.

- Icon Design Guidelines, Android 2.0: from Developer.Android.com. Includes ways to download the template pack and other useful stuff, and is quite detailed for the graphic designer, for a change (Photoshop); *http://developer.android.com/guide/practices/ui_guidelines/icon_design.html*.

- 14 Android GUI Widgets: by which they seem to mean default desktop widgets (e.g., clock) versus UI widgets such as scroll bars (Photoshop); *http://bharathp666.deviantart.com/art/Android-Widgets-182178250*.

- HTC HD2 PSD Vector: 150 layers, 60 groups, almost all vector (Photoshop); *http://delightgraphic.deviantart.com/art/HTC-HD2-Smartphone-vector-183186532*.

- HTC G2: 370 layers, almost all vector (Photoshop); *http://zandog.deviantart.com/art/HTC-G2-PSD-182148621*.

- Motorola Droid: 2,350 layers, almost all vector (Photoshop); *http://zandog.deviantart.com/art/Motorola-Droid-2-PSD-180694347*.

- Samsung Galaxy S: device not specified, 127 layers (Photoshop); *http://zandog.deviantart.com/art/Samsung-Galaxy-S-PSD-178972735*.

- Samsung Galaxy Tab P1000: 110 layers (Photoshop); *http://zandog.deviantart.com/art/Samsung-Galaxy-Tab-P1000-PSD-180537736*.

- Android Honeycomb Tablet: preview of wireframe components, sketch sheets, etc. (based on screenshots, emulators, etc., so not final); from Zurb (OmniGraffle); *http://www.zurb.com/playground/honeycomb-stencils*.

- Google Android 1.6 Wireframe: stencil for OmniGraffle, from MarlinMobile (OmniGraffle); *http://www.graffletopia.com/stencils/630*.

- Google Android 2.1/2.2 Wireframe: stencil for OmniGraffle, from MarlinMobile (OmniGraffle); *http://www.graffletopia.com/stencils/673*.

iOS

- iPhone, iPod Touch, and iPad: (Photoshop) *http://www.teehanlax.com/blog/2010/06/14/iphone-gui-psd-v4/*.

- Library of iPhone UI Elements: Photoshop library of iPhone-specific elements only, from Teehan-Lax. Updated occasionally, so if you see a new one, please update the link for us (Photoshop); *http://www.teehanlax.com/blog/2010/08/12/iphone-4-gui-psd-retina-display/*.

- Library of iPhone4 UI Elements: same as above but for the iPhone 4 resolution instead (Photoshop); *http://fantasy-apps.deviantart.com/gallery/#/d2r7sus*.

- iPhone UI Kit: lots of UI widgets like switches, buttons, etc. (Photoshop); *http://dribbble.com/shots/42977-iPhone4-Retina-GUI-PSD?offset=7*.

- Another Library of iPhone4 UI: based on the Teehan-Lax work above, but presumably different in some way (Photoshop); *http://www.usercentred.net/2010/06/28/illustrator-template-for-iphone-design/*.

- Vector Wireframe Templates for iPhone (Illustrator); *http://reddevilsx.deviantart.com/art/iPhone-4-PSD-HD-176629139*.

- iPhone 4 HD: another one, not sure who made it or if it's fresh or a derivation of the Teehan-Lax stuff again (Photoshop); *http://www.vcarrer.com/2010/09/iphone-wireframe-kit-google-docs.html*.

- iPhone Wireframe Kit: in, of all things, Google Docs. Haven't tried it, but I really need to. (Google Docs); *http://browse.deviantart.com/resources/applications/psd/?order=5&q=ipad#/d2qvrn1*.

- iPad Template: preview only has the frame, so not sure what is included (Photoshop). *http://browse.deviantart.com/resources/applications/psd/?order=5&q=ipad#/d2qvrn1*

- iPad Wall Presenter: I think just the frames with guides to put backgrounds and designs into (PNG and Photoshop); *http://mrforscreen.deviantart.com/art/iPad-Wall-Presenter-179103252*.

- Sketching the iPad from Zurb: to encourage sketching, even as far as to print empties and use your mad Sharpie® skills (OmniGraffle): *http://www.zurb.com/playground/ipad-stencils*.

- iPad and iPhone Design: by da5id, via Graffletopia (OmniGraffle); *http://graffletopia.com/stencils/570*.

MS Windows Phone 7

- Design Templates for Windows Phone 7: from Microsoft. Includes 28 layered Photoshop template files and the Segoe system font (Photoshop); *http://go.microsoft.com/FWLink/?Linkid=196225*.

- Microsoft Windows Phone 7 wireframe stencil for OmniGraffle: from MarlinMobile (OmniGraffle); *http://www.graffletopia.com/stencils/689.*

Symbian Family (S40, S60, and Maemo)

- Templates for mobile phones: meaning mostly S40 feature phones, but also some older S60 devices; *http://www.forum.nokia.com/Develop/Web/Mobile_web_browsing/ Web_templates/Templates_for_mobile_phones/.*

- Templates for smartphones: meaning S60 3rd and 5th editions, Symbian ^3, Maemo, and some newer S40 devices; *http://www.forum.nokia.com/Develop/Web/Mobile_ web_browsing/Web_templates/Templates_for_smartphones/*

- S60 Wireframing Stencils: S60 (through 5th edition) elements of all sorts (Illustrator, Fireworks); *http://www.forum.nokia.com/info/sw.nokia.com/id/cfc7b6a4-2dc5-4c91- 88a5-c35764fff8fe/S60_Wireframing_Stencils.html*

UI Guidelines

OS and device manufacturers have increasingly been providing guidelines for the implementation of their UI. This may include brief, general guidance, relatively specific format rules, or quite involved discussions of design implementations and the reasoning behind them.

All provide much more documentation for developers, so if there is insufficient information in the UI guidelines, you may wish to browse the rest of the developer resources for the answer. Some UI or UX information may be included in other sections.

Android (from Google and OEMs)

- Android User Interface Guidelines, *http://developer.android.com/guide/practices/ ui_guidelines/index.html*

- Motorola's Best Practices for Android UI, *http://developer.motorola.com/docstools/ library/Best_Practices_for_User_Interfaces/*

- Mutual Mobile Android Design Guidelines, *http://www.mutualmobile.com/wp- content/uploads/2011/03/MM_Android_Design_Guidelines.pdf*

- Listing of Android patterns, real ones, with problems and solutions as drawings, from UNITiD in Amsterdam, *http://www.androidpatterns.com/*

- Google TV Android Developer's Guide, both from code.google.com; Google TV Web Developer's Guide, *http://code.google.com/tv/web/* and *http://code.google.com/ tv/android/index.html*

Apple iOS

- iPhone Human Interface Guidelines, iPhone Human Interface Guidelines (PDF), *http://developer.apple.com/library/ios/documentation/userexperience/conceptual/mobilehig/MobileHIG.pdf*

- iPad Human Interface Guidelines, *http://developer.apple.com/library/ios/#documentation/general/conceptual/ipadhig/Introduction/Introduction.html*

- iPad Human Interface Guidelines (PDF), *http://developer.apple.com/library/ios/documentation/general/conceptual/ipadhig/iPadHIG.pdf*

HP Palm WebOS

- User Interface Summary, *https://developer.palm.com/content/api/design/mojo/user-interface-summary.html*

- Human Interface Guidelines, *https://developer.palm.com/content/api/design/mojo/hi-guidelines.html*

MeeGo

- Handset UI Design Principles, *https://meego.com/developers/ui-design-guidelines/handset*

- UX Design Principles (a brief, neat overview of how to approach the overall interaction of the OS), *https://meego.com/developers/meego-ux-design-principles*

- Hildon UI Guidelines for Nokia Maemo, *http://www.forum.nokia.com/info/sw.nokia.com/id/eb8a68ba-6225-4d84-ba8f-a00e4a05ff6f/Hildon_2_2_UI_Style_Guide.html*

- MeeGo UI Design Guidelines for Handset, *http://meego.com/developers/ui-design-guidelines/handset*

Microsoft Windows CE, Windows Mobile, and Windows Phone

- Windows Touch UI Guidelines, *http://msdn.microsoft.com/en-us/library/cc872774.aspx*

- UI Guidelines for Windows Mobile 6.5, *http://msdn.microsoft.com/en-us/library/bb158602.aspx*

- UI Design & Interaction Guide for Windows Phone 7 (v2.0 PDF, from Microsoft), *http://go.microsoft.com/FWLink/?Linkid=183218*

RIM BlackBerry Smartphone

- UI Guidelines for BlackBerry 6.0 Smartphones, *http://docs.blackberry.com/en/developers/deliverables/17965/index.jsp?name=UI+Guidelines+-+BlackBerry+Smartphones6.0&language=English&userType=21&category=Java+Development+Guidelines&subCategory=*

- UI Guidelines for BlackBerry 6.0 Smartphones (PDF), *http://docs.blackberry.com/ en/developers/deliverables/17964/BlackBerry_Smartphones-UI_Guidelines-T893501- 980426-0721013746-001-6.0-US.pdf*

- UI Guidelines for BlackBerry 4.x, 5.x Smartphones, *http://docs.blackberry.com/en/ developers/deliverables/20196/index.jsp?name=UI+Guidelines+-+BlackBerry+Smart phones2.5&language=English&userType=21&category=Java+Develop ment+Guidelines&subCategory=*

- UI Guidelines for BlackBerry 4.x, 5.x Smartphones (PDF), *http://docs.blackberry .com/en/developers/deliverables/20195/BlackBerry_Smartphones-UI_Guidelines- T1011811-1011811-0903100131-001-2.5-US.pdf*

- BlackBerry Browser Content Design Guidelines (PDF), *http://docs.blackberry.com/ en/developers/deliverables/4305/BlackBerry_Browser-4.6.0-US.pdf*

QNX/Tablet

- UI Guidelines for the BlackBerry PlayBook Tablet, online at *http://docs.blackberry .com/en/developers/deliverables/27299/* or the same thing as a PDF at *http://docs .blackberry.com/en/developers/deliverables/27298/BlackBerry_PlayBook_Tablet- UI_Guidelines--1361251-0418095918-001-1.0.1-US.pdf*

Samsung Bada

- Bada Application UI Guide, *http://dpimg.ospos.net/contents/docs/resources_1004/ com.osp.appuiguide.help/html/FramesetMain.html?menu=MC01010403&mtb1= &mtb2*

Symbian Family (S40 and S60)

- Nokia Design & User Experience Library, *http://library.forum.nokia.com/index .jsp?topic=/Design_and_User_Experience_Library/GUID-A8DF3EB8-E97C-4DA0- 95F6-F464ECC995BC_cover.html*

- Nokia Mobile Design Patterns (a series of articles, approaching a pattern library; note that some of these may be very familiar, as it is a wiki, so some older patterns from this book and other information posted to the wiki have been copied to here by the authors, and other users); *http://wiki.forum.nokia.com/index.php/Category:Mobile_ Design_Patterns*

- Forum Nokia Design Portal (PDF), *http://www.forum.nokia.com/Design/*

- Symbian UI wiki, *http://developer.symbian.org/wiki/User_Interface*

- Nokia Series 40 UI Style Guide, *http://www.forum.nokia.com/info/sw.nokia.com/ id/73e935fe-8b59-43b2-ab3e-1c5f763672db/Series_40_UI_Style_Guide.html*

- S60 Visualization and Graphic Design Guideline (PDF), *http://www.forum.nokia .com/info/sw.nokia.com/id/34762388-9434-4c42-9c5e-3e545b0975ea/S60_Platform_ Visualization_and_Graphic_Design_Guideline_v1_0_en.pdf.html*

- Symbian^3 UI Style Guidelines (PDF), *http://www.forum.nokia.com/info/sw.nokia .com/id/5c419b14-75ff-4791-b1a8-db1e0d72e36e/Symbian_3_UI_Style_Guide.html*

Feature Phones

Feature phones, by the classic definition, do not have a "named OS." More to the point, applications cannot generally be loaded in the native OS, so other methods must be used.

J2ME

The Java Platform, Micro Edition (the same thing used to have a "2" in there, and is still mostly referred to by that abbreviation) is a common platform for running applications on practically all feature phones, worldwide. It was developed by Sun, so is now owned by Oracle. Installation methods vary, and some applications will work on basically all devices, while some are targeted to specific operators, regions, manufacturers, or classes of devices. Oracle tells us that J2ME runs on more than 3 billion handsets at the time of this writing. This is a big deal.

- Sony Ericsson UI Rulebook for JavaPlatform8 and later, *http://developer.sonyericsson .com/cws/download/1/716/984/1262667210/DW-102212-UI_Rulebook.pdf*

- Java ME Landing Page (a bunch of introductions, as well as developer resources, from Oracle), *http://www.oracle.com/technetwork/java/javame/index.html*

BREW

This is a competitor to J2ME, created by QUALCOMM, and also allowing a (more or less) single piece of software to run on a range of handsets.

- AT&T UI Elements for Brew MP, *http://developer.att.com/developer/forward .jsp?passedItemId=1900002*

Native operating systems

Feature phones do have operating systems, mostly developed by the individual manufacturers, and native applications are in fact developed outside of the manufacturer's direct control for these devices. However, they generally cannot be installed by end users. These are created or specified by operators, and installed under controlled conditions. This is mostly stuff like a custom phone book. Even the default web browser is likely to be J2ME or BREW.

Although UI guidelines do exist, these are generally proprietary, so they cannot be distributed. I have a few of them, but cannot share them. They would also be generally irrelevant; if you have to develop for this situation, you will be given the most applicable version by the operator and/or manufacturer, and have to work closely with them to get the software implemented.

Emulators

Simulators and emulators help with design, development, testing, and demonstration of software, when the actual environment is unavailable or unsuitable for testing. These are particularly applicable for mobile devices, as hardware is not always even out, service contracts and the number of devices for testing makes them very expensive, and it can be slow and cumbersome to load for each incremental code change.

Though often incorrectly used as such, they are not interchangeable terms, however:

- A *simulator* is software appearing on a computer that acts like the target environment, but is technically dissimilar in some key aspect, and maybe all of them. The simulator will superficially behave like the actual device, but is driven by different code entirely.

- An *emulator* runs the actual code; just within a virtual machine which itself simulates the hardware environment. From a technical perspective you can have much more confidence in the fidelity of the experience. Emulator problems arise from the virtual machine, which may have bugs (or the lack of them) that vary from the actual device. Often, the developer is allowed to select the available memory or processor or network speed, which is useful for unit testing, but must be toned down for realistic testing or demonstrations later.

New ones are constantly being added, or replaced, so please help us keep this up to date. Visit the wiki at *http://www.4ourth.com/wiki* or contact us with updates you may encounter.

An entirely other class of resources listed here are *remote testing labs*. I only know for sure how Device Anywhere works so I cannot comment on the others, but I believe they all work the same way. Actual handset hardware is disassembled, things are soldered to it, and it's strapped to cabinets. You get to press buttons and it goes over a real network, and gives you the screen output. These are all fairly pricey fee-for-service programs, but some operator developer programs give away a few hours for free. Be careful, as they mostly charge per minute connected, not per click or based on activity; disconnect as soon as you are done.

A good place to start is with MobiForge (*http://mobiforge.com/testing/story/a-guide-mobile-emulators*), which has published a useful guide to actually getting more than a dozen emulators to run. Many of these are buried under their developer sites, so you may not have even found them. Most have some trick or other to get them running, especially if you are not steeped in technical minutiae.

Web browsers

- dotMobi Online Emulator: web simulator (despite the name) with focus on common phones. Run by YoSpace. Works in the browser, so is quick and works on all OSes. *http://mtld.mobi/emulator.php*.

- Opera Mini Simulator: Opera's J2ME browser in the Web, so it works fine for every platform. *http://www.opera.com/products/mobile/operamini/demo.dml*.

- Opera Mobile Developer Tools: complete SDK with emulators for Opera Mobile. This is not the same as Opera Mini. For Windows, Mac, and Linux. *http://www.opera.com/developer/tools*.

- Bolt Browser Simulator: intended to sell the product, but useful to check how things might work on a small-screen, scroll-and-select browser. *http://boltbrowser.com/demo*.

- iMode HTML simulator 2: from NTT DoCoMo, displays (depending on the platform) HTML, Flash, and PDF content, for devices using the i-mode Browser 2.0, released up through 2009, so certainly still in service. Windows only. *http://www.nttdocomo.co.jp/english/service/imode/make/content/browser/html/tool2*.

- Access Developer Tools: Access J2ME browser development tools. *http://www.accessdevnet.com/index.php/Downloads*.

Remote testing labs

- Device Anywhere: several services, such as automated testing. You want the developer tools, which let you click around and see what is happening. *http://www.deviceanywhere.com*.

- MDPi: offered by Keynote as a component of its whole suite of quality assurance, testing, and monitoring products. *http://www.keynote.com/products/mobile_quality/on_device/mobile-device-application-testing.html*.

- Nokia Remote Device Access: service provided for free with a membership to Forum Nokia, for all the Symbian, S^3, and Maemo devices. Limited to eight hours a day. *http://www.developer.nokia.com/Devices/Remote_device_access*.

- Samsung Lab.Dev: the same service as the Nokia RDA, just offered for Samsung Android handsets only, through the Samsung site and with a required membership to the Samsung Mobile Innovator program. Click the Android button at the top of the page to start the service. *http://innovator.samsungmobile.com/bbs/lab/view.do?platformId=1*.

OS simulators and emulators

- BlackBerry Simulators: relatively simple-to-use BlackBerry simulators, without extra programming interfaces being required. Each one is a separate package, so checking multiple devices can be tedious. Windows only. *http://us.blackberry.com/developers/resources/simulators.jsp*.

- BlackBerry Playbook Emulator: complete emulator running in Air, for Windows and Mac. *http://us.blackberry.com/developers/tablet*.

- Archived Openwave Phone Simulators: when Openwave got sold a few years back, it removed all the old development links. However, there are plenty of devices that still access the Internet via the Openwave browsers, so it might be important for you. Dennis Bournique of WAP Review has very nicely hosted the old installers, and for a while at least you can get them here. *http://blog.wapreview.com/3733*.

- iPhone/Pre Emulator: Windows-only desktop emulator, emulates both iPhone and Pre. Paid service with 16-day trial period only. *http://www.genuitec.com/mobile.*

- Palm SDK: complete development environment for Windows, Mac, and Ubuntu. Relatively easy to use and complete interface. Use a full mouse to control. Center button is the home key. *https://developer.palm.com/content/resources/develop/sdk_pdk_download.html.*

- Palm Project Ares: full web development environment for WebOS in the browser. *http://ares.palm.com/Ares/about.html.*

- iOS SDK: includes simulator for all iOS devices. Mac only. *https://developer.apple.com/devcenter/ios/index.action#downloads.*

- Android SDK: start here. Multistep install process, and the emulator is pretty barebones out of the box, so hard to use for design validation unless you are also somewhat of a developer. Numerous add-ons for neat hardware like the Galaxy Tab, Motorola Xoom, and so on. Windows, Mac, Linux. *http://developer.android.com/sdk/installing.html.*

- Nokia Tools: Nokia keeps up all its old ones, so you can get emulators and SDKs for S40, older S60, Qt, Maemo, and everything Nokia works on, right from here. Too many to list out. *http://www.developer.nokia.com/Resources/Tools_and_downloads.*

- Windows Phone 7 Emulator: generic emulator, not just Windows only, but requires Vista SP2 or Windows 7 to run. *http://www.microsoft.com/download/en/details.aspx?id=13890.*

- Access Developer Tools: Access, which I still think of as a browser maker, seems to have bought up the old Palm OS (called Garnet now) and offers a suite of OS and web products. Many are for Windows, Mac, and Linux, but lots of tools, so no promises for any individual one. *http://www.accessdevnet.com/index.php/Downloads.*

- Windows Mobile 6.1: complete SDK. Windows only. *http://www.microsoft.com/download/en/details.aspx?id=16182.*

- Windows Mobile 6: complete SDK. Windows only. *http://www.microsoft.com/download/en/details.aspx?displaylang=en&id=6135.*

- Standalone Device Emulator: the emulator from Visual Studio 2005, without installing the whole package. Comes with Windows Mobile 5 skins, but not really an emulator of anything specific. Windows only. *http://www.microsoft.com/download/en/details.aspx?id=20259.*

- LG Java (J2ME feature phone) SDK: for developing custom applications on LG feature phones. Custom-install for each and every phone LG offers. Windows only. *http://developer.lgmobile.com/lge.mdn.tnd.RetrieveSDKInfo.dev?modType= T&objectType=T&menuClassCode=&saveFileName=&resourceNo=&selectedTy pe=&tabIndex=1.*

- Sony Ericsson Java (J2ME feature phone) SDK: for developing custom applications on Sony Ericsson feature phones. Windows only. *http://developer.sonyerics son.com/wportal/devworld/downloads/download/dw-99962-semcjavamecldcsdk2 506?cc=gb&lc=en.*

- Samsung Java (J2ME feature phone) SDK: for developing custom applications on Samsung feature phones. Windows only. *http://innovator.samsungmobile.com/ down/cnts/toolSDK.detail.view.do?cntsId=7500&platformId=3.*

- Bada SDK: complete SDK including emulator for Samsung's sorta-not-a-feature phone OS. Windows only. *http://developer.bada.com/devtools;jsessionid=nlZpTq yR6lrC7rj2shKWwWW8kmQf4Cw22sfgK4vNnJvzS608lyml!-939324231.*

- SDK for Brew and Brew MP: components to create an SDK for various existing platforms (such as Eclipse), with instructions, for the Brew feature phone application environment. Windows only. *https://developer.brewmp.com/tools/brew- mp-sdk.*

Prototyping/wireframing

- iPhone screen projector: "tethers" with your Mac to display content from your desktop on your iPhone. Requires a free iPhone app. *http://www.zambetti.com/ projects/liveview.*

- iPhoneprototype: Firefox plug-in for iPhone mockups. *http://sourceforge.net/ projects/iphoneprototype.*

- Adobe Device Central: part of the Adobe CS product line, lets you see what the design might look like on various mobile device screens. Includes nonwhites, nonblacks, glare, etc. *http://www.adobe.com/products/creativesuite/devicecentral.*

Full design suite

- MIDS, Mobile Interface Design System: flow charts, navigation, wireframes, rendering, and even a physical test device. *http://www.g5e.com/games/MIDS.*

Color Deficit Design Tools

Mobile devices are used in all sorts of environments, and must be clear and easy to read when bright, dark, dirty, rain-covered, and so on. General heuristic principles should be followed to dual- (or triple-) encode information so that it is clear in all conditions.

Therefore, the design of mobile devices should not be significantly affected by considerations for users with color vision deficits. Practically, many, many interfaces are still single-coded, with (for example) red type used for important notices and no use of iconography or type weight to reinforce this.

The information and guidelines in this appendix should serve not just as a reminder of how to address the needs of a significant portion of the population, but also of how color vision can fail every user in certain conditions. If colorblind users are considered in design, the product will work better for all users.

General Information on Color Deficits

Color vision deficiency (commonly referred to as *colorblindness*) is a condition in which certain colors cannot be distinguished or can only be distinguished with difficulty. It is most commonly due to an inherited condition. There is no treatment.

Red/green colorblindness is by far the most common form. Blue/yellow and other forms also exist, but are rarer and harder to test for. Complete colorblindness (seeing only in shades of gray) is extremely rare. Color deficiencies of one sort or another occur in about 8% to 12% of males and about .5% to 1% of females (of European origin).

Clinically, disturbances of color vision will occur if the amount of pigment within a cone is reduced or if one or more of the three cone systems are entirely absent. The gene for this is carried in the X chromosome. Therefore, colorblindness occurs much more commonly in males and is typically passed to them by their mothers. Reduction in pigment in one or more channels is more common than the loss of one or more sets of cones; therefore, most have a strict color deficiency, and perceive certain color channels more poorly, versus not at all.

Types of Color Deficiency

The following are various types of color deficiency:

Protanomaly (partial red/green, 1% of males)
> Also known as "red-weak." Complete or partial inability to see using long-wavelength sensitive retinal cones making it hard to distinguish between colors in the green-yellow-red section of the spectrum. Red perception is reduced both in saturation and in brightness. Red, orange, yellow, yellow-green, and green appear somewhat shifted in hue toward the green, and all appear paler than they do to the normal observer. The redness component that a normal observer sees in a violet or lavender color is so weakened for the protanomalous observer that he may fail to detect it, and therefore sees only the blue component.

Deuteranomaly (partial red/green, 5% of males, the most common form of colorblindness)
Also known as "green weak." Complete or partial inability to see using middle-wavelength sensitive retinal cones making it hard to distinguish between colors in the green-yellow-red section of the spectrum. He makes errors in the naming of hues in this region because they appear somewhat shifted toward red for him; difficulty in distinguishing violet from blue.

Dicromasy
Individuals with the following two conditions normally know they have a color vision problem and it can affect their lives on a daily basis. They see no perceptible difference between red, orange, yellow, and green. All these colors that seem so different to the normal viewer appear to be the same color for this 2% of the population.

Deuteranopia (complete red/green, 1% of males)
The deuteranope suffers the same hue discrimination problems as the protanope, but without the abnormal dimming. The names red, orange, yellow, and green really mean very little to him aside from being different names that every one else around him seems to be able to agree on. Similarly, violet, lavender, purple, and blue seem to be too many names to use logically for hues that all look alike to him.

Protanopia (complete red/green, 1% of males)
For the protanope, the brightness of red, orange, and yellow is much reduced compared to normal. This dimming can be so pronounced that reds may be confused with black or dark gray, and red traffic lights may not appear to be illuminated at all. He may learn to distinguish reds from yellows and from greens primarily on the basis of their apparent brightness or lightness, not on any perceptible hue difference. Violet, lavender, and purple are indistinguishable from various shades of blue because their reddish components are so dimmed as to be invisible (e.g., pink flowers, reflecting both red light and blue light, may appear just blue to the protanope).

The following are some less-common color deficit conditions:

- Tritanopia (complete blue/yellow).

- Tritanomaly (partial blue/yellow). Complete or partial inability to see using short-wavelength sensitive retinal cones making it hard to distinguish between colors in the blue-yellow section of the spectrum.

- Monochromacy. Complete or partial inability to distinguish colors. Complete colorblindness is very rare.

Common usability complaints among colorblind individuals

The following are some common usability complaints among those who are colorblind:

- Weather forecasts and similar infographic maps. Certain colors cannot be distinguished in the map or legend.

- Lighted indicators (e.g., charging LED). Is the indicator light red, yellow, or green?

- Position is often the only way to decipher traffic lights. This is an important reason they are standardized with red on top. Horizontal signals, or single signals that change color, are a problem.

- Color observation by others. "Look at those lovely pink flowers on that shrub." My reply, looking at a greenish shrub: "What flowers?"

- Kids and crayons. Color vision deficiencies bother affected children from the earliest years. At school, coloring can become a difficulty when one has to take the blue crayon, and not the pink one, to color the ocean.

- Chemical test strips (e.g., for hard water, pH, swimming pools).

- When cooking, red deficient individuals cannot tell whether their piece of meat is raw or well done. Many cannot tell the difference between green and ripe tomatoes or between ketchup and chocolate syrup. They can, however, distinguish some citrus fruits. Oranges seem to be of a brighter yellow than lemons.

Design patterns for users with color deficits

Design patterns to design around users with a color deficit (e.g., colorblindness) are available at WeAreColorblind.com (*http://wearecolorblind.com/category/patterns/*).

These are heavily focused on infographics, but the principles are easy to follow.

Color deficit simulators

Here is a list of color deficit simulators:

- Sim Daltonism: color blindness simulator for OS X, shows results much like the Apple Digital Color Meter. *http://michelf.com/projects/sim-daltonism/*.

- Vischeck: color simulator plug-ins for Photoshop. *http://www.vischeck.com/downloads/*.

- Entre: web-based colorblindness simulator. Upload any graphic to test. *http://www.etre.com/tools/colourblindsimulator/*.

- Wickline: web-based colorblindness simulator. Specify a URL to see the resultant page modified. *http://colorfilter.wickline.org/*.

- Colorlab: lets you pick from a (web-safe) palette, and see the results. Seems to have a lot more information, all to help you understand how to pick colors that are safe for color deficit users, but I am not totally up on how it works yet. *http://colorlab.wickline.org/colorblind/colorlab/*.

- Ishihara Test for Color Blindness: *http://www.toledo-bend.com/colorblind/Ishihara.asp*.

Mobile Typography

Introduction to Mobile Typography

Mobile and small-screen design is largely about communicating information to the user. More often than not, regardless of how exciting and shiny the interface is, this will still be centered on the display of text content.

Mobile typography is about the selection and use of all the type elements within the design. It is only partly about the selection of the correct font and face, and has a great deal to do with selecting display technologies, understanding sizes, and applying conventional design methodologies (size, shape, contrast, color, position, space, etc.) to best employ the type elements.

Challenges of Mobile Typography

Computer-based type, especially for Internet display, has always been a challenge due to display technologies (resolution), availability of type, color and contrast reproduction variations, and size variations. Mobile devices take these issues, magnify them, and add a spate of unique environmental and use-pattern issues. The primary barrier is of technology, and the primary concern is of readability within the user's context.

Type originally existed as shapes cut out of metal. When digital typesetting first became commercially viable, this model was followed and these letterforms were turned into vector glyphs, mathematically describing the shape of the character.

Smartphones now generally support this sort of type, and may include many fonts and faces and scale uniformly, allowing almost unlimited display options. Custom type can be included with applications loaded to the device relatively easily, and web-based font embedding is now appearing as well.

Older and low-end devices, including the billions of feature phones in the world, mostly only support "bitmap" fonts. These do not use vector glyphs, but instead draw each character as a small graphic of pixels. For each size or weight, a new set of type images must be loaded. Only a small number of fonts will be loaded, and generally only three sizes will be supported by J2ME applications running on these handsets.

All digital display devices render the final shapes as pixels on a square grid. Even ePaper devices must communicate with the stochastic display device with the same technology. Vector shapes, including type, are "rasterized" to comply with this format, and turned into pixels. Subtle angles and curves can become lost, or appear jagged regardless of antialiasing techniques. See Figure C-1.

Figure C-1. *Comparing vector and bitmap (or raster) glyphs.*

Although very high-resolution devices make some of the problems hard to see, almost eliminating them, the basic issues persist, and should be considered during design, and selection of proper type.

Technology continues to improve, and both digital foundries and OS makers are regularly implementing new techniques to improve rendering and readability.

Technology

Although some devices are beginning to allow effectively unlimited type selection, support vector glyphs, and have large amounts of storage and running memory, most mobile devices are still resource- and technology-constrained. General issues of storage on the device, running memory, download times, and cost of network access limit availability of type for mobile application design. As many devices require raster (bitmap) faces, each size is loaded as a complete, different typeface. Most products end up with the device's default type, or with a very limited set of choices for their application.

Although this challenge will slowly dissolve, it will always be present to some degree. Inexpensive devices, specialist devices (youth, elderly, and ruggedized), and emerging market needs seem to indicate these issues will persist for another several decades at least.

Usability

Mobiles are used differently from desktops, and even most print use of type. They are closest, perhaps, to signage in that they must be comprehended by all user populations, under the broadest possible range of environmental conditions (e.g., poor lighting) and at a glance. The typical mobile user is working with the device in a highly interruptible manner, glancing at the screen for much of the interaction. The type elements must be immediately findable, readable, and comprehendible.

This is different from the technical challenge in that it is inherent in the mobile device. Users will always interact with their devices in this manner, so it must always be addressed, regardless of the technical implementation.

An Introduction to Typography

To understand mobile typography, you will need to learn a little of the language and principles of typography in general. Do note that this is a very cursory review of this field. Please use the terms to search for more information, and check the sources listed in the references.

Baselines and measurements

The basic building block is lines and measures (see Figure C-2).

Figure C-2. *Baselines, x-height, and some of the other basic ways of measuring type height and vertical position.*

The baseline is where all type rides. Note that some round characters can dip below this, but only very little; probably not at all at mobile screen sizes. X-height is the height of the lowercase *x*, or any lowercase character, excluding the ascenders for those characters. Cap height is the height (more or less) of the tallest ascenders, and of the uppercase characters.

Letter height and measurements

During the letterpress era, letters were created from cast *sorts*, or blocks of lead, with the letter being a raised portion in the middle. These sorts were arranged together to form words and sentences. Each metal sort was designed to have a specific measurement. The letter's height was measured from the top to the bottom of the sort, not the actual letter, which could vary. This standard measurement became known as the type's *point size*.

Today, type is still commonly referred to by its size in points. Other related measures are also useful to know, and are increasingly supported by modern mobile device OSes. These are the most important measures:

Point

> One point is 1/72nd of an inch or about .35 mm. It used to be a slightly odd number, but has been standardized with digital typesetting.

Pica

> 12 points is one pica. Picas are not used to specify type, but can be used as a larger measure for any other dimensions of layout, such as spacing or column width.

Em

> This is the height of the sort, which is still defined by an invisible box which contains the glyph shape. Although a relative measure (it depends on the typeface and the size of the type being referenced), it is a general unit of measure, and distances and spacing can be referred to in "ems." Em-dashes are very long dashes.

En

> This is simply half an em. It is mostly encountered not as a unit of measure (though technically it is) but as a definition of a shorter dash. If it takes a short dash, use an en dash, not a hyphen. There are also 1/4 and 1/8 ems, used to define spaces, but they have no special names.

Twip

> Rarely used, but encountered deep in some interactive systems as a scale measure without need of decimals, is the very small twip. This is, properly, 1/20th of a point, but sometimes has other meanings, so be careful when you encounter it.

Hairline

> This is an old printing term, meaning the smallest consistently printable element, and is always a rule such as a line to separate columns. So, it is variable depending on the printing and other reproduction technology available, and not well supported with digital typesetting systems. For digital display, this has no explicit meaning. But the concept is valid, and it would be "one pixel." Understand the limits of your technology and design to those.

Abbreviations for points and picas can be odd. When alone, "p" is for pica and "pt" is for point, but typographers have a convention (supported by all serious digital design tools still) of "picas" p "points," as in "3p6," for example. This may even be encountered as a way to express points without the preceding zero, such as "p6." Type is always specified as points, such as "72pt."

Letterforms and their parts

When choosing the appropriate mobile type, we must understand that each typeface has unique characteristics that affect its legibility across device screen technologies, reading distances, and screen sizes.

In order to create effective message displays that are legible for mobile displays, understanding the basic elements of type is important. The information that follows will assist you when choosing the appropriate typeface for your design:

Font

The physical character or characters that are produced and displayed.

Typeface

A collection of characters—letters, numbers, symbols, punctuation marks, etc.

Glyph

The smallest shape of a character that still conveys its meaning.

Baseline

The invisible guideline upon which the main body of text sits. Some letters may, of course, extend below the baseline, with descenders. Think *g*, *j*, *q*.

X-height

The height of the main lowercase body from the baseline. It is usually defined as the size of the lowercase letter *x*, hence the name. It excludes ascenders and descenders. For body copy in mobile and small-screen devices, the x-height must be between 65% and 80% of the cap height for readability.

Cap height

The distance from the baseline to the height of the capital letter (and often all ascenders). When measuring to determine a font's point size, the cap height is used.

Descender line

The part of a letter that extends below the baseline. The descender line is the guide to which all descenders within the font family rest against. For mobile and small-screen devices, do not use excessive descenders. Avoid exceeding 15% to 20% of the cap height, to avoid excessive leading.

Ascender

The part of the letter that extends above the x-height. For mobile and small-screen devices, do not have ascenders above the cap height. This is critical for non-English languages, to better support accent marks without excess leading or overlapping type.

Counters

The negative spaces formed inside characters, such as the shape in the middle of an O. Small type sizes or heavy typefaces may cause counters to appear to fill in and look solid if they are too small, or complex.

Stress

Adding curvature to the straight shapes of a letterform. This is generally not desired for mobile faces. At best, the small rendered size will simply blur out these subtleties. It could also make it impossible to render sharp letters at small sizes. See Figure C-6.

Stems

The main vertical or diagonal elements of a character.

Bowls

The main, generally curved area that forms an enclosed area for a letterform with a stem. Think of the round, enclosed areas on either type of the lowercase *a*.

Type is composed of letterforms, and very importantly to the readability, the counter forms made up of the white space inside letters (see Figure C-3).

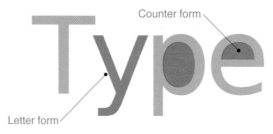

Figure C-3. *Letterforms and counters, or the negative space inside the letterforms.*

The space between letters is another sort of counter form, and we will discuss it shortly.

Every piece of the letterform also has a name. There are many more of these, but the following list discusses the most important ones for understanding and selecting type appropriate for mobile and small-screen use (see Figure C-4 and Figure C-5).

Figure C-4. *Some of the parts of a glyph (or letterform): the stem, descender, bowl, and crossbar.*

Figure C-5. *More parts of the glyph: the ascender and serif.*

Figure C-6. *Stressed versus nonstressed stems. Dashed lines are straight, so you can more clearly see the curve on the stressed type.*

Serifs

Finishing details at the ends of a character's main stroke. They extend outward. They are not solely decorative. They also help with our ability to discriminate other characters that make up lines of text. Serif faces are more readable for large blocks of type than sans-serif faces. However, in small mobile type, these may become undetectable, become blurry, and decrease legibility due to the limits of screen pixel technologies.

Sans serifs

Characters without serifs. For mobile type, sans serif is often the default choice as it works well enough for all uses, at all sizes. For users that have poor vision, you may need to use sans serifs that include more visually distinct characters in certain cases.

Square serifs

Also called *slab serifs*, these use heavy, squared-off serifs. Using these may be a good compromise, to ensure that the serifs display at the rendered sizes. Appropriate kerning is important to use for letter discrimination and legibility.

Serifs are available in a few different styles. The basic choices are shown in Figure C-7.

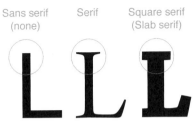

Figure C-7. *Types of serifs: sans serif without any serif of the conventional type, and a square or slab serif without notable tapers.*

For mobile type, sans serif is often the default choice as it works well enough for all uses, at all sizes. Serif faces are more readable for large blocks of type than sans serif faces. If your application is filled with news articles or something similar, consider if a serif face might work. Square serifs work well, to ensure that the serifs display at the rendered sizes and under difficult lighting conditions. Square (or slab) serif faces are the default faces used in popular eReaders.

All these letterform details matter not just for style, but for legibility and readability. Note that most of the letterform is similar to every other letterform. The bottom half is just a series of undifferentiated legs, as shown in Figure C-8.

letterform

letterform

Figure C-8. *The bottom half of each letterform is just a series of undifferentiated legs. The top is where all the readability happens.*

So, fairly small portions; mostly in the top half; their ascenders, descenders, counters, and inter-forms are the keys to readers understanding the form and reading the phrase.

Families and styles

A *font family* is a group of stylistically related fonts.

A *font* is a small group of (generally) very similar typefaces, which mostly will vary in weight. A common weight is "bold." If you have ever heard the term *boldface*, that's because it's a "bold typeface."

Black is a weight heavier than bold. Roman, Book, or Normal is the default weight. Thin or light is thinner or lighter, essentially the opposite of bold. Numerous other terms are also used, and some fonts have more than a dozen weights.

Weight will, usually, strongly influence the width of the overall type. Readability in long strings can be quite negatively impacted by large blocks of bold text. Very often, other methods (color, shape, position, whitespace) are more suitable for emphasis than bold on small screens. See Figure C-9.

Weight and width
Weight and width

Figure C-9. *Weight will usually change the width and readability of type. Be careful when using weight changes as a part of a focus change or other transient state.*

Italics are almost script-like faces, both tilted and with additional curves and more decorative serifs. Oblique is the same font more or less just slanted over as a sort of fake italic.

Italics and obliques are not suggested for any purposes on displays with less than about 150 pixels per inch. The angled forms, much like stressed verticals, cannot be rendered sharply. The stairstep effect is noticeable even with type-specific antialiasing techniques. Over about 200 ppi, this begins to become less noticeable, and such types can be used effectively.

Using large blocks of all caps, bold, or italics (or obliques) will result in reduced reading speeds, and lower comprehension. The best results are achieved with mixed case (both uppercase and lowercase) Romans. This can apply even to small passages, such as titles, so if readability is critical consider testing the use of styles such as all-caps titles to be sure they work.

All-caps is sometimes used for emphasis, but this is not always useful, and may provide effectively less emphasis due to reduced legibility.

Space: Kerning and leading

Proper spacing between characters and lines of type is crucial to readability, but must often be handled automatically for interactive systems. Even well-designed ones must take into account multiple screen sizes and aspect ratios, and sometimes even user-controlled zoom. There is little room for the designer to individually kern letter pairs.

Instead, be aware of the issue, pick fonts that are well designed, and use type display tools that respect the embedded kerning tables and work well for the devices and functions you are designing.

The key spacing attributes are:

Kerning
> The space between any two characters. Ideally, this is automatically generated and follows well-established principles laid out by the type designer, and set forth in a kerning table. See Figure C-10.

Tracking
> The overall intercharacter spacing for a block of text. Tracking has a parent relationship to kerning, if both are defined.

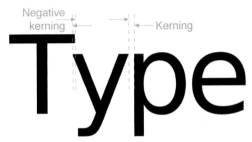

Figure C-10. *Kerning, showing space between, or space overlapping, individual letterforms.*

Do not, as a general rule, use monospace faces, where each character takes the same horizontal width, except for specialist reasons, such as examples of code or tables rendered as text, where spacing is crucial to display clarity.

Leading

Also known as line spacing, or the amount of vertical space from one baseline to the next baseline. Larger leading can often increase readability in poor conditions, such as low light or in motion, but too much can make it impossible to understand that two lines are related to each other. Larger than default leading can be used to make improperly long lines of text more readable, if they are otherwise unavoidable.

This is measured as the vertical distance between baselines (or any like-to-like line). Sufficient room must be provided for the ascenders and descenders to not collide with each other. Additional room must be provided to ensure that lines of type can be read, without forming intercharacter counters in the space between lines. See Figure C-11.

Figure C-11. *Leading is the vertical distance between lines.*

Many typefaces suitable for mobile display have very small descenders to help you eliminate extra leading, and fit more information on a page.

Most languages have some accent characters. These are usually placed above the letterform, occupying the ascender space for lowercase characters, but an above-cap-height space for uppercase characters. Typically, this will require very large leading to enable readability of the accents. See Figure C-12.

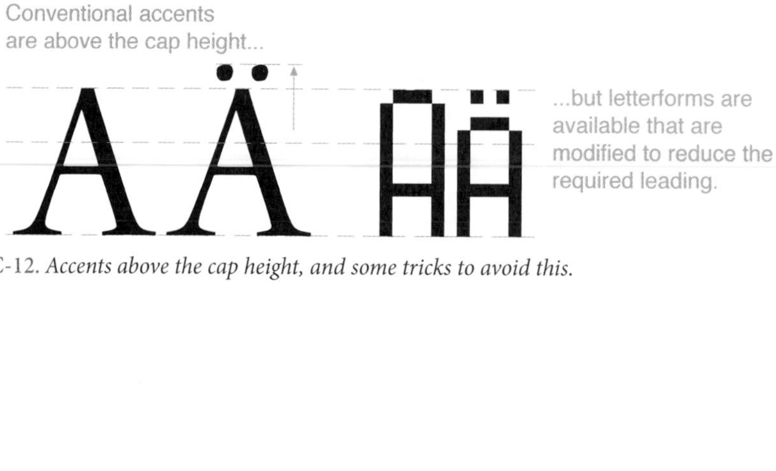

Conventional accents
are above the cap height...

...but letterforms are
available that are
modified to reduce the
required leading.

Figure C-12. *Accents above the cap height, and some tricks to avoid this.*

This generally is unacceptable for small-screen display, and some faces have special letterforms that fit the accent and character within the cap height. If your application must support accent-enabled languages, consider the effect this will have on your design.

Alignment

Alignment of type is used as a design tool, and to make the text fit appropriately. All type is "aligned." Some is justified, but this is a subset. Available types of alignment are:

Left align

The most common alignment by far. Aside from tabbed-in first lines of a paragraph, and so on, all type is aligned to a single invisible vertical guideline. The type continues to the right until no more complete words can fit, and then it breaks to the next line. The irregular right side makes it also sometimes called "flush left" and/or "ragged right." Hyphenation can be used to make the "rags" less dramatic, but interactive systems have poor or no automatic hypehnation currently.

Right align

The opposite of left align, sometimes called "flush right." Used for special design purposes or to set apart small amounts of type.

Center

Most used for titles and other items that easily fit the space. For items longer than a line, the same rule about breaking as in left align is used, but there is no straight side, and both edges are ragged. Note that the rags are perfectly symmetrical.

Justify left

Only really available when optical typesetting appeared, and most successful in the digital age. This uses line-by-line tracking (and sometimes kerning) controls to essentially remove the ragged right side from the left-aligned type by stretching and squeezing the amount of text on one line to make the lines even, or "flush" on both sides. Hyphenation is also heavily relied on to make the lines of more even length. This does not preclude first lines of paragraphs being indented and other spacing for design purposes. How last lines are handled varies stylistically, but usually is allowed to be "ragged," with the conventional tracking.

Justify right

Identical to justify left, but flipped so that the ragged last line is on the right.

For narrow columns, texts with long words (such as technical jargon), or systems with poor tracking control (such as those that only adjust interword spacing), distracting "rivers" of whitespace may appear in the text. If this is common, justification may not be suitable.

Do not use monospace fonts and additional spaces to simulate justified paragraphs. This is highly unreadable.

Guidelines for Selecting a Typeface for Mobile and Small-Screen Devices

Research shows that reading from a computer screen is about 25% slower than from paper. Many mobile devices, due to size or environmental complexities, can make this even more challenging. Understand the principles of type and the limits of mobile to help you make appropriate decisions for your content, and your users. See Figure C-13.

Must have:

- X-height between 65% and 80% of the cap height
- Strong counters (or "counter forms"); often, using squared-off shapes for small counters is a good idea
- Unstressed forms; straight, even-width lines
- No excessive descenders; avoid exceeding 15% to 20% of the cap height, to avoid excessive leading
- No ascenders above the cap height; this is critical for non-English languages

Un-stressed forms   Descender height 17%   x-height 69%   Clean, open counters

Figure C-13. *Some pointers to selecting suitable mobile type.*

Good mobile type should also:

- Be space-efficient. Generally this means narrow, to allow sufficient height for all users to read the characters.
- Not look compressed.
- Be well kerned. Letters should not run together, or have spaces that look like word breaks.
- Have the same, or similar, width for all weights and styles (there is no penalty for using oblique/italics or boldface for emphasis).
- Use subtle serifs when it can be beneficial to the form; consider them for a face, or for some characters of a face.
- Include a true italic. A sloped Roman ensures that hardly any elements are vertical; a true italic can preserve legibility, following the aforementioned rules, while also being different enough to read as "other than body."

- Be part of a complete family. Serif and sans serif can both be used (titles and body text have different needs), as well as many weights of each, if space is available on the device.

Non-Latin languages will have additional requirements. As always, be aware of regional needs and cultural distinctions.

Readability and Legibility Guidelines

Message Display Characteristics and Legibility

Mobiles are used differently from desktops, and even most print use of type. They are closest, perhaps, to signage in that they must be comprehended by all user populations, under the broadest possible range of environmental conditions (e.g., poor lighting) and at a glance. The typical mobile user is working with the device in a highly interruptible manner, glancing at the screen for much of the interaction. The message display is competing with millions of other stimuli in the visual field. Therefore, mobiles must be designed to stand out appropriately, beginning with legibility.

The following is a list of legibility guidelines for mobile devices:

- Avoid using all caps for text. Users read paragraphs in all caps about 10% slower than mixed cases (Nielsen 2000).

- With age, the pupil shrinks, allowing less light to enter the eye. By the age of 80, the pupil's reaction to dim light becomes virtually nil (Nini 2006). Use typefaces that have more visually distinct characters in certain cases, while still maintaining a desired unity of form.

- Older viewers with aging eyesight can benefit from typefaces that have consistent stroke widths, open counter forms, pronounced ascenders and descenders, wider horizontal proportions, and more distinct forms for each character (such as tails on the lowercase letters *t* and *j*).

- Try to use plain-color backgrounds with text, because graphical or patterned backgrounds interfere with the eye's ability to discriminate the difference.

- Sans serif is often the default choice as it works well enough for all uses, at all sizes. Serifs may or may not help with readability, so there is no special reason to use them. Consider using default faces that carry through the sensibility of the OS, such as Helvetica (iPhone's default typeface). Verdana is also good, as it has a larger x-height and simpler shapes, designed specifically for on-screen readability. There are numerous mobile-specific faces as well.

- Almost all text should be left-aligned, and only items such as Titles should be centered.

- Use typefaces that have strong and open counters (or counter forms). Often, using squared-off shapes for small counters is a good idea.

- Use typefaces with unstressed forms; straight, even-width lines.

- Consider how different pixel densities across mobile platforms affect physical sizes of elements. Using ems to define the size of fonts in CSS may provide ease.

Message display readability

As we discussed earlier, legibility is determined by how we can detect, discriminate, and identify visual elements. But legibility is not the end to this process of designing effective mobile displays. We must make sure they are readable. When readability is achieved, the user can then evaluate and comprehend the meaning of the display.

Readability is based on the ease and understanding of text. It is determined by whether objects in the display have been seen before. Thus, readability is affected by the message's choice of words, the sentence structure, appropriate language, and the reading goals of the user (Easterby 1984).

Reading modes

In September 2004 Punchcut worked with QUALCOMM to develop a typographic strategy with respect to its custom user interfaces within its mobile operating system and applications. In that study, Punchcut determined three modes of reading and understanding behavior that occur within the mobile context (Benson 2006).

These modes consider the duration and size of focus for textual attention:

Glance
> Users are in "Go" mode and they need quick access to frequent and/or critical tasks. Their focus is distracted and minimally on the task at hand. Users rely on shape and pattern recognition in this mode.

Scan
> Users are sorting and selecting, scrolling through lists of choices and options. They apply more focus than when glancing, but not their entire focus.

Read
> Users are reviewing and comprehending information, and apply their entire focus on desired information.

Language requires context

These three modes are based within the context of the user's goals. Just like a user's goals require context, so does the meaning of language derived from the message's choice of words. Without context, the intended message display's meaning may become misunderstood.

In our everyday context, we must derive meaning from others' intentions, utterances, eye movements, prosody, body language, facial expressions, and changes in lexicon (McGee 2001). When context uses more individual pieces, the less ambiguity we will encounter in the message's meaning.

In mobile devices today, we may not have all of those contextual aids available to help us derive meaning. The mobile user is presented only with the aids the device's OS, the technology, or the designer's intention have provided.

Readability guidelines for message displays

The following guidelines and suggestions should be considered when designing readable mobile displays:

Vocabulary

- For on-page elements—such as the text of Titles, Menus, and Notifications— use vocabulary that will be familiar to your user. Do not use internal jargon or branded phrasing only familiar to the company and its stakeholders.

- Use vocabulary specific to the task at hand. General and vague titles will create confusion.

- Strive for no help or "FAQs" documentation. With clear enough labels, you can often avoid explaining in a secondary method.

Images as aids

- Images can provide additional visual aids to reinforce a message's intended meaning. When needed, use images that are clear in meaning. Do not use images that have arbitrary meanings or require learning.

- Image meanings are socially and culturally created. Even the best icons will not be understood for some users. Using images alone without text labels may result in varying interpretations.

Overflow or truncation

- Make sure key titles, button labels, soft-key labels, and similar items fit in the available space. Do not allow them to wrap, extend off the screen, or truncate.

- For certain cases, such as lists, tables, and some other descriptive labels, the text may truncate. When possible, do not break words. To indicate the truncation, use an ellipsis character. If not available, use three periods, never two or four or any other number. Sometimes it may be suitable for the word to disappear under the edge of the viewport or some other element instead to indicate the truncate.

- Rarely, the marquee technique may be useful for list items. Text is truncated, flowing off the screen, and when in focus it scrolls slowly so that after a brief time the entire text can be read by the user.

- Long text, in paragraphs, will always wrap using one of the methods described in more in the "Alignment" section.

- If the entire text of a paragraph cannot fit, or is not needed (such as the beginning of a long review) it may begin as a paragraph, and truncate at the end, to indicate there is more content.

Line length

- From 50 to 60 characters per line is suitable. If you need longer lines, increase leading, but there are limits to how far you can go. At around 120 characters, users might find it difficult to scan that far.

- The optimal line length varies around a multitude of factors. These factors include display size, line size, type size, kerning, color, margins, columns, and visual acuity.

- Users' subjective judgments and performance do not always correlate. Studies have shown that users may prefer shorter line lengths, but may actually read faster with longer lines.

Typefaces for Screen Display

Back in 1975, AT&T wanted a new typeface to commemorate the company's 100th anniversary. AT&T indicated that the requirements for the new typeface must fit more characters per line without reducing legibility to reduce paper consumption, reduce the need for abbreviations and two-line entries, increase legibility at the smaller point sizes, and be used for the phone book directory. Matthew Carter, a type designer, got the job and created the new typeface Bell Centennial.

Carter's new sans serif typeface was more condensed, increased the x-height, and used more space in the open counters and bowls. Aware of the printing limitations, he used letters with deep "ink traps," which allowed for the counter forms to be open, making them more legible at smaller point sizes instead of becoming filled with ink; this also reduced ink usage, a serious consideration by those who buy ink by the truckload. His typeface was not effective at larger point sizes or on stock paper because the ink traps, little gaps in the corners of the counters, would not fill in and could be seen.

For its intended purpose, at small sizes, on poor paper, it was very effective. A particular typeface often must be selected, or designed, for the specific context. In 1975, available technology, the medium, and users were considered. As designers and developers today, these considerations remain throughout the design process. You need to understand that when choosing a typeface, context of use is key.

- Headlines, especially large ones, will often need to be a different typeface or font than the body copy.

- Italics have been ineffective on digital displays, but very high resolutions will allow their increasing use. Consider them carefully, and understand the difference between a simple oblique version of a font, and a true italic.

- Type can express a hierarchy beyond just size and weight. Should your captions, or bullet lists, use another size, weight, style, or font?

- As in the Bell Centennial example, legibility, readability, and appropriateness are key. If making type small enough makes it illegible, maybe this can be solved with a new font, instead of larger type.

- Show restraint. Designers mock designs with too many fonts as "ransom notes." There is no hard and fast rule on a number, though.

Challenges of Mobile Typography Technology Today

Although some devices are beginning to allow effectively unlimited type selection, support vector glyphs, and have large amounts of storage and running memory, most mobile devices are still resource- and technology-constrained. General issues of storage on the device, running memory, download times, and cost of network access limit availability of type for mobile application design. As almost all devices require raster (bitmap) faces, each size is loaded as a complete, different typeface. Most products end up with the device's default type, or with a very limited set of choices for their application.

Digital Fonts Today

Digital fonts used today are constrained by the technology display capabilities and the device OS. Most current mobile phones use antialiased fonts. Because of the display's square pixel layout, antialiasing is used to render some of the pixels' shades of gray along the edges of the letter. This helps users to perceive the letter as being smooth. Antialiased text is more legible when using larger font sizes for titles and headings; however, using antialiasing text in small font sizes tends to create a blurry image. Consider the mobile display's capabilities when choosing the font size and font family because they might not be available for that mobile device.

Companies such as Microsoft, Bitstream, Monotype, and E-Ink have introduced their own type of font display technologies to improve readability for such devices. Some of these font technologies are constrained to a specific fixed arrangement of pixel display technology. ClearType, Microsoft's subpixel rendering technology, is orientation-specific and will not work on devices whose display orientation can change. However, ClearType works very well on LCD displays because of the fixed pixel layout. On some OSes, there may be an "automatic" font setting which detects if the display is an LCD or CRT to turn on or off subpixel rendering.

Font Rendering Technologies

The following are examples of some of the available rendering technologies used to improve type legibility on digital displays:

- ESQ Mobile Fonts by Monotype Imaging (*http://www.monotypeimaging.com*) are optimized for smartphone and feature phone displays in addition to other consumer electronics and embedded systems. ESQ Fonts also include WorldType, which offer wide language support on mobile devices.

- Bitstream Font Technology by Bitstream (*http://www.bitstream.com*) uses Font Fusion and Panorama technologies optimized for resource-constrained mobile devices.

- Apple Advanced Typography, or AAT, is Apple's (*http://developer.apple.com/fonts*) advanced font rendering software. It is a set of extensions to the TrueType outline font standard, with similar smart-font features to the OpenType font format.

- ClearType by Microsoft (*http://www.microsoft.com/typography/default.mspx*) has greater pixel control by turning on and off each of the colors in the pixel. This makes on-screen text more detailed and increases legibility.

- TrueType is an outline font standard originally developed by Apple Computer in the late 1980s. TrueType is common for fonts on both the Mac OS and Microsoft Windows operating systems. TrueType was developed to ensure high control of how fonts are displayed down to individual pixels.

Be aware of display technology. ePaper, for example, is a series of technologies used for electronic digital displays such as eReaders (note that "E-Ink" is but one of many brands, and is not a generic label). ePaper generally relies on reflected, not emitted, light, by suspending particles in a liquid; a charge causes the dark particles to rise and be visible, or bicolor particles to rotate from light to dark (technologies vary) and display dark areas on a lighter surface, much like ink on paper.

Although the stochastic nature of the display elements, contrast and low speed of the display change, make ePaper design very different, current display technologies can only drive them with conventional backplane technology, so pixels are effectively square, as with all other display types.

Although ePaper is a dramatic departure, every display technology has its unique attributes. OLED and AMOLED displays, for another example, are lit-pixel displays (without a backlight). White text on black will, unlike a backlit display, use much less power than black text on white. This may be a key design consideration for those devices.

Greeking

Greeking is a design term used to refer to using placeholder text or images instead of the real, final text. Sometimes this is important or unavoidable as the real text is not available, or would be irrelevant such as with a sample document.

Design documentation must by nature represent content. Many wireframe techniques use boxes or other greeking methods, instead of using real content, even when it is available. This is a somewhat contentious issue, but our experience in design indicates that the proper choice is crucial in several different facets. See Figure C-14.

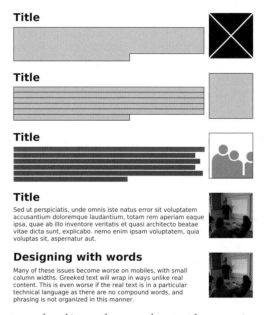

Figure C-14. *The many types of greeking, and some real content for comparison.*

Aside from graphical representations of type, another common representation is the classic typographer's greeking. The entire, original version of it is reproduced here:

Sed ut perspiciatis, unde omnis iste natus error sit voluptatem accusantium doloremque laudantium, totam rem aperiam eaque ipsa, quae ab illo inventore veritatis et quasi architecto beatae vitae dicta sunt, explicabo. nemo enim ipsam voluptatem, quia voluptas sit, aspernatur aut odit aut fugit, sed quia consequuntur magni dolores eos, qui ratione voluptatem sequi nesciunt, neque porro quisquam est, qui dolorem ipsum, quia dolor sit, amet, consectetur, adipisci velit, sed quia non numquam eius modi tempora incidunt, ut labore et dolore magnam aliquam quaerat voluptatem. ut enim ad minima veniam, quis nostrum exercitationem ullam corporis suscipit laboriosam, nisi ut aliquid ex ea commodi consequatur? quis autem vel eum iure reprehenderit, qui in ea voluptate velit esse, quam nihil molestiae consequatur, vel illum, qui dolorem eum fugiat, quo voluptas nulla pariatur?

At vero eos et accusamus et iusto odio dignissimos ducimus, qui blanditiis prae-sentium voluptatum deleniti atque corrupti, quos dolores et quas molestias excepturi sint, obcaecati cupiditate non provident, similique sunt in culpa, qui officia deserunt mollitia animi, id est laborum et dolorum fuga. et harum quidem rerum facilis est et expedita distinctio. nam libero tempore, cum soluta nobis est eligendi optio, cumque nihil impedit, quo minus id, quod maxime placeat, facere possimus, omnis voluptas assumenda est, omnis dolor repellendus. temporibus autem quibusdam et aut officiis debitis aut rerum necessitatibus saepe eveniet, ut et voluptates repudiandae sint et molestiae non recusandae. Itaque earum rerum hic tenetur a sapiente delectus, ut aut reiciendis voluptatibus maiores alias consequatur aut perferendis doloribus as-periores repellat.

Hanc ego cum teneam sententiam, quid est cur verear, ne ad eam non possim ac-commodare Torquatos nostros? quos tu paulo ante cum memoriter, tum etiam erga nos amice et benivole collegisti, nec me tamen laudandis maioribus meis corrupisti nec segniorem ad respondendum reddidisti. quorum facta quem ad modum, quaeso, interpretaris? sicine eos censes aut in armatum hostem impetum fecisse aut in libe-ros atque in sanguinem suum tam crudelis fuisse, nihil ut de utilitatibus, nihil ut de commodis suis cogitarent? at id ne ferae quidem faciunt, ut ita ruant itaque turbent, ut earum motus et impetus quo pertineant non intellegamus, tu tam egregios viros censes tantas res gessisse sine causa?

Quae fuerit causa, mox videro; interea hoc tenebo, si ob aliquam causam ista, quae sine dubio praeclara sunt, fecerint, virtutem iis per se ipsam causam non fuisse. — Torquem detraxit hosti. — Et quidem se texit, ne interiret. — At magnum periculum adiit. — In oculis quidem exercitus. — Quid ex eo est consecutus? — Laudem et cari-tatem, quae sunt vitae sine metu degendae praesidia firmissima. — Filium morte mul-tavit. — Si sine causa, nollem me ab eo ortum, tam inportuno tamque crudeli; sin, ut dolore suo sanciret militaris imperii disciplinam exercitumque in gravissimo bello animadversionis metu contineret, saluti prospexit civium, qua intellegebat contineri suam. atque haec ratio late patet.

John McWade wrote in *Before and After Magazine* 4(2):

After telling everyone that Lorem ipsum, the nonsensical text that comes with Page-Maker, only looks like Latin but actually says nothing, I heard from Richard McClintock, publications director of The Garnet at Hampden-Sydney College in Virginia, who had enlightening news:

"Lorem ipsum" is Latin, slightly jumbled, the remnants of a passage from Cicero's De Finibus 1.10.32, which begins thus: "Neque porro quisquam est qui dolorem ipsum quia dolor sit amet, consectetur, adipisci veldt…" [There is no one who loves pain it-self, who seeks after it and wants to have it, simply because it is pain.]. De Finibus Bonorum et Malorum, written in 45 BC, is a treatise on the theory of ethics; it was very popular in the Renaissance.

What I find remarkable, is that this text has been the industry's standard dummy text ever since some printer in the 1500s took a galley of type and scrambled it to make a type specimen book; it has survived not only four centuries of letter-by-letter resetting but even the leap into electronic typesetting, essentially unchanged except for an occasional 'ing' or 'y' thrown in. It's ironic that when the then-understood Latin was scrambled, it became as incomprehensible as Greek—hence, the term "Greek" for dummy text.

This term has since been expanded to the broader meaning of any false content, even to include simple gray lines or boxes representing type.

Designing with Words

There are several issues with the use of greeking:

- It is a specialized design language. The styles using boxes or lines especially will not always be understood by consumers of the document. Any confusion, even if solved before implementation, can cause delays, or induce errors in estimation.

- Far too often, the exact same fake content is used over and over again. Every bullet starts "Lorem ipsum," for example. This does not accurately represent the variety of content that would be encountered, so it is not an adequate representation of the final design.

- Text greeking of the "lorem ipsum" variety is not universally recognized as fake content, so it may have to be explained.

- Latin is not English, and does not have the same cadence, or average length of words or distribution of those words. Letters are also used at different rates, adding to the lack of fidelity with real content.

Many of these issues become worse on mobiles, with small column widths. Greeked text will wrap in ways unlike real content. This is even worse if the real text is in a particular technical language as there are no compound words, and phrasing is not organized in this manner.

The suggestion, therefore, is to use real copy whenever possible. Content either already exists, or will be created for your project. Either get this copy as early as possible (even if it is in draft format) or work closely with the content team to create enough for your mockups.

In the same way you are a design or software professional, writers have a specific skill set that cannot be simulated. I rely on writers whenever possible to help create everything from page titles to button labels. They will do a better job, in less space, than you can do.

Respect Names and Languages

Although already mentioned within the Ordered Data pattern, it is worth restating here that real names must always be displayed. Use caution when picking names of individuals, products, and locales when providing placeholder or sample content. Make sure there is room for as many of the likely results as possible.

It may be useful to specifically keep long names for various data fields. For names, for example, I have long used a former coworker (who has since returned to Thailand and teaches HCI), Narin Jaroensubphayanont. When you include hyphenated last names, this is really not that overwhelmingly long, even for the United States.

This also points out a crucial lesson in regionally specific information. Different languages have different cadences, so if you're designing a global application, periodically load a language other than your default. Check local address and name formats to make sure they fit.

Seriously, no jokes

If you have to make up content for temporary, internal use, there is often an inclination to use pop culture references, or make internal jokes. If you are still doing this, you will be surprised to find out what a bad sense of humor many, many people have.

Designers have been fired and customers have been lost over placeholder copy. Although guidelines could be provided, even around the edges of these there is room to make costly mistakes.

Even seemingly innocuous fake content can cause immense trouble. There are incidents where placeholder phone numbers have made it into production. To look real, they were not 555 numbers or anything else clearly fake, so ended up not being fake but dialing real locations. Two real-world cases:

- The phone listed for a retail store instead resolves to a house. The retired residents, having lived there for decades, received dozens of calls a day and do not want to change their long-held number that has now been published in numerous locations. Lawyers got involved, and it took years to sort out.

- The phone number listed to contact customer care dials an "adult chat" number. This was easily rectified, but offended customers left, or had to be offered apologies from senior management and expensive enticements to remain customers.

Yes, really. These both happened.

Human Factors

Human Factors and Physiology

Sensation: Getting Information into Our Heads

Your mind is like a leaky bucket. It holds plenty of information, but can easily let information slip away and spill out. If you can understand how visual information is processed and collected, you can create effective visual interactive displays that resemble the way the mind works.

This can help to limit the cognitive load and risks of information loss during decision-making processes. Your perception model is complex, and there are many theories explaining its structure which is beyond the scope of this book. A general description of visual sensation and perception is described in this appendix.

Sensation is a process referring to the capture and transformation of information required for the process of perception to begin (Bailey 1996). Each of our sensors (eyes, ears, nose, skin, mouth) collects information, or stimuli, uniquely, but all will transform the stimulus energy into a form the brain can process.

Collecting visual stimuli: How the eye works

The eye is an organ responsible for vision. Many people use the analogy that our eye works like a camera. Both eye and camera have a lens, an aperture (pupil), and a sensor (retina). However, the manner in which sensing and processing occurs is very different. This should be understood at least a little by the designer in order to create displays that are easy to see and understand.

The eye collects, filters, and focuses light. Light enters through the cornea and is refracted through the pupil. The amount of light entering the lens is controlled by the iris. The lens focuses the beam of light and then projects it onto the back part of our retina where it contacts the photoreceptors known as rods and cones.

These receptors are light-sensitive and vary in relative density; there are about 100 million rods and only 6 million cones. The cones are used for seeing when there is bright light; three kinds of cones, each with their own pigment filter, allow perception of color. The rods are sensitive to dim lighting and are not color sensitive. These receptors convert light into electrochemical signals which travel along the optic nerve to the brain for processing.

Visual acuity and the visual field

Visual acuity is the ability to see details and detect differences between stimuli and spaces. Inside our eye, at the center of our retina, lies our fovea. The fovea is tightly packed only with cones (approximately 200,000) and it is here where our vision is most focused. The fovea is the central 1 to 2 degrees of our eye, and the last .5 degree is where we have our sharpest vision. The farther away objects extend beyond our fovea range, the lower the resolution and color fidelity. We can still detect items peripherally, but with less clarity. Types of color perception vary by their location as well; blue can be detected about 60 degrees from our fixed focal point, while yellow, red, and green are only perceptible within a narrower visual field.

Factors affecting visual acuity depend on many things, including the size of the stimulus, the brightness and contrast of the stimulus, the region of the retina stimulated, and the physiological and psychological condition of the individual (Bailey 1996).

Size of the stimulus: Visual angle

The actual size of an object is basically unimportant as far as how easy it is to perceive. Instead, it is the *visual angle* or the relative size to your eye. This takes into account both size and distance from the viewer. Discussions of this in various technical fields often discuss the angular resolution, as true resolution is unimportant.

The visual angle can be calculated using the following formula:

Visual Angle (minutes of arc) = (3438)(length of the object perpendicular to the line of sight)/distance from the front of the eye to the object

Visual angle is typically measured in much smaller units than degrees such as seconds or minutes of arc (60 minutes in a degree, 60 seconds in a minute). Other specialized units may also be encountered such as milliradians, or may simply be in degrees with annoyingly large numbers of decimal places.

With an understanding of visual angle, we can determine the appropriate size of visual elements including character size viewed at specific distances. According to the Human Factors Society (1988), the following visual angles are recommended for reading tasks:

- When reading speed is important, the minimum visual angle should not be less than 16 minutes of arc (moa) and not greater than 24 moa.

- When reading speed is not important, the visual angle can be as small as 10 moa.

- Characters should never be less than 10 moa or greater than 45 moa.

So, let's assume you are designing text that is to be read quickly on a mobile device, with a viewing distance of 30 cm (11.8 in). The equation would look like this:

Length = 16 minutes of arc (30)/3438

The smallest acceptable character height would be 0.14 cm, or about 10 pt. Remember, all this exists in the real world; you will have to take into account real-world sizes, never pixels, when designing for perception.

Visual Perception

After our senses collect visual information, our brain begins to perceive and store the information. Perception involves taking information that was delivered from our senses and interacting it with our prior knowledge stored in memory. This process allows us to relate new experiences with old experiences. During this process of visualization of perception, our minds look to identify familiar patterns. Recognizing patterns is the essential for object perception. Once we have identified an object, it is much easier to identify the same object on a subsequent appearance anywhere in the visual field (Biederman and Cooper 1992).

The Gestalt School of Psychology was founded in 1912 to study how humans perceive form. The Gestalt principles that were developed can help designers create visual displays based on the way our minds perceive objects. These principles, as they apply to mobile interactive design, are:

Proximity
> Objects that are close together are perceived as being related and grouped together. When designing graphical displays, having descriptive text close to an image will cause the viewer to relate the two objects together. This can be very effective when dual-coding graphical icons.

Similarity
> Objects sharing attributes are perceived to be related, and will be grouped by the user. Navigation tabs that are similar in size, shape, and color will be perceived as a related group by the viewer.

Continuity
> Smooth, continuous objects imply they are connected. When you design links with nodes or arrows pointing to another object, viewers will have an easier time establishing a connected relationship if the lines are smooth and continuous and less jagged.

Symmetry
> Symmetrical relationships between objects imply relationships. Objects that are reflected symmetrically across an axis are perceived as forming a visual whole. This can be bad more easily than good. If a visual design grid is too strict, unrelated items may be perceived as related, adding confusion.

Closure

A closed entity is perceived as an object. We have a tendency to close contours that have gaps in them. We also perceive closed contours as having two distinct portions: an inside and an outside. When designing list patterns, such as the grid pattern, you should use closure principles to contain either an image or a label.

Relative size

Smaller components within a pattern are perceived as objects. When designing lists, entities such as bullets, arrows, and nodes inside a group of information will be viewed as individual objects that our eyes will be drawn to. Therefore, make sure these objects are relevant to the information that it is relating to. Another example of relative size is a pie with a missing piece. The missing piece will stand out and be perceived as an object.

Figure and ground

A figure is an object that appears to be in the foreground. The ground is the space or shape that lies behind the figure. When an object uses multiple Gestalt principles, figure and ground occur.

Visual Information Processing

The information that is visually collected begins early in our visual perception process. In a parallel, bottom-up, top-down process, neural activity rides two information-driven waves concurrently. The first wave occurs within the bottom-up process. Information collected by the retinal image passes to the back of our brain along the optic nerve in a series of steps that begin pattern recognition:

1. Features in our visual field, such as size, orientation, color, and direction, are processed by specific neurons. Millions of these features are processed and used to construct patterns.

2. Patterns are formed from processed features depending on our attention demands. Here, visual space is divided by color and texture. Feature chains become connected and form contours. Many of these cognitive pattern recognitions are described through Gestalt principles.

3. Objects most relevant to our current task are formed after the pattern-processing stages filter them. These visual objects are stored in our working memory, which is limited in its capacity. Our working memory holds only about three visual objects in attention at one time. These visual objects are linked to other various kinds of information that we have previously stored.

While the first bottom-up wave is processing patterns, the second top-down wave is processing which information is relevant to us at that moment and is driven by a goal. In addition, we associate actions that are then primed for our behaviors. So, through a series of associated visual and nonvisual information and action priming, we can perceive the complex world around us.

Articulating Graphics

Now that we have an understanding that visual object perception is based on identifying patterns, we must be able to design visual displays that mimic the way our mind perceives information. Stephen Kossyln states, "We cannot exploit multimedia technology to manage information overload unless we know how to use it properly. Visual displays must be articulate graphics to succeed. Like effective speeches, they must transmit clear, compelling, and memorable messages, but in the infinitely rich language of our visual sense" (Kossyln 1990).

Display elements are organized automatically

This follows Gestalt principles. Objects that are close by, are collinear, or look similar tend to be perceived as groups. So when designing information displays, such as maps, adding indicators, landmarks, and objects that are clustered together, will appear to be grouped, and share a relationship. This may cause confusion when the viewer needs to locate his exact position.

Perceptual organization is influenced by knowledge

When we look at objects in a pattern for the first time, we may not fully understand or remember the organization of those objects. However, if we see this pattern again over time, we tend to chunk this pattern and store it in our memory. Think of a chessboard with its pieces played out. A viewer who has never seen this game before will perceive the board as having many objects. However, an experienced chess player will immediately identify the objects and the relationships they have with one another and the board. So when designing visual displays, its essential to know the mental model of your user so that she may quickly identify and relate to the information displayed.

Images are transformed incrementally

When we see an object move and transform its shape in incremental steps, we have an easier time understanding that the two objects are related or identical. However, if we only see the object's beginning state and end state, our minds are forced to use a lot of mental processing and load to understand the transformation. This can take much more time and also increase errors or confusion. When designing a list where items move, such as a Carousel, make sure the viewer can see the incremental movement.

Different visual dimensions are processed by separate channels

Object attributes such as color, size, shape, and position are processed with our minds using separate processing channels. The brain processes many individual visual dimensions in parallel at once, but can only deal with multiple dimensions in sequence. For example, when designing a bullet list with all black circles, we can immediate identify all of them. However, if we add a bullet that is black, is the same size, but is in diamond shape, our minds have to work harder to perceive them as being different.

Color is not perceived as a continuum

Designers often use color scales to represent a range, such as red is hot, blue is cold. Temperatures in between will be represented by the remaining visual spectrum between them. The problem is that our brains do not perceive color in a linear dimension as it physically exists like this. We view color based on the intensity and amount of light. So a better way of showing this temperature difference would be to use varying intensity and saturation.

If a perceptually orderable sequence is required, a black to white, red to green, yellow to blue, or saturation (dull to vivid) sequence can be used (Ware 2000).

When high levels of detail are to be displayed, the color sequence should be based mostly on luminance to take advantage of the capacity of this channel to convey high spatial frequencies. When there is little detail, a chromatic sequence or a saturation sequence can be used (Rogowitz and Treinish 1996).

In many cases, the best color sequence should vary through a range of colors, but with each successive hue chosen to have higher luminance than the previous one (Ware 1988).

Hearing

How Our Hearing Works

Our sense of hearing plays a critical role in our ability to perform daily activities. Before our ears can detect sound, acoustic energy begins in the form of sound waves traveling through the air. These sound waves strike our eardrum, which sends them into our inner ear. Inside the inner ear are nerve impulses that are then transmitted to and perceived by the brain.

Measuring sound

Sound waves are measured in two ways: frequency and intensity.

Frequency is the number of cycles of pressure change occurring in one second and is measure in hertz (Hz). We perceive frequency as pitch. Humans can hear sound wave frequencies ranging from 20 to 20,000 Hz. When designing for the general population, expect them to detect frequencies from 1,000 to 4,000 Hz.

Intensity is determined by the amount of pressure with which a sound wave strikes the eardrum and is measured in decibels (dB). We perceive intensity as loudness. Frequency ranges from 1,000 to 8,000 Hz require the least intensity to be heard. And tones at lower frequencies must have larger intensities to be heard. As people age, their ability to hear higher frequencies is greatly reduced. By the time most people reach age 65, very few can still detect frequencies greater than 10,000 Hz.

Decibels are a logarithmic measure, not a linear one. Note that sound, power, and voltage decibels are all different, and many of those use varying reference levels. For sound pressure levels, a 20 dB difference is a 10-to-1 SPL change; a 40 dB change is a 100-fold increase. For sound pressure, the common reference level of 0 dB is about the threshold of human hearing, at 1,000 Hz.

One decibel is, conveniently, around the smallest perceptible difference to human hearing. It is also interesting that sound is perceived differently than its actual pressure values. As a rule of thumb, an increase of 10 dB in measured sound pressure is perceived to be only about twice as "loud." A 20 dB increase sounds about four times as loud (instead of 10 times) and a 40 dB increase is perceived to be about 16 times as loud, instead of the 100 times it actually is.

Here are some typical sound pressure levels, and their perceived levels:

Event	Sound pressure	Relative perceived loudness
Rustling leaves	10 dB	1/32nd
Whispered conversation	20 dB	1/16th
Quiet office interior	30 dB	1/8th
Quiet rural area	40 dB	1/4th
Dishwasher in next room	50 dB	1/2
Normal conversation	60 dB	Baseline
Dial tone	70 dB	2 times
Car passing nearby	80 dB	4 times
Truck or bus passing nearby	90 dB	8 times
Passing subway train	100 dB	16 times
Loud nightclub	110 dB	32 times
Threshold of pain	120 dB	64 times

Note that hearing damage begins to be possible at 140 dB.

Understanding how our auditory sense works can give us greater insight into designing audio alters and notifications. Be mindful of how increased age affects sound wave detection, as well as what levels of loudness are appropriate in our devices.

Brightness, Luminance, and Contrast

The terms *brightness*, *luminance*, and *contrast* are confusing to almost everyone, largely as a result of hardware manufacturers mislabeling display controls since the dawn of television. Apparently, this was to make it easier on the general public, but the result is an improper mental model, making it hard to adjust or design for electronic displays properly.

Brightness

Brightness refers to our subjective perception of how bright an object is. Therefore, what may seem very bright to you may be less bright to me. We can use subjective words such as *dim* and *very bright* to describe our perceptions of brightness.

When mobile device displays provide controls to adjust screen brightness, they're really controlling the amount of light emitted from the device. But when we are controlling the amount of light on a display, we're really concerned with how bright it feels to us, and how comfortable we are at that level of luminance.

Luminance

Luminance is the measure of light an object gives off or reflects from its surface. Luminance is measured in different units such as candela (cd/m^2), footlambert (ftL), mililambert (mL), and Nit (nt).

When a brightness control is available, the actual function being controlled is the display luminance, either via control of backlight brightness for LCDs, or of the "white point" setting for LEDs. Since luminance values are not commonly encountered, some example figures may be usefully illustrative:

- A typical computer display emits between 50 and 300 cd/m^2.

- Some mobile devices are now capable emitting up to 300 cd/m^2 of luminance.

- Riggs (1971) notes that in starlight (luminance of .0003 cd/m^2) we can see the white pages of a book but not the writing on them.

- The recommended luminance standard for measuring acuity is 85 cd/m^2 (Olzak and Thomas 1996).

- For text contrast, the International Standards Organization (ISO 9241, part 3) recommends a minimum of a 3:1 luminance ratio of text and background, though a ratio of 10:1 is preferred (Ware 2000).

Remember that luminance and brightness are not measured in the same manner. For example, if you lay out a piece of black paper in full sunlight on a bright day, you may measure a value of 1,000 cd/m^2. If you view a white piece of paper in an office light, you will probably measure a value of only 50 cd/m^2. Thus, a black object on a bright day outside may reflect 20 times more light than white paper in the office (Ware 2000).

Contrast

Contrast is the difference in visual properties that makes an object stand apart from other objects or backgrounds. Generally, this is the difference between the perceived brightness values of the highest white level compared to the darkest black level. High ambient light levels can reduce the perceived (or, depending on the display technology, actual) contrast.

Functional contrast can be strongly influenced by the designer. If multiple gray tones are used adjacent to each other, they may be perceived easily under ideal conditions, but blend together in poor conditions.

Black Level

Contrast is to black level as brightness is to luminance. Generally, contrast control is performed by adjusting the black level, or the amount of light emitted or transmitted by the low end of the display output.

Note that these controls work differently on other display types, such as CRTs. Some display types (most ePapers) have no meaningful control over contrast (or black level) at all.

General Touch Interaction Guidelines

The minimum area for touch activation, to address the general population, is a square 3/8 of an inch on each side (10 mm). See Figure D-1. When possible, use larger target areas. Important targets should be larger than others.

There is no distinct preference for vertical or horizontal finger touch areas. All touch can be assumed to be a circle, though the actual input item may be shaped as needed to fit the space, or express a preconceived notion (e.g., button). Due to reduced precision and poor control of pressure, but smaller fingers, children who can use devices unassisted have the same touch target size.

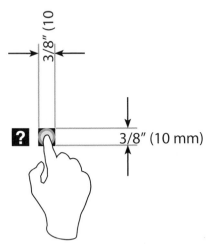

Figure D-1. *Minimum area for touch activation. Do not rely on pixel sizes to measure touch targets. Pixel sizes vary based on device and are not a consistent unit of measure.*

Targets

The visual target is not always the same as the touch area. However, the touch area may never be smaller than the visual target. When practical (i.e., there is no adjacent interactive item), the touch area should be notably larger than the visual target, filling the "gutter" or whitespace between objects. Some dead space should often be provided so that edge contact does not result in improper input.

In the example shown in Figure D-2, the orange dotted line is the touch area. It is notably larger than the visual target, so a missed touch (as shown) still functions as expected.

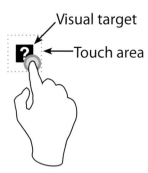

Figure D-2. *Visual target compared to the touch area. The touch area should never be smaller than the visual target.*

Touch Area and the Centroid of Contact

The point activated by a touch (on capacitive touch devices) is the centroid of the touched area; that area where the user's finger is flat against the screen.

The centroid is the center of area whose coordinates are the average (arithmetic mean) of the coordinates of all the points of the shape. This may be sensed directly (the highest change in local capacitance for projected-capacitive screens) or calculated (center of the obscured area for beam sensors).

A larger area will typically be perceived to be touched by the user, due to parallax (advanced users may become aware of the centroid phenomenon, and expect this). See Figure D-3.

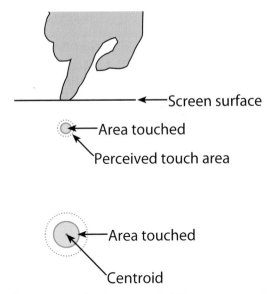

Figure D-3. *The centroid area compared to the area touched. Due to screen parallax, we typically perceive a larger area exists to touch.*

Bezels, Edges, and Size Cheats

Buttons at the edges of screens with flat bezels may take advantage of this to use smaller target sizes. The user may place her finger so that part of the touch is on the bezel (off the sensing area of the screen). This will effectively reduce the size of her finger, and allow smaller input areas.

This effective size reduction can only be about 60% of normal (so no smaller than 0.225 inch or 6 mm) and only in the dimension with the edge condition. This is practically most useful to give high-priority items a large target size without increasing the apparent or on-screen size of the target or touch area. See Figure D-4.

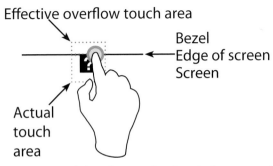

Figure D-4. *By using the space provided on the screen bezel, or the frame around the screen, the actual target size can be slightly reduced and speed of interaction can be increased.*

Fitts's Law

Paul M. Fitts (1912–1965) was a psychologist at both Ohio State University and the University of Michigan. In 1954, he created a mathematical formula to determine the relationship for how long it takes a user to either select an object on the screen or physically touch it, based on its target size and distance from the selector's starting point.

Fitts's Law is widely used today by UX designers, human factors specialists, and engineers when designing graphical user interfaces and comparing performance of various input devices. Fitts's Law finds that:

- The time required to move to a target is a function of the target size and distance to the target.

- The farther a target object is from the initial starting position will require a longer time to make that successful selection.

- That time can be increased when the target size is too small.

In mobile devices, we know that screen display size is limited and its space is valuable. In addition, mobile users require quick access to the content they are looking for. Using Fitts's Law together with these constraints can improve the user experience:

- Buttons and selectable controls should be an appropriate size because it is relatively difficult to click on small ones. Using the screen bezel overflow can provide a trick for allowing smaller, yet highly selectable targets . For mouse or other pointer-driven systems, the edge of a selectable area becomes "infinitely" deep, so the selectable area is functionally much larger. For touch devices with a flat bezel, the user may place part of his finger off the screen and activate a much smaller target than usual. Refer to the section "General Touch Interaction Guidelines" earlier in this appendix.

- Pop ups and tooltips can usually be opened or activated faster than pull-down menus since the user avoids travel.

- Reduce the number of clicks to access content by providing surface-level sorting and filtering controls to access indexed information quickly.

References

Works Cited

Bailey, R. W. (1996). *Human Performance Engineering: Designing High Quality Professional User Interfaces for Computer Products, Applications and Systems (3rd Edition)*. Upper Saddle River, NJ: Prentice Hall.

Benson, J., Olewiler, K., & Broden, N. (2006). "Typography for Mobile Phone Devices: The Design of the QUALCOMM Sans Font Family." AIGA Design Archives.

Biederman, I., & Cooper, E. E. (1992). "Size Invariance in Visual Object Priming." *Journal of Experimental Psychology: Human Perception and Performance* 18(1): pp. 121–133.

Briggs, A., & Burke, P. (2009). *A Social History of the Media: From Gutenburg to the Internet*. Cambridge, England: Policy Press.

Dyson, M. C. (2004). "How physical text layout affects reading from screen." *Behaviour & Information Technology* 23(6): pp. 377–393.

Easterby, R. (1984). Processes and Display Design. In R. Easterby, R. Easterby, & H. Zwaga (Eds.), *Information Design: The Design and Evaluation of Signs and Printed Material* (pp. 19–36). Chichester, England: John Wiley and Sons.

Human Factors and Ergonomics Society. (1988). *American National Standard for Human Factors Engineering of Visual Display Terminal Workstations*. Santa Monica, CA: HFES.

Kosslyn, S. M., Chabris, C. F., & Hamilton, S. (1990). "Five Psychological Principles of Articulate Graphics." *Multimedia Review*, pp. 23–29.

Kryter, K. D. (1972). Speech communication. In H. P. Van Cott & R. Kinkade, *Human Engineering Guide to Equipment Design*. Washington, DC: US Government Printing Office.

Lynch, K. (1960). *The Image of the City*. Cambridge, MA: MIT.

McGee, D., Pavel, M., & Cohen, P. (2001). "Context Shifts: Extending the Meanings of Physical Objects with Language." *Human Computer Interaction*, Vol. 16.

Nielson, J., & Pernice, K. (2010). *Eyetracking Web Usability*. Berkeley, CA: New Riders.

Nini, P. (2006, January 23). "Typography and the Aging Eye: Typeface Legibility for Older Viewers with Vision Problems." Retrieved May 15, 2011, from AIGA: *http://www.aiga .org/content.cfm/typography-and-the-aging-eye*.

Nooteboom, S. G. (1983). "The temporal organization of speech and the process of spoken-word recognition." IPO Annual Progress Report.

Norman, D. A. (1988). *The Design of Everyday Things*. New York: Basic Books.

Olzak, L. A., & Thomas, J. P. (1986). "Seeing Spatial Patterns." *Handbook of Perception and Human Performance* 1(7): pp. 7–56.

Parkinson, R. (1972). "The Dvorak Simplified Keyboard: Forty Years of Frustration." *Computers and Automation*, pp. 18–25.

Rogowitz, B., & Treinish, L. (1996). "How Not to Lie with Visualization." *Computers in Physics* 10(3): pp. 268–273.

Saffer, D. (2009). *Designing Gestural Interfaces (http://oreilly.com/catalog/9780596518394)*. Sabastopol, CA: O'Reilly.

Saffer, D. (2005, March 6). *The Role of Metaphor in Interaction Design*. Retrieved from Slideshare: *http://www.slideshare.net/dansaffer/the-role-of-metaphor-in-interaction-design*.

Visser, F. S., Stappers, P. J., & Sanders, E. B. (2005). "Contextmapping: Experiences from Practice." *CoDesign: International Journal of CoCreation in Design and the Arts*, Vol. 1, pp. 119–149.

Ware, C. G. (1988). "Dynamic Adjustment of Stereo Display Parameters." *IEEE Transactions on Systems, Man and Cybernetics* 28(1): pp. 56–65.

Ware, C. (2000). *Information Visualization: Perception for Design*. San Diego: Academic Press.

Ware, C. (2008). *Visual Thinking for Design*. Burlington, MA: Morgan Kaufmann Publishers.

Wroblewski, L. (2010, September 13). *Data Monday: Input Matters on Mobile*. Retrieved May 3, 2011, from LukeW: *http://www.lukew.com/ff/entry.asp?1188*.

Additional Resources

Wiki

We put together this wiki as a resource for this book. This site has an extensive number of mobile resources, including research-based frameworks, pattern libraries, drawing tools and templates, and plenty more enlightening mobile stuff. Please check it out!

4ourth Mobile
http://4ourth.com/wiki

Books

Here is a list of books that inspired us while we were writing this book. If you're new to the field of mobile UI design, or you want to expand on your interests, these books can offer you an excellent source of background knowledge.

- *A Pattern Language* by Christopher Alexander, Sara Ishikawa, Murray Silverstein, Max Jacobson, Ingrid Fiksdahl-King, and Shlomo Angel (Oxford University Press, 1977)

- *About Face 3: The Essentials of Interaction Design* by Alan Cooper, Robert Reinmann, and David Cronin (Wiley, 2007)

- *Blink: The Power of Thinking Without Thinking* by Malcolm Gladwell (Time Warner Book Group, 2005)

- *Cockpit Engineering* by D. N. Jarrett (Ashgate Publishing Limited, 2005)

- *Designing by Drawing: A Practical Guide to Creating Usable Interactive Design* by Steven Hoober (Little Springs Design, 2009)

- *Designing Interfaces*, Second Edition (*http://oreilly.com/catalog/0636920000556*) by Jenifer Tidwell (O'Reilly, 2011)

- *Envisioning Information* by Edward R. Tufte (Graphics Press, 1990)

- *Emotional Design: Why We Love (or Hate) Everyday Things* by Donald Norman (Basic Books, 2005)

- *Gesture and Thought* by David McNeill (The University of Chicago Press, 2005)

- *Information Architecture for the World Wide Web: Designing Large-Scale Web Sites* (*http://oreilly.com/catalog/9780596527341*) by Peter Morville and Louis Rosenfeld (O'Reilly, 2006)

- *Information Design* by Robert Jacobson (MIT, 1999)

- *Mobile Design and Development* (*http://oreilly.com/catalog/9780596155452*) by Brian Fling (O'Reilly, 2009)

- *The Visual Display of Quantitative Information* by Edward R. Tufte (Graphics Press, 1983)

- *Universal Principles of Design: 125 Ways to Enhance Usability, Influence Perception, Increase Appeal, Make Better Design Decisions, and Teach through Design*, Second Edition, by William Lidwell, Katrina Holden, and Jill Butler (Rockport Publishers, 2010)

- *Web Form Design: Filling in the Blanks* by Luke Wroblewski (Rosenfeld Media, 2008)

Index

O

OLED (organic light-emitting diodes), 44–45, 422

one-axis interaction, 322

on-screen buttons, 256

On-screen Gestures pattern
about, 320
cross-referencing patterns, 57, 58, 161, 162, 273, 277, 447
pattern description and usage, 347–352

open lists, defined, 151

operating systems
drawing tools by, 477–479
simulators and emulators, 484–486

option menus, 34

Ordered Data pattern
about, 228
cross-referencing patterns, 276, 277
greeking considerations, 512
pattern description and usage, 229–232

ordering data
challenges in, 230–231
information design and, 66–68

ordinal classification scheme, 64, 169

organic light-emitting diodes (OLED), 44–45, 422

organic pagination control, 189, 191–192

Orientation pattern
about, 425
cross-referencing patterns, 447, 450
pattern description and usage, 433–438

Other Hardware Keys pattern
about, 320
cross-referencing patterns, 447, 450
pattern description and usage, 333–337

output and input. *See* input and output

overlay method, 215

P

page layouts
display principles, 2–3
guidelines for mobile users, 3–4

page patterns. *See also* composition patterns
about, 1–5, 173
composition principles, 8–9
defined, 1
history of composition, 7

Pagination pattern
about, 175
cross-referencing patterns, 58, 160, 163, 274, 275, 277
pattern description and usage, 188–192

paging channel, defined, 455

paired buttons for directional entry, 321

PAN (personal area network), 471

parallax phenomenon, 296–297, 423

paths, defined, 167

pattern reference charts
for audio and vibration patterns, 448–449
for composition patterns, 56–59
for control and confirmation patterns, 161–162
for display of information patterns, 159–161
for drilldown patterns, 274–275
for general interactive controls patterns, 446–448
for information control patterns, 277–278
for input and selection patterns, 448–449
for labels and indicators patterns, 276–277
for lateral access patterns, 273
for revealing more information patterns, 162–163
for screens, lights, and sensors patterns, 449–450
for text and character input control patterns, 445–446

patterns. *See also* specific patterns
about, xix–xx
for audio and vibration, 389–415
authors' development of, xx–xxii
common practice versus best practice, xxii–xxiii
for composition, 10–60
for control and confirmation, 110–136
design concepts to consider, xxvi–xxviii
for displaying information, 68–105
for drilldown, 197–224
for general interactive controls, 313–364
as guidelines, xix
for information controls, 251–278
for input and selection, 365–388
for labels and indicators, 225–249
for lateral access, 169–195
organization of information about, xxiii–xxvi
for revealing more information, 141–163
for screens, lights, and sensors, 417–452
for text and character input controls, 281–312
visual perception of, 515–517

PCS (Personal Communications Service), 462

PDC network technology, 466

Peel Away pattern
about, 175
cross-referencing patterns, 273
pattern description and usage, 180–184

About the Authors

Steven Hoober has been documenting the design process for all of his 15 year design career, and entered mobile full time in 2007 when he joined Little Springs Design. His work includes *Designing by Drawing*, templates for mobile design, and he frequently blogs on design and UX topics. Steven has led projects on security, account management, content distribution, and communications services for numerous products, from construction supplies to hospital recordkeeping. Steven's mobile work has included design of browsers, e-readers, search, NFC, mobile banking, data communications, location, and OS overlays. Steven spent eight years at US mobile operator Sprint and has also worked with AT&T, Qualcomm, Samsung, Skyfire, Bitstream, VivoTech, The Weather Channel, Lowe's, and Hallmark Cards.

Eric Berkman is an Interaction Designer and Experience Architect at Digital Eskimo, a leading user-centered design agency whose projects involve inspiring change. Eric's design career has included developing mobile UI experiences for global telecommunications companies; branding and packaging design for Coca-Cola, Miller Brewing Company, and Bristol-Meyers Squibb; and interactive museum exhibitions. His expertise and interests focus on a user-centric, participatory design approach to create meaningful individual, social, and cultural interactions. He has both a bachelor's degree in industrial design and a master's in interaction design from the University of Kansas. He currently resides in Sydney, Australia.

Colophon

The image on the cover of *Designing Mobile Interfaces* is a lovebird.

The name "lovebird" refers generally to any of nine species of the genus *Agapornis* (from the Greek *agape*, meaning "love," and *ornis*, meaning "bird"). More commonly, they're known as small parrots. They're named for their monogamous pair bonding and their tendency to spend long periods of time sitting with their partners. When kept singly as pets, lovebirds will often bond with their human owners. Despite their small size, these affectionate birds are just as intelligent and colorful as their larger parrot cousins, although they're not considered to be as great of talkers, as some never learn to "speak," or mimic humans.

Eight of the nine species of lovebird are native to continental Africa, while the ninth species is native to Madagascar. They live in small flocks, and most eat grass, vegetables, seeds, and fruit, although the black-winged lovebird eats insects and figs, and the black-collared lovebird eats only figs native to its area, which makes it difficult to keep in captivity. Lovebirds have stocky builds and are usually between five and six inches long. They have short, blunt tails and long, sharp beaks. Most lovebirds have green plumage on their lower bodies; the coloring of their upper bodies depends on the species. They usually live between 10 and 15 years.

Lovebirds are popular as pets, due in part to their capacity for affection. If they bond with their human owner early on, they can be trained to do tricks and show great loyalty, so much so that they can become aggressive to other birds or humans. When kept paired in captivity, it's important that the members of the pair get along with each other. Pairs that are truly bonded can be seen feeding and grooming each other, while mismatched pairs do not display such affection.

The cover image is from Johnson's *Natural History*. The cover font is Adobe ITC Garamond. The text font is Adobe Minion Pro, and the heading and note font is Adobe Myriad Pro Condensed.

Get even more for your money.

Join the O'Reilly Community, and register the O'Reilly books you own. It's free, and you'll get:

- $4.99 ebook upgrade offer
- 40% upgrade offer on O'Reilly print books
- Membership discounts on books and events
- Free lifetime updates to ebooks and videos
- Multiple ebook formats, DRM FREE
- Participation in the O'Reilly community
- Newsletters
- Account management
- 100% Satisfaction Guarantee

Signing up is easy:

1. **Go to: oreilly.com/go/register**
2. **Create an O'Reilly login.**
3. **Provide your address.**
4. **Register your books.**

Note: English-language books only

To order books online:
oreilly.com/store

For questions about products or an order:
orders@oreilly.com

To sign up to get topic-specific email announcements and/or news about upcoming books, conferences, special offers, and new technologies:
elists@oreilly.com

For technical questions about book content:
booktech@oreilly.com

To submit new book proposals to our editors:
proposals@oreilly.com

O'Reilly books are available in multiple DRM-free ebook formats. For more information:
oreilly.com/ebooks

O'REILLY®

Spreading the knowledge of innovators   oreilly.com

Have it your way.